SAGE was founded in 1965 by Sara Miller McCune to support the dissemination of usable knowledge by publishing innovative and high-quality research and teaching content. Today, we publish over 900 journals, including those of more than 400 learned societies, more than 800 new books per year, and a growing range of library products including archives, data, case studies, reports, and video. SAGE remains majority-owned by our founder, and after Sara's lifetime will become owned by a charitable trust that secures our continued independence.

Los Angeles | London | New Delhi | Singapore | Washington DC | Melbourne

A Bridge to Mathematics

A Bridge to Mathematics

Shobha Bagai
Professor, Cluster Innovation Centre, University of Delhi

Amber Habib
Professor, Department of Mathematics, Shiv Nadar University

Geetha Venkataraman
Professor, School of Liberal Studies, Ambedkar University Delhi

SAGE | TEXTS

Los Angeles | London | New Delhi
Singapore | Washington DC | Melbourne

First published in 2017 by

SAGE | TEXTS

SAGE Publications India Pvt Ltd
B1/I-1 Mohan Cooperative Industrial Area
Mathura Road, New Delhi 110 044, India
www.sagepub.in

SAGE Publications Inc
2455 Teller Road
Thousand Oaks, California 91320, USA

SAGE Publications Ltd
1 Oliver's Yard, 55 City Road
London EC1Y 1SP, United Kingdom

SAGE Publications Asia-Pacific Pte Ltd
3 Church Street
#10-04 Samsung Hub
Singapore 049483

Published by Vivek Mehra for SAGE Publications India Pvt Ltd, typeset in 11/13pts Times New Roman by Archetype, and printed at Saurabh Printers Pvt Ltd, Greater Noida.

Library of Congress Cataloging-in-Publication Data Available

ISBN: 978-93-864-4612-1 (PB)

SAGE Team: Amit Kumar, Indrani Dutta, Vandana Gupta and Ritu Chopra

Brief Contents

Detailed Contents

List of Tables

List of Tables

List of Figures

Preface

Liberal arts students, the world over, are expected to develop reasoning and data analysis skills as part of their undergraduate repertoire. These involve the ability to analyse and reason, to be able to use the analysis to model situations/problems and finally to be able to infer, present and communicate their analysis. From the time of Pythagoras, mathematics has been seen as essential to reasoning. It provides a model of perfect reasoning as well as essential tools for understanding information and reaching conclusions.

Yet many of these students have a fear associated with numbers and mathematics. In India, they may be hampered by not having studied mathematics beyond Grade X, while also carrying an abiding scar from their actual engagement with school mathematics. Most texts that are available under the category of 'Mathematics for the Liberal Arts' engage with topics designed to allay these fears as well as build some skills. They are, however, typically designed for a terminal course in mathematics for students who will otherwise have little to do with the subject. Our intent, on the other hand, is to provide a bridge back into the world of mathematics for students who have felt isolated from it. We hope that some of our readers will be enabled to take further courses in mathematics, while others will gain a good foundation for carrying out or understanding basic mathematical or statistical work in various disciplines.

Among others, this book aims at:

1. Increasing the appreciation of mathematics as an art and a human endeavour.
2. Motivating students towards the study and use of mathematics by providing them basic tools to understand critical issues.
3. Making the reader understand the handling of information and integrating it into graphs and figures that capture its essence.

One way of providing a link with mathematical reasoning for students is to present the ideas in a real-life or everyday context. This not only enables readers to identify with the mathematics that they encounter but also shows the further necessity to build up on mathematical skills. Often, a historical viewpoint also tells the story of how humankind has developed and effectively utilised mathematical ideas. This is the theme taken up in the first chapter.

In order to be able to understand certain real-life problems and solve them, expressing the problems mathematically is essential. This helps clarify and refine the problem so as to be able to find a solution. It is, therefore, imperative to learn the basic language for expressing mathematical ideas, namely that of sets. The second chapter helps the reader with the concept of a set and the language of set theory. The third chapter deals with analysing arguments. This begins with the idea of what statements are and then builds up to a systematic method of figuring out what valid arguments are.

Being able to interpret and handle data is a necessity of our times. This is dealt with in the fourth chapter, along with material on computer-based practicals for handling data. The fifth chapter continues the study of this aspect of mathematics by taking up the topic of probability and the first ideas of mathematical statistics.

The book also has chapters on symmetry, perspective and art. These topics allow a student to reason, model and analyse situations that arise in an everyday context. They also increase an awareness of patterns and the mathematics that pervades the world around us.

The treatment of all parts is gentle, non-calculus-based and example-oriented. Solved examples, exercises and suggestions for projects provide the reader with hands-on experience in developing their analytic and quantitative skills. We further this practical aspect by providing computer-based lab work which will help students to experience handling data and the use of computational tools to analyse and present data.

The topics covered in the book would not only be useful to liberal arts students but also to practitioners of social sciences, humanities and life sciences. For many, this book would help bridge the gap that would have emerged between mathematical skills picked up at school and the quantitative and analytical skills that are required as part of their profession or academic lives. The material covered would enable these practitioners with much needed foundational skills in reasoning, data analysis, measuring chance, spatial awareness and quantitative methods that are a requisite for progressing to higher levels.

We also intend that the book will not only have a remedial aspect, but also an inspirational one. School teachers at the higher secondary level as well as their students will find it an accessible source of further mathematics and of projects they can carry out. This is, however, only the first bridge that a reader would need to traverse to connect back to mathematics. Other roads would be needed to take the readers further into their mathematical journey. Each chapter of the book ends with a list of references —books, articles and websites—that would provide the reader with further direction and assistance.

Acknowledgements

The authors are deeply grateful to Ambedkar University Delhi (AUD) for supporting this book-writing project in many ways, not least of which was through financial grants, administrative support and encouragement.

This book has its inception in a course designed for liberal arts students of Ambedkar University Delhi (AUD). We were unable to find a single book that would serve as a textbook for the course. Thus began a textbook-writing project supported by the University. But administrative and academic commitments of the authors intervened to extend the writing period. During the writing, we had support from our respective institutions and colleagues. In particular, the authors wish to thank Asha Devi D, of AUD, for secretarial help on the book-writing project. Thanks are also due to our students Akhil Puthiyadhatu Veetil, Sameer Mallik, Asmita Punia and Divyansh Dev who read early drafts and gave us useful suggestions. We would like to acknowledge the support of our friends and families for the myriad and unobtrusive ways in which they provided support and succour.

Shobha Bagai would like to thank her husband Anshu Bagai and son Sukul Bagai who supported and encouraged her in her writing endeavour, though it meant time spent away from them; her parents Maj S. K. Sood (late) and mother Sushma Sood who have always allowed her to follow her ambitions throughout her journey so far. The book would have taken longer to complete without the support of her colleagues with a special mention of her dear friend Jyoti Sharma.

Amber Habib expresses his gratitude to the memory of his uncle Professor Mohd Mohsin for showing him the path to becoming a mathematician.

Geetha Venkataraman would like to thank her partner Mahesh Rangarajan, her parents Visalakshi Venkataraman and WgCdr PS Venkataraman (late), and her mother-in-law Shantha Rangarajan for their patience, encouragement and unstinting support. A special thanks goes to her daughter Uttara Rangarajan for keeping her mother grounded and her 'almost' daughter Neelambari Bhattacharya. Many thanks are particularly due to her young friend Sheuli Chander who was delightful company whenever a break was needed from mathematics. Thanks are also due to the female members of the WhatsApp 'women' group Sumangala Damodaran, Smita Bannerjee, Shobha Talengala, Shakun Srivastava and Asmita Kabra for their wonderful company. Thank you Ishani Bannerjee for creating the WhatsApp group for us. The writing of this book went parallel with the period when birdwatching and photography began consuming large chunks of difficult to spare time. So thanks are also due to birding friends, Savithri Singh, Sudeshna Dey and Subramanian Venkataramani, who helped to prolong the writing process by organising many a long weekend of birdwatching and photography.

About the Authors

Shobha Bagai is currently a Professor at the Cluster Innovation Centre, University of Delhi. Professor Bagai completed her masters in 1989 and PhD in Mathematics in 1993 from Indian Institute of Technology (IIT) Delhi. She was a gold medallist from Panjab University in 1987 scoring 100% marks in Mathematics.

Having taught a myriad of courses—calculus, algebra, analysis, differential equations, mechanics, probability, statistics, discrete mathematics, linear programming—to undergraduate and postgraduate students for the past 22 years, she has been invited by a number of institutions to talk on various topics and applications of mathematics. She had also been a Visiting Assistant Professor at IIT Bombay (2002–2004) and a fellow in mathematics at the Institute of Lifelong Learning, University of Delhi (2008–2009). Professor Bagai has published a number of research papers in reputed journals and has edited and co-authored a book on building mathematical ability. Along with her teaching and administrative assignments, she is actively involved in various projects with her students, many of which have resulted in undergraduate research papers.

During her free time, she enjoys doing puzzles that keeps her mentally active, dancing that keeps her physically fit, and nature photography that lets her enjoy the beauty of the nature.

Amber Habib is currently a Professor, Department of Mathematics, School of Natural Sciences, at Shiv Nadar University (SNU), Uttar Pradesh. Professor Habib did his MS (Integrated) in Mathematics at IIT Kanpur and PhD at University of California, Berkeley. These were followed by postdoctoral positions at the Harish-Chandra Research Institute, Allahabad, and the Indian Statistical Institute, Bangalore Centre. His research interests include representation theory and mathematical finance.

For over 15 years, he has been involved in efforts to make mathematics education at the school and college levels more interesting and fulfilling through special topics, projects and an appreciation of the myriad links of mathematics with other disciplines. This began during his time as a teacher at St. Stephen's College, University of Delhi, and was his main focus during the years he spent with the Mathematical Sciences Foundation. He has also served several times as a resident faculty for the Mathematical Training and Talent Search Programme sponsored by the National Board for Higher Mathematics.

He has received the SNU Faculty Excellence Award for Excellence in Teaching during 2011–2014. Apart from his academic interests and responsibilities, he finds fascination and recreation in nature photography. One can find a rich repository of the same in https://www.flickr.com/photos/amberhabib/.

Geetha Venkataraman is a Professor of Mathematics at Ambedkar University Delhi (AUD). She did an MA and DPhil (doctorate) in Mathematics at the University of Oxford. Her area of research is finite group theory. She is a co-author of a research monograph titled *Enumeration of Finite Groups*, published in 2007.

Prior to joining AUD as a Professor in 2010, Professor Venkataraman taught at St. Stephen's College, University of Delhi. She has published several research papers related to group theory and articles related to education, with an emphasis on undergraduate education. Apart from her interest in group theory and related areas, she is deeply interested in popularising mathematics, mathematics education and issues related to women in mathematics. She has given several research talks and popular talks on mathematics in India and other countries to a varied audience ranging over middle- and high school children, schoolteachers and mathematicians.

Professor Venkataraman did a BSc (Honours) in Mathematics at St. Stephen's College, University of Delhi. She stood first in Delhi University in BSc (H) Mathematics. The Inlaks Scholarship and the Burton Senior Scholarship from Oriel College, University of Oxford, enabled her to complete a masters and a DPhil respectively from the University of Oxford.

She is a keen birdwatcher and photographer of birds. Recently, four of her bird photographs have been published in a bird book for children, written in Manipuri, titled *On the Lap of Nature: Let's See Let's Find*, by Dr Kh. Shamugnou. Some of her photographs have been chosen for the database Oriental Bird Images and can be seen at http://orientalbirdimages.org/photographers.php?action=birder&Birder_ID=1291

1

CHAPTER

Why to Solve It?

In this chapter, we illustrate how mathematics has grown out of human needs and our attempts to understand and influence the world around us. We use examples from history to highlight different aspects and applications of mathematics, beginning from the dawn of human civilization and working up to the present day. Later chapters will show us how to think mathematically. Here, we are concerned with the even more basic question, "Why should we care about mathematics?"

1.1 Marking Time

Our ancestors must have remarked at a very early time the prevalence of cyclic phenomena in nature, especially the changing of seasons. They would have soon connected the seasonal weather changes with other events of importance to them such as the availability of food and the migration of animals. As culture and technology grew more complex, they would have felt an ever greater need to accurately anticipate these changes so that they could plan the planting of seeds, the harvesting and storage of their crops and the travel to trade with other communities.

Predicting seasonal changes is difficult because of irregularity in the aspects which affect us most directly, such as temperature and rainfall. A sudden warm day in winter does not necessarily mean that spring is near, nor does a summer drizzle always portend the coming of the monsoon. Luckily, there are also changes which are regular, if less noticed, and these can be used for more reliable forecasting. The most striking of these seasonal changes is the shrinking of the duration of the day as summer changes to winter, followed by a lengthening as summer returns. This change makes us consider the motion of the Sun in the sky, and we see that in the long term the seasonal changes can be correlated with how high it rises. Other heavenly bodies have their own roles to play. The Moon, through the changing of its phases, provides an easily used clock and the months are a convenient intermediate unit between days and years.

It is one thing to crudely observe these broad connections, quite another to understand the patterns precisely enough for accurate prediction. We must first acquire the ability to count and measure, and then to calculate. Thus, the need for a useful calendar was a prominent driving force for the birth and development of mathematics. On the one hand, the need to measure time through clocks created a need for the study of numbers and arithmetic. On the other hand, describing the motion of the Sun and the Moon required the development of concepts related to space and geometry.

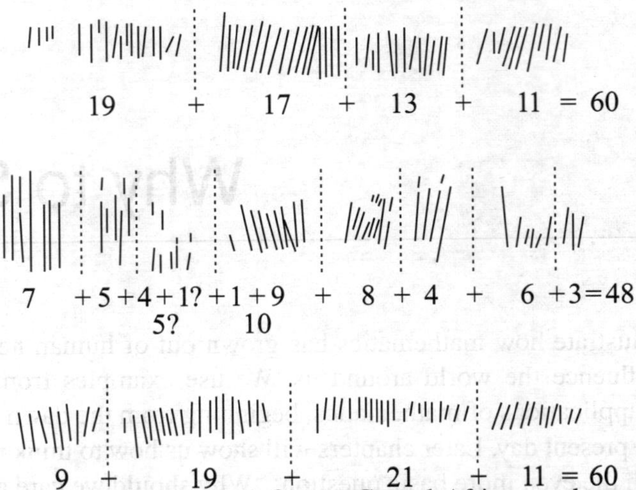

The *Ishango bone* is a mammal bone from a Central African settlement dated to over 20,000 years ago. Etched on it are three rows of marks which are further grouped in various ways. If the marks were intended for counting, then their distribution shows interesting mathematical patterns. For example, the number of marks in two of the rows adds up to 60, perhaps indicating a connection with monthly cycles. The numbers in one row are grouped to represent doubling, and so on. [C1]

The movement of the Sun across the sky allows us to measure time through changes in the shadows it casts. This is the principle underlying sundials. For a sundial, the basic unit of time is one day-length, and it can be calibrated to any fraction of this unit. The drawback is that since a sundial measures fractions of a day, it cannot detect variation in day-lengths. We can use a sundial to define an hour as one-twelfth of a day, but then summer hours will be longer than winter hours. And, of course, sundials are useless at night.

These drawbacks are avoided by the water clock and the sand glass which are based on terrestrial rather than heavenly events. The simplest water clock is just a jar of water with a small outlet. Water flows continually out through the outlet, and the changing level of water indicates how much time has passed. Such water clocks have been found from ancient Egypt and Mesopotamia, dating back to 3,000 or 4,000 years ago. And from Mesopotamia, we even have tables of day-lengths at different times of the year (Figure 1.1.1).

The ancient Mesopotamians seem to have first estimated the longest day as being twice as long as its night. Consequently, night watchmen were paid twice as much in the summer than in winter![1] An interesting aspect is that the night watchmen's wages had just two levels: one for the six months of summer and another for the six months of winter.

---

[1] Neugebauer, 'The water clock in Babylonian astronomy' [6].

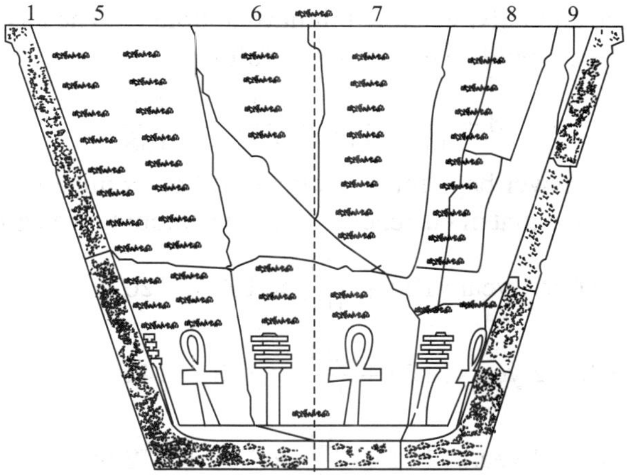

Figure 1.1.1 Section of a water clock found at the Karnak temple in Egypt and dated to about 1,400 BC. Water dripped out from a hole in the base, and the elapsed time was read from the water level, using marks on the inside wall. The Egyptians defined an hour as one-twelfth of the night, so this water clock has a separate column for each month to account for the monthly variation in hour-lengths. [C2]

Later, this ratio was corrected to a more accurate 3:2. Moreover, the Mesopotamians moved to a more refined system of day-lengths, gradually changing from shortest to longest. They created tables in which the day-length changed uniformly with time. Similar ideas are recorded in India in the *Jyotisa Vedanga* which describes the division of a day and night into 30 *muhurta*s. The longest day was stated to be 18 *muhurta*s and the corresponding night 12 *muhurta*s. The day-length was assumed to change at a constant rate of one *muhurta* per month.[2]

Thus, the need for a calendar contributed to the development of two fundamental concepts in mathematics. One is periodicity, as in recurring cycles of seasons and day-lengths. The other is linearity, equal changes in one quantity corresponding to equal changes in another. Although we shall not go into further details, it is worth noting that this study led to the development of astronomy and geometry, and the need for a practical geometry led to trigonometry. In school, we learn of trigonometry as a tool for measuring heights of buildings or distances between mountains. Yet that is not how it began. It began from the attempt to understand the motion of the heavenly bodies and our place in the universe.

1.2 Measuring Value

We have seen how mathematics received an impetus from attempting to track the flow of time. Another early source of stimulation was commerce or the flow of money. Returning again to the Mesopotamia of four millennia ago, we find a clay tablet with the following query: If a loan earns

[2] For the history of mathematics in India up to about the seventeenth century, see Plofker, *Mathematics in India* [7].

20% interest, compounded annually, when will it have doubled?[3] The answer is given in a base 60 system, and, in our notation, would be expressed as follows:

$$3 + \frac{47}{60} + \frac{13}{60^2} + \frac{20}{60^3} = 3\frac{85}{108}$$

This very precise-looking answer begs for investigation! Let us imagine that the loan was of 100 mina (a unit of weight and also that of currency). Then the amount owed grows annually as follows:

$$\text{After 1 year}: 100 + \frac{20}{100} \times 100 = 120$$

$$\text{After 2 years}: 120 + \frac{20}{100} \times 120 = 144$$

$$\text{After 3 years}: 144 + \frac{20}{100} \times 144 = 172.80$$

$$\text{After 4 years}: 172.80 + \frac{20}{100} \times 172.80 = 207.36$$

Thus, the precise time of doubling is somewhere between 3 and 4 years, and likely closer to 4 than to 3. But what justifies an answer as detailed as $3\frac{85}{108}$?

Here, we have to imagine possible conventions for calculating interest at intermediate times. If you were going to earn 60 mina interest over a year but you took your money back after just six months, what interest should you earn? One plausible answer is that over half the time you should earn half the interest, that is, 30 mina. Taking this logic further, over three months, you should earn 15 mina; over one month, 5 mina and so on. Let us apply this convention to the Mesopotamian problem:

- The interest earned over the full fourth year would be $207.36 - 172.80 = 34.56$.
- To achieve doubling, the actual interest earned should be $200 - 172.80 = 27.20$.
- The time taken to earn an interest of 27.20 is $\frac{27.20}{34.56}$ years.

And $\frac{27.20}{34.56} = \frac{85}{108}$ exactly matches the Mesopotamian answer! Here too they have used the concept of equal growth in equal time, or what we would call linear growth. Figure 1.2.1 illustrates the two different ways in which they used linear growth in keeping a calendar as against keeping accounts.

1.3 Making the Best Selection

One of the challenges in counting or measuring is dealing with extremes. Suppose you are asked to count a number of marbles spread on the floor. Doing an ordinary count of 1, 2, 3, . . . will be time-consuming, and it will be easy to lose your place and have to start over. In this case, you may

[3] Boyer and Merzbach, *A History of Mathematics* [2].

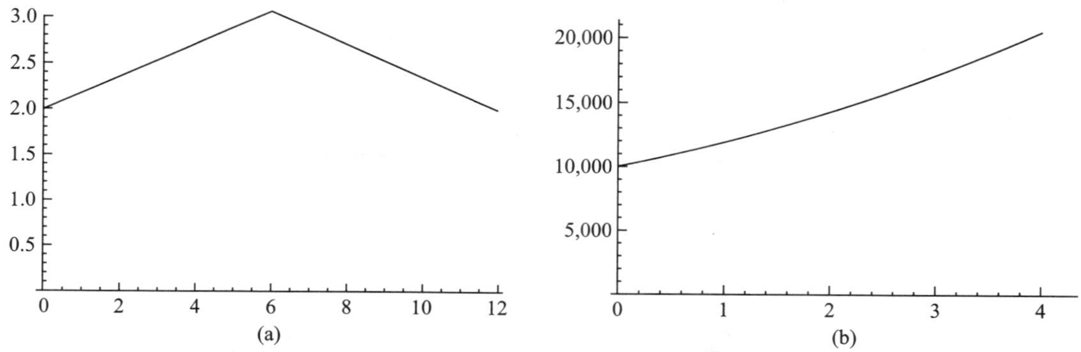

Figure 1.2.1 (a) The first graph represents the Mesopotamian technique of increasing the day-length by constant amounts till it reached its greatest value, and then decreasing it at the same rate to return to its lowest value a year later. (b) The second graph represents their method of calculating interest, with linear pieces of an ever increasing slope. The same convention is found in modern books on finance.

prefer to first count in larger groups, say of 5, and resort to counting in ones only when a very few are left. Thus, you may count 5, 10, 15, 20, 21, 22, 23 instead of 1, 2, 3, . . . , 23.

A similar situation prevails in weighing, with the added complication of needing to account for fractional amounts. No set of weights will allow us to *exactly* weigh any given object, but by having a variety of weights we can hope to measure accurately enough for our needs. Sets of weights have been recovered from the earliest civilizations and it is evident that careful thought has gone into devising them.

Cubes of chert (a sedimentary rock) have been found in large numbers at Harappa and Mohenjo-Daro. If we use 13.6 g as a unit, their weights run as

$$\frac{1}{16}, \frac{1}{8}, \frac{1}{4}, \frac{1}{2}, 1, 2, 4, 10, 20, 40, 100, 200, 400, 500 \text{ and } 800$$

We do not know what weights the people of this civilization took as units. The 13.6 g weights are usually taken as the base unit because they are very common and are also reasonably central. A case could also be made for the sixth weight, of 27.2 g, which in some places is the most common. If we take 27.2 g as unit, we have the values

$$\frac{1}{32}, \frac{1}{16}, \frac{1}{8}, \frac{1}{4}, \frac{1}{2}, 1, 2, 5, 10, 20, 50, 100, 200, 250 \text{ and } 400$$

whose integer values from 1 to 100 match the common choices for currency notes in our time. It is also interesting to see the variation from the smallest weight of 0.85 g to the highest weight of 10.9 kg. The presence of such small weights indicates that there were items considered precious enough to be weighed to high precision.

The main quality of a good set of weights (apart from accuracy) is that we should be able to weigh an object with a small number of weights. In the Indus system, with 27.2 g as unit, we can cover all (whole number) weights from 1 to 10 with just one weight of 5, two weights of 2 and one weight of 1. By adding higher weights in a similar pattern—one weight of 50, two of 20, one of

Figure 1.3.1 Weights from the Indus Valley Civilization, exhibited at the National Museum, Delhi. [C3]

10—we can cover all whole weights from 1 to 100. As an illustration, suppose we need a weight of 79. We can obtain it using one weight of 50, one of 20, one of 5, and two of 2 (Figure 1.3.1).

If we remove either the weight of 200 or 250 (with 27.2 g as unit), then the Indus system has an additional attractive property: the quickest way to weigh an object on a balance is to fill the other side with the heaviest weights that do not tip the balance.

For example, suppose we have to weigh an object with a weight of 136. We put it on one pan of the balance. On the other pan, we see that 100 is too low and 200 is too high, so we put the 100 weight. Next, we add 50 and find that it is too heavy, so we leave it out and try 20. At this point, we have 120 units on the other side. We similarly add further weights of 10, 5 and 1 to reach 136 and achieve a balance. This process of measuring by using the heaviest weight that fits is called the **greedy algorithm.**

The greedy algorithm is not optimal for every set of weights so the Indus choice is no accident. Indeed, these observations hold also for the fractional weights of 1/2, 1/4, 1/8, 1/16 and 1/32. All multiples of 1/32 can be weighed optimally by the greedy algorithm.

In Mesopotamia, a much richer historical record shows a multiplicity of weight units and systems. One of the more curious finds is a collection of tiny weights, initially mistaken for beads, whose weights can be expressed as 1, 3, 6, 9, 15, 24 in units of 0.14 g.[4] (The Mesopotamians counted in multiples of 60 and 0.14 g is one-sixtieth of the weight unit named shekel.) If we

[4] W B Hafford, Hanging in the balance: precision weighing in antiquity [5].

ignore the smallest weight, these numbers grow in an interesting way: $3+6=9$, $6+9=15$, $9+15=24$, each number being the sum of the two preceding ones.

Consider a set of such weights with members weighing 1, 1, 3, 6, 9, 15 and 24 in units of one-sixtieth of a shekel. These can be used to measure any weight up to 59/60 of a shekel using the greedy algorithm.

The process of weighing by gradually filling up one side of a pan is a special instance of a more general problem that we call the **Knapsack problem**. The Knapsack problem is typically posed as follows. Consider a bag (knapsack) that can hold up to a certain volume. Consider also a bunch of objects with various values. Which of these should we put in the bag (their total volume cannot exceed that of the bag) so as to have as much value as possible? If the bag is large and the objects are varied in their sizes and values, it may take a long time to figure out the answer! The cleverness of the Indus and Mesopotamian approaches is that the weights have been chosen so that the problem can always be quickly solved.

The Knapsack problem crops up everywhere in the modern world—loading cargo planes, choosing which assets to invest in, determining the product mix for a production line and even deciding which questions to attempt on an exam! As mentioned above, the Knapsack problem can take a very long time to solve, and it is not known if there is a fast way to solve such problems. The expectation is that there is no method which will always be fast!

1.4 Finding the Path

You may remember the story of Alexander Fleming's discovery of penicillin in 1928. He noticed that a bacteria culture left open by mistake had been infected by a mold that appeared to inhibit the growth of the bacteria. Thus began our age of antibiotics. This episode demonstrates how a small event, when viewed in a new way by an alert mind, can lead to dramatic advances in our knowledge.

The history of mathematics too has a famous story of this type, connected to the city of Kaliningrad in Russia. Kaliningrad lies between Lithuania and Poland and is at some distance from the rest of Russia. In fact, it was originally a German town and was called Königsberg. The river that runs through this town was then called the River Pregel. The Pregel branched and looped through Königsberg, as shown in Figure 1.4.1, and in the eighteenth century, there were seven bridges across it.

A challenge took shape around the river and the bridges—it was to find a path that would let one walk across all seven bridges exactly once. No bridge could be missed or crossed twice and, of course, there was to be no swimming across the river! This puzzle was eventually brought to the attention of the mathematician Leonhard Euler with the information that no such path had yet been found and opinion was divided about the possibility of proving its absence.

Euler noted that one could attempt to solve the problem by carefully listing all possible paths, but rejected this as being too tedious and also because it would not lead to a general method applicable to similar problems with more complicated configurations. Instead, he noted that while this problem was clearly related to geometry, it was different from the usual geometry problems in that the actual lengths and angles of the configuration were irrelevant. This reminded him of the

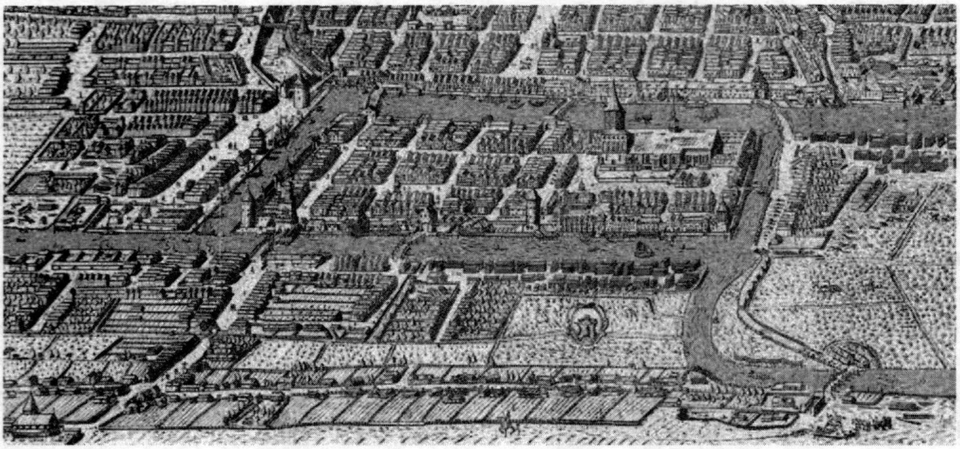

Figure 1.4.1 Königsberg in the seventeenth century. The Pregel river has been shaded to make the bridges more visible. [C4]

desire expressed some 60 years earlier by Leibniz for a new kind of mathematics that would deal directly with position instead of magnitude.

Euler began his solution by stripping away all the irrelevant and distracting details and just briefly sketching the river, the land areas and the bridges. He named the four land areas A, B, C and D; and the bridges a, b, c, d, e, f and g (as shown in Figure 1.4.2).

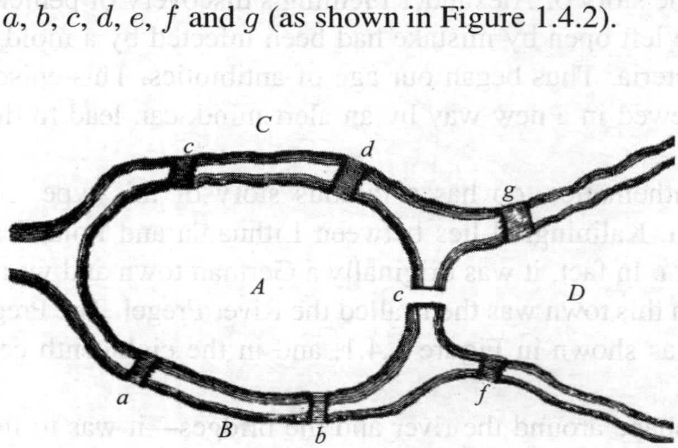

Figure 1.4.2 Euler's simplified representation of the bridges of Königsberg. [C5]

He represented the movement of a person from area A to B by AB and from B to A by BA. Thus, the string of letters $ABDCAB$ represents a path which that starts in A, crosses over to B (say by bridge a), then to D (by bridge f), then to C (by bridge g), then to A again (say by bridge c) and finally ends at B (by bridge b). At this stage, we see that there are no unused bridges connecting B and so this path has reached a dead end. And the bridges d and e have not been crossed!

Now, let us count how often each letter should show up in a string that uses up all the bridges exactly once. As an example consider the area marked D which has three bridges. If we start in D, it gives us one occurrence. At some later stage, we must re-enter and then immediately leave D, giving one more occurrence and thus a total of 2. If we start outside D, at some point, we must pass through it, using up two of its bridges and creating one occurrence of D in the string. Eventually, we must use the remaining bridge to enter D creating a second occurrence. So we see that, wherever we start, the letter D occurs twice.

The same logic applies to areas B and C as they also have three bridges. Area A has five bridges and so the letter A must occur thrice (twice from passing through and once from starting or ending). This leads to the following table of how often each letter must occur:

A	B	C	D
3	2	2	2

This means that the path must be represented by a total of nine letters. But this total of nine letters implies eight crossings, and since there are only seven bridges, we are not allowed eight crossings. Hence, such a path is impossible. (This is an example of a *proof by contradiction*. You have probably seen another famous proof by contradiction in school: the proof that there is no rational number whose square is 2. If you had trouble with this argument, please revisit it after reading Chapter 3.)

Euler's approach was later sharpened as follows. Since the exact shape of the land areas is not relevant, we can reshape them as we like, as long as we keep the right number of connecting bridges. The river itself is also not relevant! In fact, we can shrink the four land areas to points and use curves joining them to represent the bridges. The resulting diagram is called a **graph** (Figure 1.4.3).

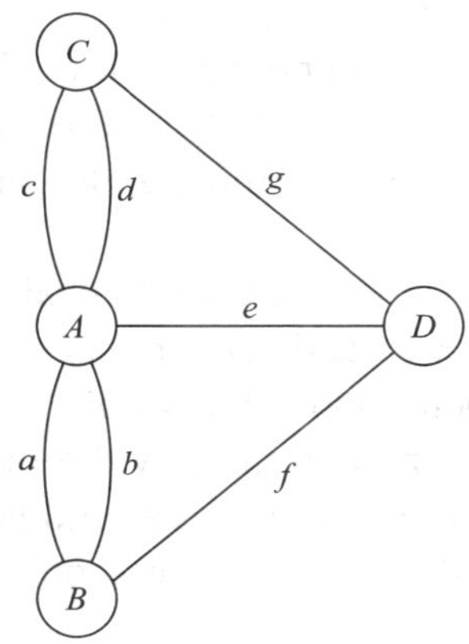

Figure 1.4.3 The graph depicting the bridges of Königsberg.

Graphs have become an essential tool for understanding connections and relationships. Important examples of such graphs are railway or airline networks, the web with hyperlinks connecting webpages, and even the networks of 'friends' on social media. The success of Google relative to earlier search engines was founded on its use of the graph created by hyperlinks to decide which pages were likely to have more information on the search phrase.

1.5 Learning from Nature

The topics that we have taken up so far involved humans trying to bring order to their concepts of the world around them. We look for patterns and analogies, and proceed on that basis. We explore in our minds the consequences of certain assumptions and then apply the results to actual situations where those assumptions seem reasonable. Among the questions we ask repeatedly is: *What is the best way?* What is the best choice of standard weights so that weighing will be easy? Given a network of roads connecting certain destinations, what is the shortest route that passes through all of them? Which questions should one attempt first in an exam? When should a power company turn off a generator when demand is low, keeping in mind the extra costs of restarting it?

The area of mathematics that deals with such problems is called **optimization**, and when it deals specifically with problems of industry, it is called **operations research**. An example is the Knapsack problem mentioned earlier. Optimization problems are typically hard and do not have good solutions. When solutions exist, they tend to work well in practice when only a few factors have to be taken into consideration. As the number of factors grows, so does the time taken to compute the solution, and soon that time becomes so long that the result is useless by the time it is computed!

In recent years, mathematicians and scientists have discovered new ways to tackle optimization problems. First, we can save time by not aiming for the best way alone, but being satisfied with just a good enough way that can be found in time to be useful. Second, we can find such good practices by looking to nature for examples. Third, possible approaches can be tested by using computers to run through a very large number of possible scenarios.

In an optimization scenario, we often do not have all the information at hand. Picture the task of finding your destination in the dark with the help of a torch that can only illuminate up to a few meters (Figure 1.5.1). Locally, we see the best thing to do, but always doing the best thing locally may not be the best thing to do overall.

A new approach is to look at how various life forms deal with such problems in their lives. Of course, we do not expect that they set up and solve mathematical models for their problems. But we can expect that over the ages, their approaches would have become more and more efficient, and under the pressure of constant competition, only the most efficient approaches would have survived. We can also expect that these approaches would have simple decision rules. If we can discover these and create corresponding axioms, we should have found simple approaches that work very well on average.

For example, it was found that some species of ants, faced with the previous path choice problem, show the following behaviour. Initially, ants set out on both paths to a food source in about equal numbers. After reaching the food, each ant returns by the same path that it took to

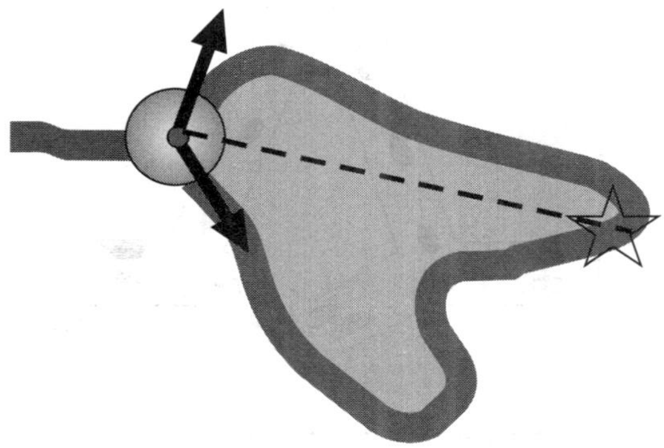

Figure 1.5.1 Suppose we wish to take the shortest path to a destination on the other side of a lake. It is night, our destination is well lit and visible, and we can use a torch to find our way and move along the lake's shore. However, we cannot see the path beyond our torch's range. Standing at the fork, we are faced with a decision: Which path to take? The available information is in favour of turning right as the visible part of that path seems to head more directly to the destination. But it turns out that this path is longer!

get there. As ants start reaching back, subsequent ants start moving over to the shorter path, and eventually almost all the ants use the shorter path.

How does this happen? Do the returning ants compare notes with the ones about to set out? Do they time each other? It turns out that much simpler rules can create the ants' behaviour:[5]

1. The ants lay a trail of chemicals (pheromones) at a constant rate as they walk along a path.
2. Given a choice of two paths, an ant is more likely to take the one on which the pheromone trail is stronger.
3. After successfully finding food, an ant returns to the nest by the same route (Figure 1.5.2).

Suppose one ant sets out on the shorter path and one on the longer. When the ant on the shorter path returns to the nest, the other one is still out on the path. At this point, the pheromone trail on the shorter path is twice as strong as on the longer one, so the next ant is more likely to take the shorter path! This tendency becomes stronger as more ants switch to the shorter path. Eventually, almost all ants move on the shorter path. The ants' memory and communication ability do not need to be invoked. (This does not mean that they are absent.)

Note that the ants are only more likely, not certain, to take the path with more pheromone. This element of randomness is key to their success in the long run. For example, suppose the environment changes and the less used path becomes better. The fact that some ants are still randomly using it can enable the ants to discover this change. Randomness also allows escape from traps like walking in circles.

[5] See Dorigo and Stützle, *Ant Colony Optimization* [3] .

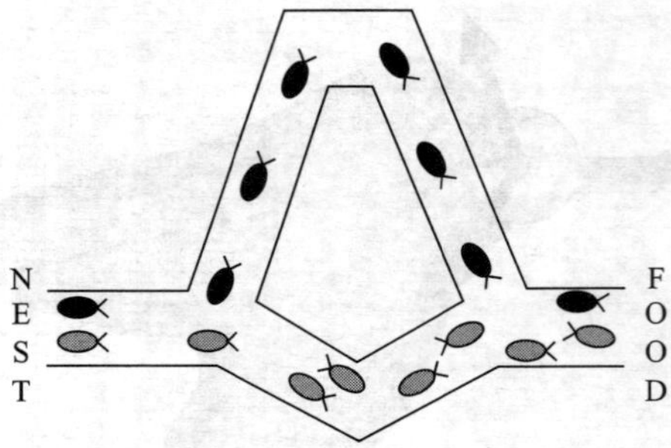

Figure 1.5.2 The two-bridge experiment. As the ants on the shorter path return earlier, the pheromone trail thickens faster on that path.

Mathematicians have made computer-based optimization techniques by mimicking the ants' behaviour. In these **ant colony optimization** techniques, 'synthetic ants' are sent out to explore and leave numerical trails. The 'synthetic ants' follow probabilistic axioms inspired by the real ants. A subsequent approach called **swarm optimization** is inspired by bees. These techniques have turned out to be very practical and improve on earlier deterministic ones in many ways. These developments bring us full circle—from developing mathematics to understand nature, we now try to understand nature to develop mathematics!

Bibliography

Books and Articles

[1] A R Angel, C D Abbott and D C Runde. 2013. *A Survey of Mathematics with Applications*. 9th edition. London: Pearson.

[2] Carl Boyer and Uta Merzbach. 2010. *A History of Mathematics*. 3rd edition. New York: John Wiley and Sons.

[3] Marco Dorigo and Thomas Stützle. 2004. *Ant Colony Optimization*. Cambridge, MA: MIT Press.

[4] Howard Eves. 1977. *Foundations and Fundamental Concepts of Mathematics*. 3rd edition. *Dover Books on Mathematics*. Mineola, NY: Dover Publications.

[5] William B Hafford. 2005. Hanging in the balance: Precision weighing in antiquity. *Expedition* 47(2), 35–37.

[6] Otto Neugebauer. 1947. The water clock in Babylonian astronomy. *Isis*: 37(1/2), 37–43.

[7] Kim Plofker. 2008. *Mathematics in India*. Princeton, NJ: Princeton University Press.

[8] Karl J Smith. 2015. *The Nature of Mathematics*. 13th edition. Boston, MA: Cengage Learning.

[9] Glen Van Brummelen. 2009. *The Mathematics of the Heavens and the Earth: The Early History of Trigonometry*. Princeton, NJ: Princeton University Press.

[10] W D Wallis. 2015. *Mathematics in the Real World*. Basel: Birkhäuser.

Website

[W1] J J O'Connor and E F Robertson. *MacTutor History of Mathematics Archive*. http://turnbull.mcs.st-and.ac.uk/history/ (accessed 27 January 2017).

Image Credits

[C1] Albert1ls/Wikimedia Commons. CC-BY-SA-3.0 license. `https://commons.wikimedia.org/wiki/File:IshangoColumnA.png; https://upload.wikimedia.org/wikipedia/commons/0/06/IshangoColumnB.png; https://upload.wikimedia.org/wikipedia/commons/d/d1/IshangoColumnC.png` (accessed 27 January 2017).

[C2] Ludwig Borchardt, 1920. Bakha~commonswiki.Wikimedia Commons. Public domain. `https://commons.wikimedia.org/wiki/File:Clepsydres-sections-borchardt.jpg` (accessed 27 January 2017).

[C3] Photo by Amber Habib.

[C4] Ivan Pozdeev. Wikimedia Commons. Public domain. `https://commons.wikimedia.org/wiki/File:Image-Koenigsberg,_Map_by_Merian-Erben_1652.jpg` (accessed 27 January 2017). The original map was published by von Merian Erben in 1652.

[C5] Ushakaron. Wikimedia Commons. Public domain. `https://commons.wikimedia.org/wiki/File%3AThe_Seven_Bridges_of_K%C3%B6nigsberg%2C_Fig._1.png` (accessed 27 January 2017).

2

CHAPTER

Sets

Forget everything you know about numbers.

In fact, forget you even know what a number is.

This is where mathematics starts.

Instead of math with numbers, we will now think about math with 'things'.

www.mathisfun.com

Mathematics is an ancient subject—human beings were concerned with counting long before they took up other intellectual pursuits such as writing, recording history or developing theories about the nature of our universe. In this long history, there have been many periods when mathematics has gone through intense and sustained development. The best-known one was Ancient Greece, highlighted by the geometry of Euclid that we study in school. India had its own, with its most spectacular achievements occurring in medieval times and ranging from geometry, number theory and algebra to the beginnings of calculus. The Arab world absorbed, combined and extended many of these ideas from both India and Greece, and laid the foundation for a fresh flowering in Renaissance Europe. Other parts of the world—the Middle East, China, Africa, the Americas—have their achievements as well.

In spite of the tremendous achievements from long ago, *our* age—starting from about 200 years ago—can be considered *the* Golden Age of Mathematics. For this, Golden Age is not restricted to one part of the world, or one section of society. During it, not only has mathematics greatly expanded its own frontiers, but also it has contributed fundamentally to expansions in other disciplines. From being considered the language of science, it is on the way to becoming the language of knowledge.

What has made this possible? A key factor is the achievement of a manner of expression which has almost entirely removed ambiguity. This has led to an ability to describe and classify relations in ways that are both clearer and more subtle than was possible earlier. In this chapter, we take up the language which makes this possible. It is called **set theory.**

Set theory, as we know it today, was founded by the German mathematician **Georg Cantor**[1] at the end of the nineteenth century. Cantor created his theory while attempting to clarify certain issues in the study of trigonometric series, which are useful in many areas of science.

In this chapter, we will take up the basic elements of set theory, such as various ways in which to describe sets, the laws governing their combinations, and their use in classifying and understanding complex situations.

2.1 Describing a Set

What is a set? If we look up the meaning of *set* in a dictionary, we may find a definition that resembles the following[2]:

> **Set**—a group or collection of things belonging or used together or resembling one another:
>
> (a) A group of people with common interests or occupations
> (b) (In tennis and other games) a group of games counting as a unit towards a match
> (c) (In popular music) a sequence of songs or pieces performed together and constituting or forming part of a live show or recording
> (d) (Mathematics and logic) a collection of distinct entities regarded as a unit, being either individually specified or (more usually) satisfying specified conditions

Thus, in common usage, a set seems to be a collection of objects that have some connection with each other. We will be particularly concerned with the definition given in part (d) under 'Mathematics and logic'. But first let us consider examples of sets arising from the common usage.

(i) The fingers on your left hand
(ii) The residents of your apartment building
(iii) The elected members of the Indian Parliament
(iv) The prescribed textbooks for your course of study

Each of these examples consists of objects which are clearly related to each other. But this need not be the only way of creating a collection. Imagine, if you will, a rose, a mango, the city of Delhi and the colour red. This too is a collection, even though there is no particular connection between its members, other than your imagination placing them together.

Let us pause and take stock of our encounter with sets. We have come across two types of examples of sets: one where the objects in the collection have a prior relation or common property and the other where they do not. But we still have not come to the point—we have not given the mathematical definition of 'set' and how its use differs from our everyday use of the word 'collection'.

[1] You can find out about the life and work of Georg Cantor in [W3].
[2] Modified from the *Concise Oxford English Dictionary*.

Most books on sets, at this level, give the following definition:

Definition 2.1.1 A **set** is a **well-defined collection** of objects.

This is, in fact, a good working definition as long as we understand what 'well-defined' means here. The example given below will provide a start in this direction.

Example 2.1.1 As part of a class project, the teacher asked the students of her class to list the best five books.

Mary's list consisted of *2001: A Space Odyssey, Rendezvous with Rama, A Meeting with Medusa, The Fountains of Paradise and The Songs of Distant Earth.*

Amar's list consisted of *The Story of My Life, Dreams from My Father, Wings of Fire, The Diary of a Young Girl and My Experiments with Truth.*

Thus, Mary's understanding of 'the collection of the best five books' was not the same as that of Amar. There is an ambiguity in the description of this collection. □

This example allows us to elaborate on the concept of 'well-defined'.

When we say that a set is **well defined** or a set is a **well-defined collection**, we mean the following: given any object, we should be able to decide, without any doubt, whether it belongs to the given set or not. The decision should not depend on our personal taste or prejudice.

As discussed in Example 2.1.1, it is not possible to decide, without subjectivity, whether a book such as *Dreams from My Father* belongs to the collection of the best five books or not. Hence, the phrase *the collection of the best five books* does not define a set. The problem arises from the fact that the word 'best' is subjective and can have different meanings for different people. The next example again illustrates the notion of 'well-defined'.

Example 2.1.2 *The collection of rich people residing in New Delhi* does not define a set. *'Rich'* can have different connotations for different people. On the other hand, *the collection of all individuals in New Delhi whose income tax return for 2012–13 showed a gross income over ₹1 crore* does describe a set. □

The reader should not assume that the only way in which 'well-defined' fails is due to the use of a subjective term in the definition. There are other situations too, but these are beyond the scope of this book.[3]

We will now take up two convenient methods for describing particular sets.

Roster Method

The objects belonging to a set are called its **elements** or **members.** The simplest way of describing a set is to list its elements inside curly brackets { }.

---

[3] If you are intrigued by this statement and wish to explore this topic further, find out about Russell's paradox. For example, see Section 1.10 of Hammack [4].

Example 2.1.3 Suppose that the only varieties of fruits that Arjun likes are apple, grape, orange, banana and guava. Then the set of the fruits that Arjun likes can be written as

$$\{\text{apple, grape, orange, banana, guava}\}$$

Each of these five varieties of fruits constitutes an element of this set. □

Sets are usually named using capital letters. For example, the set in the above example can be given the name A by writing

$$A = \{\text{apple}, \text{grape}, \text{orange}, \text{banana}, \text{guava}\}$$

This technique of describing a set by just explicitly writing out all its elements is called the **Roster Method.** It is a convenient way of depicting sets with a small number of elements.

It is important to note that the order in which we list the elements within the curly brackets does not matter: the only issue is whether an object is an element of the set or not. There is no fixing of 'first element', 'next element', etc. And so we could just as well write

$$A = \{\text{apple}, \text{banana}, \text{grape}, \text{guava}, \text{orange}\}$$

for the set named above.

When a set has a large number of elements, it is no longer practical to write out every element. For example, consider the set M consisting of the first one thousand natural numbers. Instead of writing each element, we just write

$$M = \{1, 2, 3, \ldots, 1000\}$$

The sequence of three dots or ellipsis ' ... ' in the middle tells us that the pattern set by the first few elements is to be continued until the last element is reached. The order stays irrelevant and so we could also write

$$M = \{1000, 999, 998, \ldots, 1\}$$

This notation can be extended to cover cases when the list of elements of a set does not terminate. Thus, the set of all natural numbers could be written as

$$\mathbb{N} = \{1, 2, 3, \ldots\}$$

The lack of an element written after the ellipsis conveys the instruction that the pattern is to be continued forever, without any end.

The use of ellipsis is permissible when the elements of a set follow a clear pattern that can be indicated without ambiguity by giving a few of them. For example, consider the following set:

$$S = \{1, 4, 9, 16, \ldots\}$$

It is easily apparent that the listed numbers are the squares of the first few natural numbers: $1 = 1^2$, $4 = 2^2$, $9 = 3^2$ and $16 = 4^2$. Thus, S consists of all the squares of the natural numbers.

Set-builder Notation

Consider the set C whose members are 1, 2, 3, 4 and 5. In the Roster method, we would write

$$C = \{1, 2, 3, 4, 5\}$$

Another way of representing the set C is by writing down the property that is common to its elements (and to no others):

$$C = \{x \mid x \text{ is a natural number, } x < 6\}$$

This is read as 'C is the set of all x such that x is a natural number and is less than 6'. This manner of describing a set according to a common property or rule is called **set-builder notation.** The key part of this notation is the symbol '|' which stands for 'such that'. We may use : instead of |, so that the set C may also be written as

$$C = \{x : x \text{ is a natural number, } x < 6\}$$

The most important sets in mathematics are the ones consisting of various kinds of numbers. We list the most commonly encountered ones below, along with the standard symbols that represent them:

- The set of **natural numbers** is
$$\mathbb{N} = \{1, 2, 3, 4, \dots\}$$
- The set of **whole numbers** is
$$\mathbb{W} = \{0, 1, 2, 3, 4, \dots\}$$
- The set of **integers** is
$$\mathbb{Z} = \{\dots, -3, -2, -1, 0, 1, 2, 3, \dots\}$$
- The set of **rational numbers** is
$$\mathbb{Q} = \left\{ x \;\middle|\; x = \frac{p}{q}, \text{ where } p \text{ and } q \text{ are integers and } q \neq 0 \right\}$$

We were able to list the elements of the sets $\mathbb{N}$, $\mathbb{W}$, $\mathbb{Z}$ and to express them by the Roster method. (Note the double use of ellipses in representing $\mathbb{Z}$ to indicate the continuation in both the positive and negative directions.) The set of rational numbers $\mathbb{Q}$ is more complicated and so, rather than trying to list its elements, we preferred to use the set-builder notation and just put down their defining property.

In middle school or even earlier, one is introduced to the concept of a number line. Each point on the number line represents a 'number':

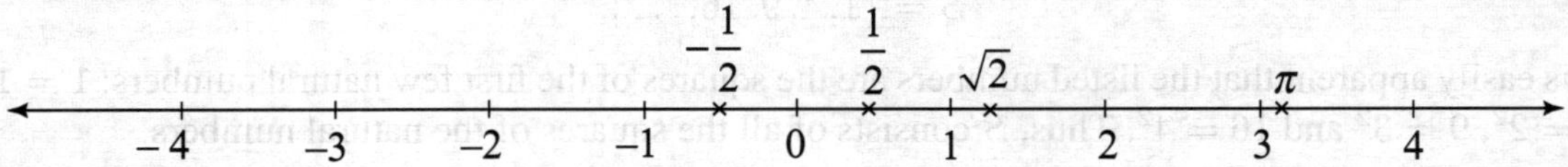

- The set of **real numbers** is
$$\mathbb{R} = \{x \mid x \text{ is a number on the number line}\}$$

We notice that all the numbers mentioned earlier are present on the number line. The integers are marked at equal intervals on both sides of zero. Between the integers are distributed the remaining rational numbers, like $1/2$ and $-1/2$. But there are also new numbers, such as $\sqrt{2}$ (the square root of 2) and π (the ratio of a circle's circumference to its diameter). These, it turns out, cannot be expressed as ratios of integers and hence are not rational numbers. Real numbers which are not rational numbers are called **irrational numbers.**

We will now look at a few examples of sets described by the Roster method as well as set-builder notation.

Example 2.1.4 Let D be the set of real numbers which satisfy the equation $x^2 = x$. In set-builder notation, we have:

$$D = \{x \mid x \text{ is a real number, and } x^2 = x\}$$

This describes the set formally but doesn't really give us a handle on what the set looks like. To do that we have to actually find all the solutions of the equation $x^2 = x$. To do so, we first rearrange it:

$$x^2 - x = 0$$

Then we factor the left-hand side:

$$x(x-1) = 0$$

The product on the left can only be zero if either of the two factors is zero, that is, if either $x = 0$ or $x - 1 = 0$. So the only elements of D are 0 and 1, and it can be expressed using the Roster method as

$$D = \{0, 1\}$$ □

Example 2.1.5 Let E be the set of the letters in the word MISSISSIPPI. We can write this set in two ways:

$$\begin{aligned} E &= \{x \mid x \text{ is one of the letters M, I, S, P}\} \\ &= \{\text{M, I, S, P}\} \end{aligned}$$

Although the alphabets I, S and P are repeated several times in the word, we list them only once when we write the set in the Roster method. □

We have seen in dealing with a set that the main thing is identifying which object is an element and which is not. So it is useful to set up some notation for this.

- If an object p is an element of a set A, we write $p \in A$ and read this as 'p belongs to A' or 'p is in A'.
- If p is not an element of A, we write $p \notin A$ and read this as 'p does not belong to A' or 'p is not in A'.

Example 2.1.6 Consider the sets $\mathbb{N}$ and $\mathbb{W}$ of the natural and whole numbers, respectively. The number 0 belongs to $\mathbb{W}$ but not to $\mathbb{N}$. So we write

$$0 \in \mathbb{W} \quad \text{but} \quad 0 \notin \mathbb{N}$$

Similarly, we have the following statements:

$$-1 \in \mathbb{Z} \text{ but } -1 \notin \mathbb{N}$$

$$\frac{1}{2} \in \mathbb{Q} \text{ but } \frac{1}{2} \notin \mathbb{Z}$$

$$\sqrt{2} \in \mathbb{R} \text{ but } \sqrt{2} \notin \mathbb{Q}$$

□

A set is an object in its own right and can become an element of another set.

Example 2.1.7 Consider the set $F = \{1, \{2,3\}\}$. This set has two elements: the number 1 and the set $\{2,3\}$. Note that while the number 2 is a member of $\{2,3\}$ which in turn is a member of F, we do not consider 2 itself to be a member of F.

A 'real-life' analogue is as follows. The authors of this book are citizens (members) of India which is a member of the United Nations. This does not give the authors membership in the United Nations for that can only be held by countries. □

This example also shows the importance of distinguishing between an object, say 2, and the set whose only member is that object. They are not the same. For instance, $\{2\}$ is a member of $\{1, \{2\}\}$ but 2 is not.

Equality of Sets

Two sets are said to be **equal** if they have exactly the same elements. In other words, sets A and B are said to be equal if every element of A is an element of B and every element of B is an element of A. We denote this by $A = B$.

Example 2.1.8 Let $A = \{a,b,c\}$ and $B = \{c,a,b\}$. Then, $A = B$ as the two sets have the same elements, just written in a different order. □

It follows from our definition of equality that two sets A and B are not equal, denoted by $A \neq B$, if there is any object which is an element of the one but not of the other.

Example 2.1.9 Consider the following sets:

$$A = \{1,2,3,4,6,12\}$$

$$B = \{x \mid x \in \mathbb{N} \text{ and } x \text{ is a factor of } 12\}$$

$$C = \{x \mid x \in \mathbb{Z} \text{ and } x \text{ is a factor of } 12\}$$

Then, $A = B$ since A is indeed just the list of all natural numbers which are the factors of 12. But C is not equal to these sets, since it also contains numbers such as -2 and -6. □

Example 2.1.10 Let $A = \{a,b\}$ and $B = \{a,\{b\}\}$. Then, we see that the element b belongs to A but not to B. So $A \neq B$. □

Quick Review

After reading this section, you should be able to:

- Decide whether a collection has been defined clearly enough to be called a set.
- Write a set formally using either the Roster method or set-builder notation.
- Identify and distinguish between the various number systems: $\mathbb{N}$, $\mathbb{W}$, $\mathbb{Z}$, $\mathbb{Q}$ and $\mathbb{R}$.
- Express symbolically the relationship of an object with a set.
- Decide if two given sets are equal.

EXERCISE 2.1

1. Determine which of the following defines a set.
   (a) The collection of intelligent students in your class.
   (b) The collection of natural numbers divisible by 7.
   (c) The collection of 10 most populated cities in the world.
   (d) The collection of rivers in India.
   (e) The collection of top 20 Hollywood movies.
2. Express the following sets in set-builder notation.
   (a) $A = \{2, 3, 5, 7, 11, 13, \ldots\}$
   (b) $B = \{$Red, Orange, Yellow, Green, Blue, Indigo, Violet$\}$
   (c) $C = \{-4, -3, -2, -1, 0, 1, 2, 3, 4\}$
   (d) $D = \{$Babur, Humayun, Akbar, Jahangir, Shah Jahan, Aurangzeb$\}$
   (e) $E = \{\ldots, -9, -6, -3, 0, 3, 6, 9, \ldots\}$
   (f) $F = \{$a, e, i, o, u$\}$
3. Express the following sets using the Roster method.
   (a) $G = \{x \mid x$ is a letter of the word MATHEMATICS$\}$
   (b) $H = \{x \mid x$ is a solution of the equation $x^2 = 2x - 1\}$
   (c) $I = \{x \mid x$ is a woman and President of India$\}$
   (d) $J = \{x \mid x$ is a prime factor of $510510\}$
   (e) $K = \{x \mid x = n^2 + 1, n \in \mathbb{N}\}$
4. With reference to the sets A to K listed in the previous two questions, state whether the following statements are true or false.
   (a) $21 \in A$
   (b) $5 \notin K$
   (c) $4 \in C$
   (d) $7 \in J$
   (e) $-7 \notin J$
   (f) $-12 \notin E$
   (g) $\{T, H\} \in G$
   (h) $-1 \in H$
   (i) $101 \in J$
   (j) $21 \notin A$
   (k) $65 \in K$
   (l) $10 \in J$
5. Do the following pairs consist of equal sets? Give reasons.
   (a) $A = \{x \mid x$ is a letter in the word FOLLOW$\}$
   $B = \{x \mid x$ is a letter in the word WOLF$\}$

(b) $A = \{x \in \mathbb{R} \mid x^2 + 5x + 6 = 0\}$
$B = \{-2, 3\}$
(c) $A = \{x \mid x$ is a colour in the flag of USA$\}$
$B = \{x \mid x$ is a colour in the flag of Great Britain$\}$
(d) $A = \{1, 2, 3, 5\}$
$B = \{x \mid x$ is a prime number less than $7\}$

2.2 A Variety of Sets

In the previous section, we learnt a few things about sets: from a working definition to how we can write or describe sets and their elements. In this section, we concern ourselves with furthering our knowledge of sets. Are there types of sets that are important enough to have a special name? Can we measure the size of a set? What is the relation of a set to its parts?

Sets of Different Sizes

The two fundamental numbers are 1 and 0. One is the basic building block for all counting. We can do without zero for simple counting, but introducing it to the number system adds tremendous flexibility. Indeed, the invention of this number is considered one of the highlights of the history of mathematics in India. It turns out that sets with 0 or 1 element have similar roles in set theory.

If a set has no elements, then it is called an **empty set** or a **null set.** In the Roster method, we can write such a set as $\{\ \}$. More commonly, an empty set is denoted by the symbol $\emptyset$.[4]

An empty set is created when we consider a property (or combination of properties) that is not satisfied by any object.

Example 2.2.1 Let A be the set of women elected as the President of the United States of America prior to 2016. There were none! So this set is empty, and we write $A = \emptyset$. □

Example 2.2.2 $\{x \mid x \in \mathbb{N},\ x \leq 0\} = \emptyset$. This is because there is no natural number which is less than or equal to 0. □

Task 2.2.1 Is $\{0\}$ an empty set?

After the empty set, we have sets with just one element. Such a set is called a **singleton.**

Example 2.2.3 Consider the set of women who have served as the Prime Minister of India prior to 2016. This set is a singleton as it contains only one element—Indira Gandhi. □

We shall soon take up the ways in which sets can be combined with each other to create new sets. At that stage, we shall see that singletons are the basic building blocks for all sets. Meanwhile, we note that we can have sets whose size is any whole number of our choice. For example,

[4] The symbol $\emptyset$ should not be confused with the Greek letter phi (ϕ).

$$
\begin{aligned}
\emptyset &\text{ has 0 elements}\\
\{1\} &\text{ has 1 element}\\
\{1,2\} &\text{ has 2 elements}\\
\{1,2,3\} &\text{ has 3 elements}\\
&\cdots\cdots\\
\{1,2,\ldots,k\} &\text{ has } k \text{ elements}
\end{aligned}
$$

The number of elements in a set A is called its **cardinality** and is denoted by $|A|$. If the cardinality of a set is some whole number k, the set is called a **finite set;** otherwise, it is called an **infinite set.** Examples of infinite sets are $\mathbb{N}$, $\mathbb{W}$, $\mathbb{Q}$ and $\mathbb{R}$—if we count the members of these sets, we never reach a last element and the counting continues forever. One of Georg Cantor's great achievements was to show that there is a hierarchy of sizes among infinite sets. In this hierarchy, $\mathbb{N}$, $\mathbb{W}$, $\mathbb{Z}$ and $\mathbb{Q}$ have the same size, whereas $\mathbb{R}$ is larger.[5]

Example 2.2.4 This example shows how powerful the empty set is. We can use it to construct finite sets of any size as follows:

$$
\begin{aligned}
A_0 &= \emptyset\\
A_1 &= \{\emptyset\} \;=\; \{A_0\}\\
A_2 &= \{\emptyset,\{\emptyset\}\} \;=\; \{A_0, A_1\}\\
A_3 &= \{\emptyset,\{\emptyset\},\{\emptyset,\{\emptyset\}\}\} \;=\; \{A_0, A_1, A_2\}\\
&\vdots\\
A_k &= \{A_0, A_1, \ldots, A_{k-1}\}\\
&\vdots
\end{aligned}
$$

We can even collect all of these to create an infinite set: $\{A_0, A_1, A_2, \ldots\}$! □

If two sets have the same cardinality, they are said to be **equivalent.** It is obvious that if two sets are equal, then they will have the same cardinality. On the other hand, two equivalent sets need not be equal.

Example 2.2.5 Consider the following sets:

$$
\begin{aligned}
A &= \{\text{red, yellow, blue, green}\}\\
B &= \{\text{red, yellow, blue, orange}\}
\end{aligned}
$$

Then, A and B are equivalent since $|A| = |B| = 4$. On the other hand, $A \neq B$ since 'green' is in A but not in B. □

[5] It is beyond the scope of this chapter to elaborate on this any further. You can refer to Chapter 13 of Hammack [4].

Subsets

A set A is called a **subset** of a set B if every element of A is an element of B. This situation is represented by the notation $A \subseteq B$. If A is not a subset of B, we write $A \nsubseteq B$. If there is even *one* element in A that is not in B, then $A \nsubseteq B$.

Example 2.2.6 Consider the sets

$$A = \{\text{index finger, middle finger, ring finger}\}$$
$$B = \{\text{thumb, index finger, middle finger, ring finger, little finger}\}$$

We go through the elements of A one by one and check that each is in B:

A = {**index finger**, *middle finger*, ring finger}
B = {thumb, **index finger**, *middle finger*, ring finger, little finger}

This shows that $A \subseteq B$. □

Task 2.2.2 Consider the set $C = \{1, \{2\}\}$. Which of the following statements are correct?

1. $1 \in C$
2. $1 \subseteq C$
3. $\{1\} \in C$
4. $\{1\} \subseteq C$
5. $2 \in C$
6. $\{2\} \in C$
7. $\{2\} \subseteq C$
8. $\{\{2\}\} \subseteq C$

The following are two extreme cases:

- Every set A is a subset of itself: $A \subseteq A$.
- The empty set is a subset of every set A: $\emptyset \subseteq A$.

We visualise a subset as being obtained by removing certain elements from the original set A. If the number of elements removed is 0, we obtain A itself as the remaining subset. If we remove all elements, we obtain the empty set.

Since a subset is either the original set or is created by removing elements, it is clear that its size cannot increase: If $A \subseteq B$, then $|A| \leq |B|$.

The notion of a subset is also useful in figuring out when two sets A and B are equal:

Let A and B be sets. Then, $A = B$ if and only if $A \subseteq B$ and $B \subseteq A$.

This simple statement is surprisingly useful for the reason that it breaks a proof of equality into two parts. We can compare it to the analogous statement for numbers: If a and b are two real numbers, then $a = b$ if and only if $a \leq b$ and $b \leq a$.

The set-builder notation has a convenient modification for subsets. Suppose we start with a set A and then form a subset B from those elements which have an extra property denoted by P. Then we write

$$B = \{x \in A \mid x \text{ has property } P\}$$

and read this as 'B is the set of those elements x of A which also have the property P'.

Example 2.2.7 In the set-builder notation, we can express the natural and whole numbers as subsets of the set of integers $\mathbb{Z}$ as follows:

$$\mathbb{N} = \{x \in \mathbb{Z} \mid x \geq 1\}$$
$$\mathbb{W} = \{x \in \mathbb{Z} \mid x \geq 0\}$$

And the set of irrational numbers can be expressed as a subset of real numbers:

$$\{x \in \mathbb{R} \mid x \notin \mathbb{Q}\}$$ □

Power Set

The set whose elements are all the subsets of a set A is called the **power set** of A. We shall represent it by $\mathcal{P}(A)$.

Let us see if we can describe the power set for various sets. We begin with the smallest set: $\emptyset$. Obviously, the only possible subset of $\emptyset$ is itself. So we have

$$\mathcal{P}(\emptyset) = \{\emptyset\}$$

Now let us consider a singleton set, say $A = \{1\}$. If we retain 1, we obtain the subset A. If we drop 1, we obtain the subset $\emptyset$. So

$$\mathcal{P}(\{1\}) = \{\emptyset, \{1\}\}$$

If we use similar arguments for a set $B = \{1, 2\}$ consisting of exactly two elements, we find that there is one subset with no element, namely $\emptyset$. There are exactly two subsets that are singletons, namely, $\{1\}$ and $\{2\}$. Finally, B itself is the unique subset with two elements. Thus,

$$\mathcal{P}(\{1,2\}) = \{\emptyset, \{1\}, \{2\}, \{1,2\}\}$$

What happens if we carry out such calculations for bigger sets? Is there a pattern? The next example shows some further calculations:

Example 2.2.8 This example shows the relation between the number of elements in a set and the number of subsets it has.

A	$\lvert A\rvert$	Members of $\mathcal{P}(A)$	$\lvert\mathcal{P}(A)\rvert$
$\emptyset$	0	$\emptyset$	1
$\{1\}$	1	$\emptyset$ **{1}**	2
$\{1, 2\}$	2	$\emptyset$ {1}, **{2}** **{1, 2}**	4
$\{1, 2, 3\}$	3	$\emptyset$ {1}, {2}, **{3}** {1, 2}, **{1, 3}**, **{2, 3}** **{1, 2, 3}**	8
$\{1, 2, 3, 4\}$	4	$\emptyset$ {1}, {2}, {3}, **{4}** {1, 2}, {1, 3}, {2, 3}, **{1, 4}**, **{2, 4}**, **{3, 4}** {1, 2, 3}, **{1, 2, 4}**, **{1, 3, 4}**, **{2, 3, 4}** **{1, 2, 3, 4}**	16

Note: The bold values are the new additions at that stage.

When we add a new element to a finite set, what happens to its subsets? The old subsets are still there, and further ones are created by adding the new element to each of the old subsets. This doubles the total number of subsets. This observation explains the growth of the size of the power set as 1, 2, 4, 8, 16, ... □

The pattern illustrated above leads to the following general claim:

If A has k elements, then the number of subsets is 2^k.
Or: $|A| = k$ implies $|\mathcal{P}(A)| = 2^k$.

The next example illustrates a use of power sets.

Example 2.2.9 A power set can be used to list all the factors of a number. Suppose we wish to find all the factors of 210. The set of prime factors of 210 is

$$A = \{2, 3, 5, 7\}$$

Any factor of 210 is a product of some of these numbers. So the possible factors correspond to choices of subsets of A. The power set of A has 16 elements:

$$\mathcal{P}(A) = \{\emptyset, \{2\}, \{3\}, \{5\}, \{7\}, \{2,3\}, \{2,5\}, \{2,7\}, \{3,5\}, \{3,7\}, \{5,7\}, \{2,3,5\}, \{2,3,7\}, \{2,5,7\}, \{3,5,7\}, \{2,3,5,7\}\}$$

We can let $\emptyset$ represent the trivial factor 1; this corresponds to not picking any of the prime factors. Hence, the factors of 210 are 1, 2, 3, 5, 7, 6, 10, 14, 15, 21, 35, 30, 42, 70, 105 and 210. □

Task 2.2.3 Will the method of the previous example always work? (Hint: Try to use it to get all the factors of 8.)

If a set A is a subset of B but *is not equal to* B, then A is called a **proper subset** of B. This situation is denoted by $A \subsetneq B$.

In Example 2.2.6, $A \subsetneq B$.

Task 2.2.4 Let A be a set with k elements. Show that the number of proper subsets of A is $2^k - 1$.

Quick Review

After reading this section, you should be able to:

- Recognise sets of different sizes or cardinalities: empty set, singletons, finite sets, infinite sets.
- Determine if two finite sets are equivalent.
- Decide if a given set is a (proper) subset of another.
- List and count the (proper) subsets of a finite set.

EXERCISE 2.2

1. Which of the following sets are finite and which are infinite? If a set is finite, what is its cardinality?
   (a) The months in a year.
   (b) The lines parallel to the axis in the XY-plane.
   (c) The natural numbers divisible by 10.
   (d) The natural numbers dividing 10.
   (e) The letters of the word ARITHMETIC.
2. Identify the singleton sets, if any, among the sets listed in Exercise 2.1, Question 3.
3. Refer to the sets in Questions 2 and 3 of Exercise 2.1. Then state whether the following are True or False.
   (a) $J \subsetneq A$
   (b) $K = A$
   (c) $|C| = |J|$
   (d) B and D are equivalent sets.
4. Which of the following are empty sets?
   (a) A = the set of vowels in the word NYMPH
   (b) $B = \{x \in \mathbb{N} \mid x + 3 = 3\}$
   (c) $C = \{x \in \mathbb{R} \mid x^2 + 4 = 0\}$
   (d) D = the set of natural numbers which are factors of 7 and 5
   (e) $E = \{x \mid x \text{ is an even prime number greater than } 1000\}$
5. State whether $A \subsetneq$ (or $\subseteq$)B or $B \subsetneq$ (or $\subseteq$)A.
   (a) $A = \{x \mid x \text{ is an integer}\}$
   $B = \{x \mid x \text{ is an even number}\}$
   (b) $A = \{x \mid x \text{ is an equilateral triangle}\}$
   $B = \{x \mid x \text{ is an isosceles triangle}\}$
   (c) $A = \{x \mid x \text{ is a square}\}$
   $B = \{x \mid x \text{ is a rectangle}\}$
   (d) $A = \{x \mid x \text{ is a multiple of } 3\}$
   $B = \{x \mid x \text{ is a multiple of } 9\}$
   (e) $A = \{x \mid x \text{ is a factor of } 1024\}$
   $B = \{x \mid x \text{ is a factor of } 64\}$
6. List and count all the subsets of
   (a) $\{c, r, y\}$
   (b) The letters in the word *racer*

2.3 Set Operations

We have described different types of sets in the previous section. Given two sets, we have learnt the criteria for when two sets are equal, when one is a subset of the other and how to create a new set—the power set—out of an existing set. What are the other types of relations we might want to explore between sets? We can add, subtract, multiply and divide two numbers. Are there, similarly, operations that we can do with two sets?

The following are two obvious operations we can do with sets:

1. Given two sets, collect all the elements of each of them to form a new set.
2. Given two sets, create a new set by collecting all the elements which are common to both of them.

Union

The **union** of two sets A and B is the collection of all elements that are either in A or in B. The union of A and B is denoted by $A \cup B$.

$$A \cup B = \{x \mid x \in A \text{ or } x \in B\}$$

When we say 'all elements that are either in A or in B', we are using *or* in its inclusive sense: We are also including those elements which are in both A and B.

Example 2.3.1 Let A and B be the sets described as follows. The elements of A and B are the capitals of some countries.

$$A = \{\text{New Delhi, Islamabad, Prague}\}$$
$$B = \{\text{New Delhi, Islamabad, Kabul}\}$$

To create the union of these sets, we bring together the elements of A and B. You could imagine that we create a box and label it $A \cup B$. In this box, we first put all the elements of A: New Delhi, Islamabad, Prague. (It's a huge box!) We now start adding the elements of B. New Delhi and Islamabad are already there, so the only actual addition is that of Kabul. This gives us:

$$A \cup B = \{\text{ New Delhi, Islamabad, Prague, Kabul}\}$$

□

Now, let us consider an example involving several sets.

Example 2.3.2 Let U, A, B and C be the sets described as follows.

$$U = \{1,2,3,4,5\}, \quad A = \{1,3\}, \quad B = \{2,4\}, \quad C = \{2,3\}$$

Note that A, B and C are subsets of U. Further,

$$A \cup B = \{1,2,3,4\}$$
$$A \cup C = \{1,2,3\}$$
$$B \cup C = \{2,3,4\}$$
$$A \cup U = B \cup U = C \cup U = U$$

□

Now that we have some practice in calculating unions, we can consider some general patterns. Let X and Y be any two sets. Then:

1. $X \cup Y = Y \cup X$ (Either way, we pool all the contents of both X and Y.)
2. $X \subseteq X \cup Y$ (Since every element of X is put inside $X \cup Y$, it is a subset.)
3. $X \subseteq Y$ if and only if $X \cup Y = Y$ (If all elements of X are already in Y, they contribute nothing new to $X \cup Y$. Similarly, if X contributes nothing new to $X \cup Y$, it can only be because all its elements were already in Y.)
4. $X \cup X = X$ (If nothing enters from outside X, the union cannot grow.)
5. $\emptyset \cup X = X$ (The empty set makes no contribution to the union.)

You might have noted that the fourth and fifth properties follow from the third. (If you didn't, check it out now!) Another interesting property is:

$$(X \cup Y) \cup Z = X \cup (Y \cup Z)$$

This is certainly a reasonable looking identity. It says that when we combine the elements of three sets, the order in which we carry out the operations does not matter. We can first mix X and Y, and then add Z. Or we can first mix Y and Z, and then add X. Now, since the order in which we take the unions does not matter, we can drop the parentheses altogether and just write

$$X \cup Y \cup Z$$

for the act of pooling together all the elements of X, Y and Z.

Task 2.3.1 In Example 2.3.2, verify that $(A \cup B) \cup C = A \cup (B \cup C)$.

Let us sum up the basic properties of unions of sets.[6] Let A, B and C be the sets. Then,

1. $A \cup B = B \cup A$	4. $A \cup A = A$
2. $A, B \subseteq A \cup B$	5. $A \cup \emptyset = A$
3. $A \subseteq B$ if and only if $A \cup B = B$	6. $(A \cup B) \cup C = A \cup (B \cup C)$

Intersection

The **intersection** of two sets A and B is the collection of all elements that are in both sets. It is denoted by $A \cap B$:

$$A \cap B = \{x \mid x \in A \text{ and } x \in B\}$$

[6] Some of these properties are demonstrated using Venn diagrams in a later section.

Example 2.3.3 We consider again the sets A and B from Example 2.3.1:

$$A = \{\text{ New Delhi, Islamabad, Prague}\}$$

$$B = \{\text{ New Delhi, Islamabad, Kabul}\}$$

We look at each element of A and check if it is also in B. We find only two common elements. Therefore the intersection is

$$A \cap B = \{\text{ New Delhi, Islamabad}\}$$

□

Task 2.3.2 Let the sets U, A, B and C be defined as in Example 2.3.2. Then, describe the following sets:

1. $A \cap B$
2. $A \cap C$
3. $B \cap C$
4. $A \cap U$
5. $B \cap U$
6. $C \cap U$

As in the case of union, we can easily identify elementary properties of intersections. For example, $A \cap B = B \cap A$, since in each case we are finding the common elements of A and B. As another instance, we have $\emptyset \cap A = \emptyset$, since the empty set has no elements and therefore has nothing in common with another set.

The basic properties of intersections of sets are listed below.[7] You must compare them with the properties of unions listed earlier, and note both similarities and differences.

1. $A \cap B = B \cap A$	4. $A \cap A = A$
2. $A \cap B \subseteq A,\ B$	5. $A \cap \emptyset = \emptyset$
3. $A \subseteq B$ if and only if $A \cap B = A$	6. $(A \cap B) \cap C = A \cap (B \cap C)$

Task 2.3.3 Give justifications for the second, third and fourth identities in the above list.

The sixth identity, $(A \cap B) \cap C = A \cap (B \cap C)$, says that the order in which we take common elements does not matter. As we did for union, we can now drop the parentheses and just write

$$A \cap B \cap C$$

for the set consisting of those elements which are in A and in B and in C.

Two sets X and Y are called **disjoint** if they have no common element, that is, if $X \cap Y = \emptyset$.

Task 2.3.4 Consider the sets in Example 2.3.2. Are any of them disjoint with each other?

Complement

Let U be a set and A any subset of U. Then, the **complement** of the set A in U is the collection of all the elements of U that are not in A. The complement of A is usually denoted by A' or A^c or $\overline{A}$:

$$A' = \{x \in U \mid x \notin A\}$$

[7] Some of these properties will also be demonstrated using Venn diagrams in the next section.

The set U has not been mentioned in the symbolic representation of complement. This is because we usually use the complement in contexts where the larger set U is fixed and all complements are with respect to U alone.

Note that by definition, $A' \subseteq U$.

Task 2.3.5 Consider the sets U, A, B and C given in Example 2.3.2. Verify that the complements of A, B and C in U are $A' = \{2,4,5\}$, $B' = \{1,3,5\}$ and $C' = \{1,4,5\}$.

We list the basic results concerning complements below. Let X be a subset of U and X' denote the complement of X in U. Then,

1. $X' \subseteq U$	4. $X \cap X' = \emptyset$
2. $(X')' = X$	5. $U' = \emptyset$
3. $X \cup X' = U$	6. $\emptyset' = U$

So far we have considered each operation (union, intersection, complement) on its own. We also need to know how they are related to each other. As an analogy, the commutative and associative laws in arithmetic ($a+b=b+a$, $a+(b+c)=(a+b)+c$) only deal with one operation at a time, but the relationship between addition and multiplication is captured by the distributive law $a \cdot (b+c) = a \cdot b + a \cdot c$. In Set Theory, the relationship between union and intersection involves several laws as follows:

1. $X \cup (X \cap Y) = X$	3. $X \cup (Y \cap Z) = (X \cup Y) \cap (X \cup Z)$
2. $X \cap (X \cup Y) = X$	4. $X \cap (Y \cup Z) = (X \cap Y) \cup (X \cap Z)$

There are also a special set of rules that govern the complements of intersections and unions. These rules are called **De Morgan's laws** and they are listed below:

1. $(X \cup Y)' = X' \cap Y'$	2. $(X \cap Y)' = X' \cup Y'$

Task 2.3.6 Verify that the sets A and B in Example 2.3.2 satisfy De Morgan's laws:

$$(A \cup B)' = A' \cap B'$$
$$(A \cap B)' = A' \cup B'$$

Quick Review

After reading this section, you should be able to:

- Carry out the operations of union, intersection and complementation of sets.
- State–and justify–the properties of union, intersection and complement.

EXERCISE 2.3

1. Let $A = \{1, 2, 3\}$, $B = \{3, 4, 5\}$, $C = \{4, 5, 6\}$ and $D = \{5, 6, 7\}$. Find

   (a) $A \cup B$ (c) $(B \cup C) \cup D$ (e) $(B \cap C) \cup (B \cap D)$
   (b) $A \cap D$ (d) $B \cup (B \cap D)$ (f) $B \cap (C \cup D)$

2. Let $A = \mathbb{N}$, $B = \{x \in \mathbb{N} \mid x = 2n, n \in \mathbb{N}\}$, $C = \{x \in \mathbb{N} \mid x = 2n - 1, n \in \mathbb{N}\}$ and $D = \{x \in \mathbb{N} \mid x \text{ is a prime number}\}$. Then describe the following combinations using both the Roster and set builder notations:

   (a) $A \cap B$ (c) $A \cap D$ (e) $B \cap D$
   (b) $A \cap C$ (d) $B \cap C$ (f) $C \cap D$

3. Let

$$U = \{x \mid x \text{ is a letter in the word CONSTANTINOPLE}\}$$
$$A = \{x \mid x \text{ is a letter in the word STATION}\}$$
$$B = \{x \mid x \text{ is a letter in the word INSTANT}\}$$
$$C = \{x \mid x \text{ is a letter in the word CONSTANT}\}$$
$$D = \{x \mid x \text{ is a letter in the word STAPLE}\}$$

   Then, find the following complements in U:

   (a) A' (d) $(B \cup C)'$
   (b) B' (e) $(A')'$ and verify that $(A')' = A$
   (c) $(A \cap B)'$ (f) $(A \cup B)' \cap C'$

2.4 Euler Diagrams

The number line helps us visualise the location of numbers. It is also effective in helping understand the operations of addition and subtraction. Similarly, it is possible to visualise sets in a manner which aids our understanding of relations between sets and also operations on sets.

In this section, we will learn how to make diagrams which help us understand the relationship between given sets, and the results of applying various operations to these sets. In the next section, we will see how these diagrams can be used to justify the various laws concerning the set operations.

We shall use closed geometrical figures such as rectangles, circles and triangles to represent sets. These figures will generally be enclosed in a rectangle. This rectangle represents a set which contains all those being discussed and is treated as the arena in which the action happens. The example below should help to make this abstract description concrete.

Example 2.4.1 Consider the sets from Example 2.3.2:

$$U = \{1, 2, 3, 4, 5\}, \quad A = \{1, 3\}, \quad B = \{2, 4\}, \quad C = \{2, 3\}$$

Here, the 'arena of action' is the set U, since A, B and C are all subsets of U. So we first create a rectangular box, label it U and fill it with the elements of U:

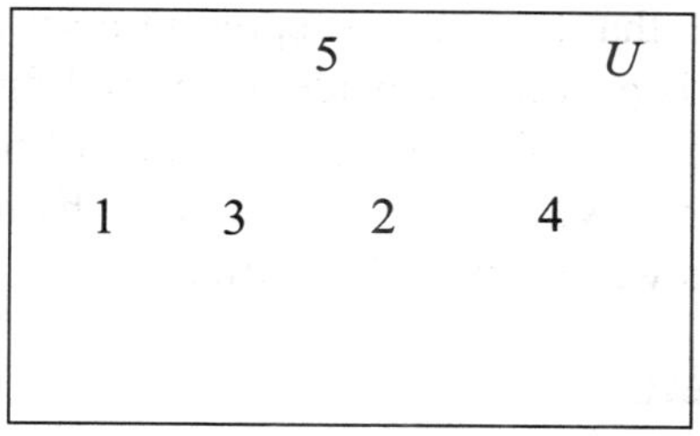

Next, we create figures inside the box U to represent the other sets. The first thing we do is to draw a circle which contains only 1 and 3—this represents A.

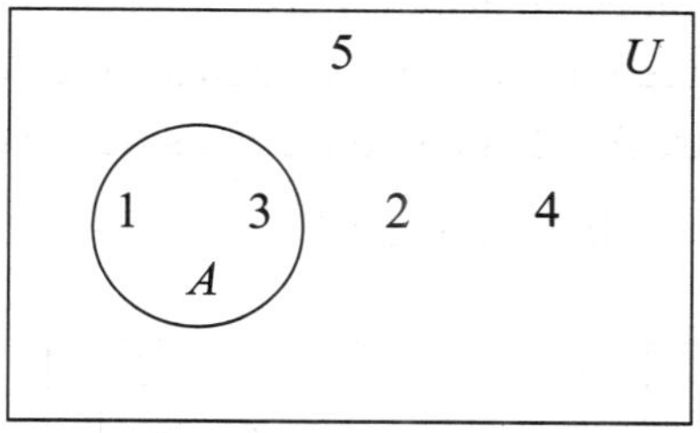

We show B by drawing a circle around 2 and 4. Since A and B have no common elements, we make sure that we draw the corresponding circles so that they do not touch or overlap.

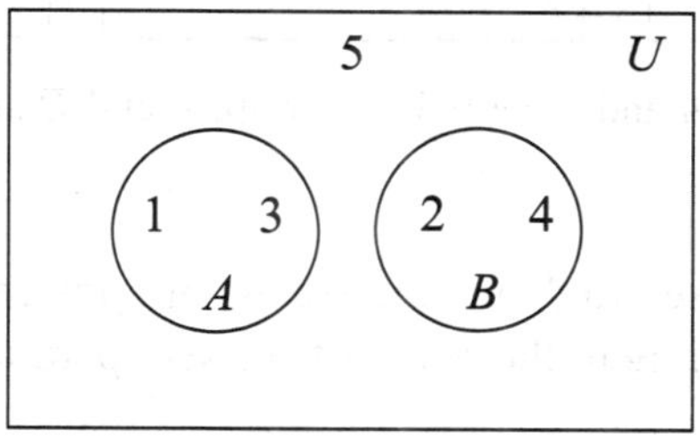

The figure is completed by drawing a circle around 2 and 3 to represent C.

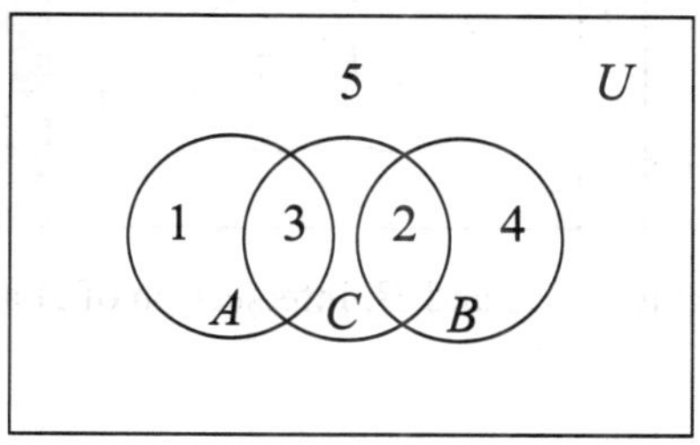

This diagram makes it easy to see the various facts and relationships. For example, we see that 5 is the only element of U which is not contained in any of the circles. In other words, $(A \cup B \cup C)' = \{5\}$, where the complement is with respect to U. Similarly, we see that no element is contained in all the circles. Thus, $A \cap B \cap C = \emptyset$. □

We see through this example, that the way to represent a set is to draw a closed figure like a circle. Every element of the set is positioned inside the closed figure. Everything *not* in the set is positioned outside the closed figure. This schematic method of representing collectives and their relations was made popular by the Swiss mathematician Leonhard Euler in the eighteenth century. These diagrams are therefore known as **Euler diagrams.**

Configurations of Two Sets

In general, if we have two subsets A and B of a set U, we can represent the relationship between them by drawing closed figures based on the following rules:

1. If A and B have some common elements, they are represented by overlapping figures.
2. If A and B are disjoint, they are represented by figures which do not touch or overlap.
3. If A is a subset of B, the figure for A is completely inside the one for B.

We have illustrated these variations in Figure 2.4.1.

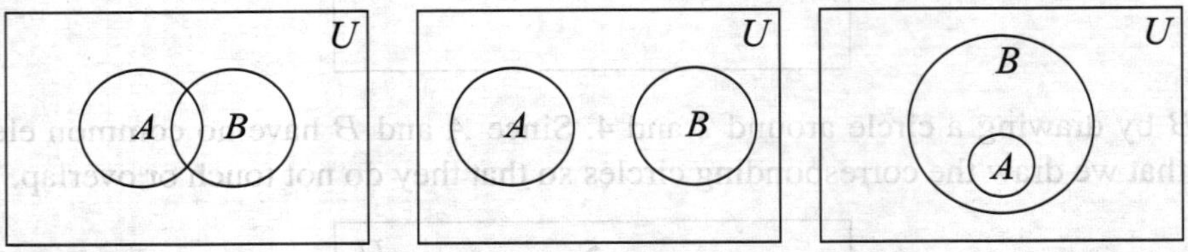

Figure 2.4.1 Left to right: A and B partially overlap, A and B are disjoint, A is a subset of B.

Euler diagrams can also be used to depict the results of applying the various set operations. We just shade the region corresponding to the result of the set operation (Figure 2.4.2).

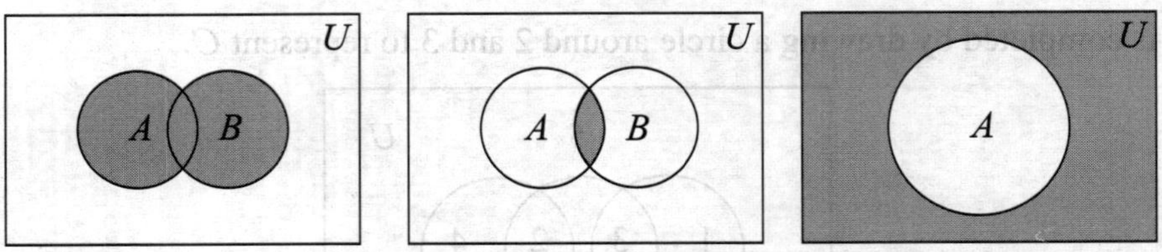

Figure 2.4.2 Left to right: union of A and B, intersection of A and B, complement of A in U.

Now we take up an example involving multiple operations. This example also shows how we can use Euler diagrams to depict abstract relationships between sets.

Example 2.4.2 Our task is to show the region represented by $A \cap B'$. The sets A and B' are represented in the two diagrams given below.

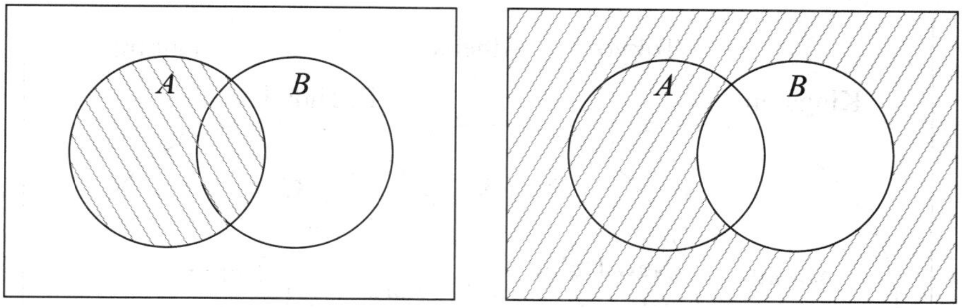

In order to represent $A \cap B'$, we need to find the 'common area' in the two diagrams that is shaded. So we overlay these two diagrams to obtain:

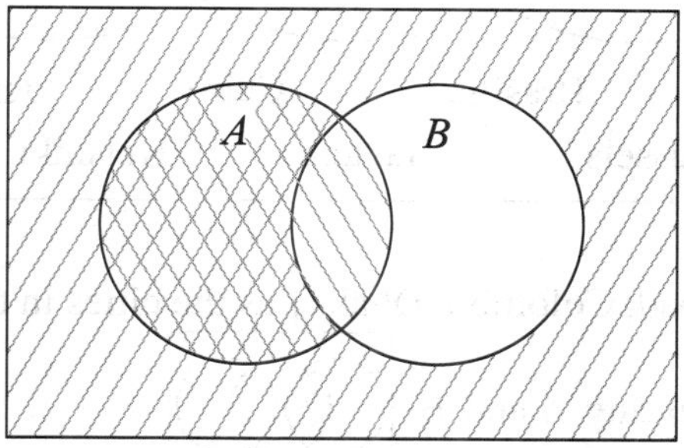

The portion shaded by checks represents the common area we were interested in. Thus, $A \cap B'$ is represented by the shaded area in the diagram below.

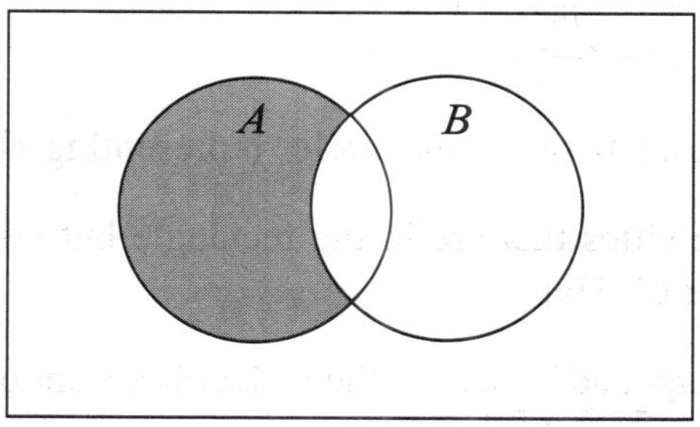

□

Configurations of Three Sets

Example 2.4.3 Consider the Euler diagram given below. Using the information given in the diagram, describe the sets $A \cap B$, $A \cap C$, $B \cap C$ and $(A \cup B \cup C)'$.

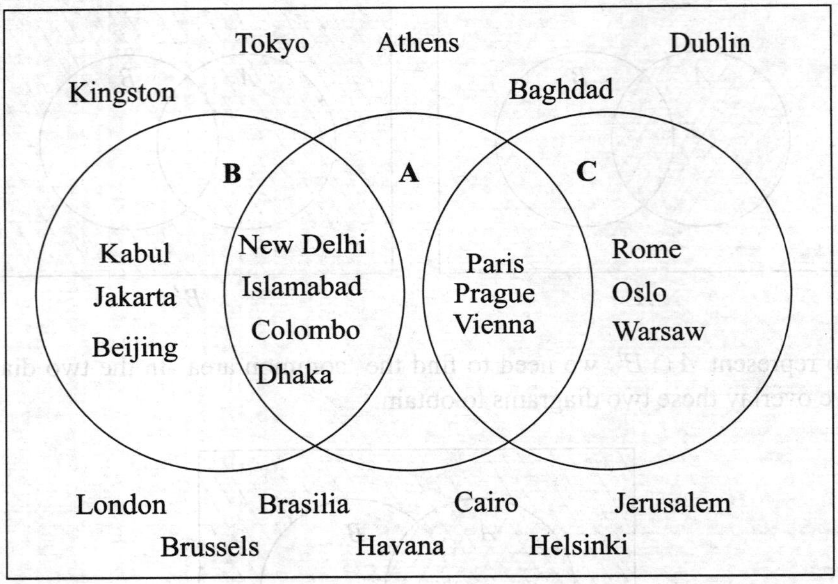

$A \cap B$ ={New Delhi, Islamabad, Colombo, Dhaka} as the cities in this set are the ones in the area common to the circles representing A and B. 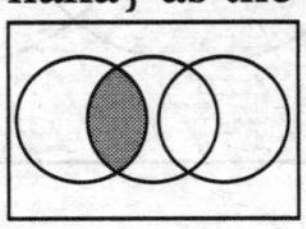

$A \cap C$ ={Prague, Paris, Vienna} as the cities in this set are the ones in the area common to the circles representing A and C. 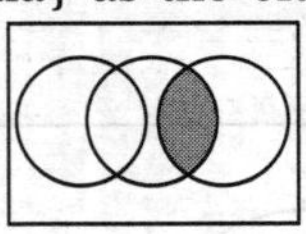

$B \cap C = \emptyset$ as there is no overlap between the circles representing B and C.

$(A \cup B \cup C)'$ consists of the cities that are in the rectangle but not in the areas covered by the circles representing A or B or C. Thus,

$(A \cup B \cup C)'$= {Kingston, Baghdad, Tokyo, Athens, Dublin, London, Brussels, Brasilia, Havana, Cairo, Helsinki, Jerusalem} 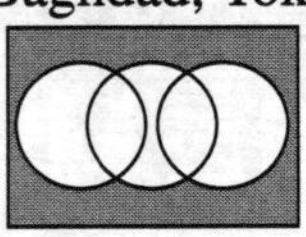

□

We have just worked through an example of how to read an Euler diagram and extract information from it. Now we shall see how to create an appropriate Euler diagram when the sets are given. We begin with the general configuration of three overlapping subsets, which is represented by the following Euler diagram:

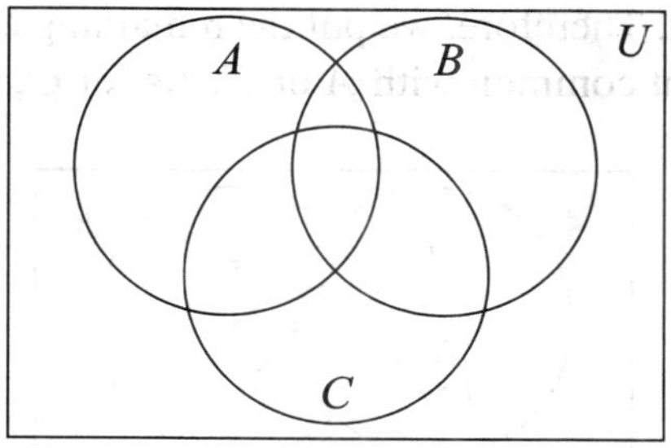

Example 2.4.4 Consider the following sets:

$$U = \{n \in \mathbb{N} : n \leq 9\}$$
$$A = \{1, 3, 5, 7, 9\}$$
$$B = \{1, 2, 3, 4, 6, 8\}$$
$$C = \{2, 3, 5, 7\}$$

We first calculate $A \cap B \cap C = (A \cap B) \cap C = \{3\}$. Therefore, we write 3 in the region common to all of A, B and C.

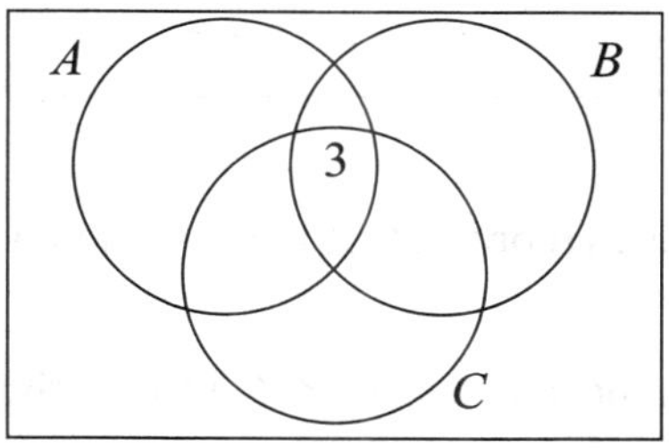

The next set of calculations is $A \cap B = \{1, 3\}$, $A \cap C = \{3, 5, 7\}$ and $B \cap C = \{2, 3\}$. Since the element 3 is already written down, we write the element 1 in the region common to A and B, excluding the region of the set C. Similarly, we write the elements 5 and 7 in the region common to A and C, excluding the region of the set B and the element 2 in the region common to B and C, excluding the region of the set A.

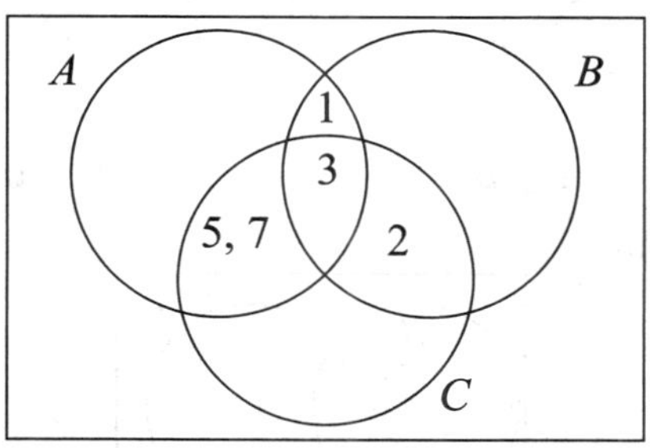

For the set A, the elements $1, 3, 5$ and 7 are already listed out. We put the remaining element 9 in the region depicts the part of A not common with B or C. Similarly, for the set B, the elements

1, 2 and 3 are already listed out. Therefore, we put the remaining elements 4, 6 and 8 in the region that represents the part of B not common with A or C. As for C, we find that all its elements are already accounted for.

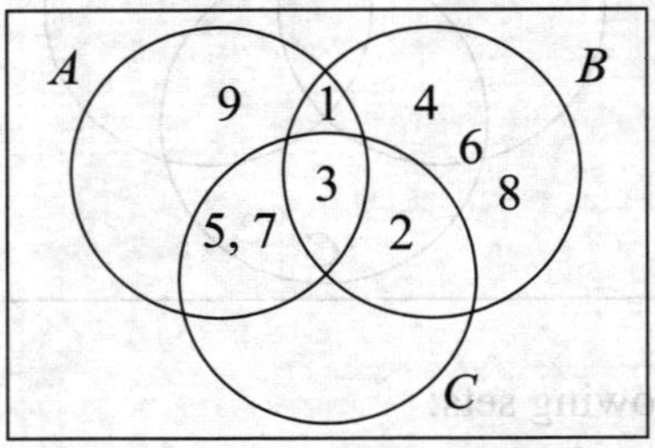

In this diagram, the part of C which lies outside A or B is empty, which means that we have the inclusion $C \subseteq A \cup B$. We can redraw the Euler diagram to show this more directly:

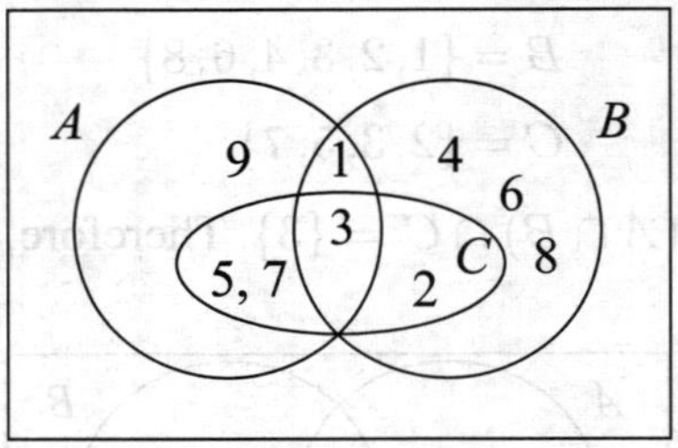

□

We can also use Euler diagrams to represent situations when we have partial knowledge of some sets.

Example 2.4.5 We are given that the subsets A, B, C of U satisfy the following relationships:

$$A \subsetneq B \quad \text{and} \quad A \cap C = \emptyset$$

Our task is to represent these in a single Euler diagram. We start by making a diagram for each relation:

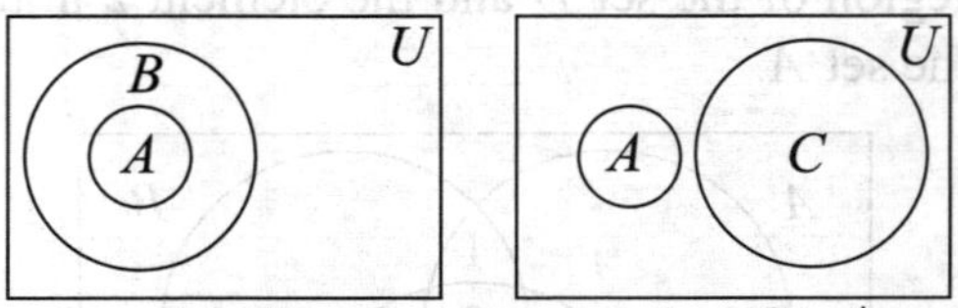

$A \subsetneq B$ $\qquad$ $A \cap C = \emptyset$

Combining the two, we obtain

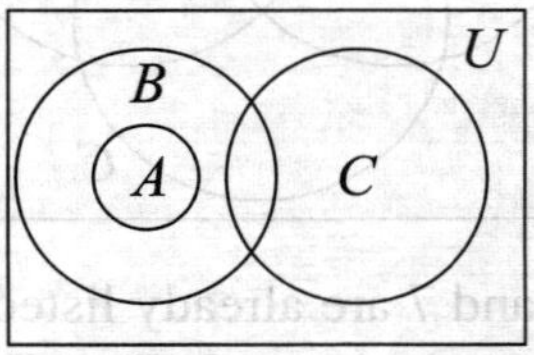

We allow C to overlap B since nothing is given about their relationship. □

Quick Review

After reading this section, you should be able to: After reading this section, you should be able to:

- Draw an Euler diagram for two or more subsets.
- Use an Euler diagram to carry out set operations.
- Represent relationships between subsets by an Euler diagram.

EXERCISE 2.4

1. Name all the set relationships among A, B and C as illustrated in the figures below:

(a)
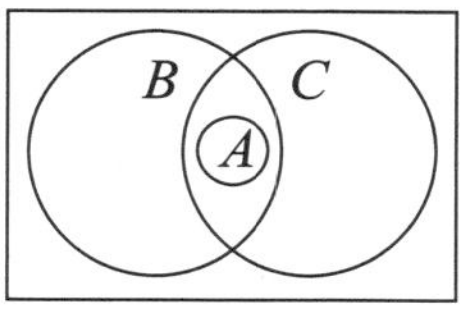

(b)
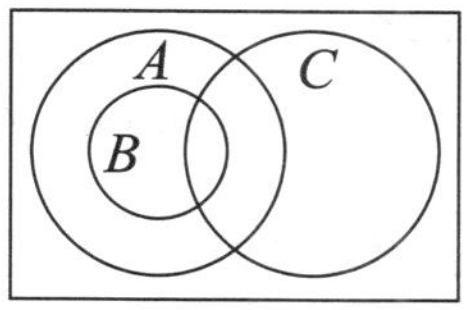

(c)
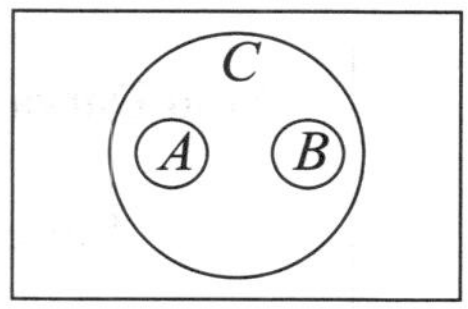

2. Draw the most general Euler diagram for the following relationships among the sets A, B and C:

(a) $B \subsetneq C$

(b) $B \subsetneq C$ and $A \cap C = \emptyset$

(c) $B \subsetneq C$ and $A \cap B = \emptyset$

(d) $A \cap C = \emptyset$ and $B \cap C = \emptyset$

(e) $A \cap B = \emptyset$, $A \cap C = \emptyset$, $B \cap C = \emptyset$

(f) $A \subsetneq B$ and $B \subsetneq C$

3. Represent the following sets with the help of an Euler diagram:

$$U = \{1, 2, 3, 4, 5, 6, 7, 8, 9\},\ A = \{1, 2, 3, 4\},\ B = \{2, 4, 6, 8\},\ C = \{3, 4, 5, 6\}$$

4. Consider the set of all living things. Draw Euler diagrams to represent the relations amongst the following combinations of its subsets:
   (a) The set of all sparrows and the set of all birds.
   (b) The set of all pigeons, the set of all crows, the set of all parrots and the set of all birds.
   (c) The set of all goats, the set of all mammals and the set of all carnivores.
   (d) The set of all mammals, the set of all herbivores and the set of all plants.

5. Consider the set of all human beings. Draw Euler diagrams to represent the relations amongst the following combinations of its subsets:
   (a) The set of all fathers, the set of all brothers and the set of all men.
   (b) The set of all men, the set of all women and the set of all teachers.
   (c) The set of all mothers, the set of all women and the set of all men.
   (d) The set of all doctors, the set of all professionals and the set of all women.
   (e) The set of all girls, the set of all dancers and the set of all students.

6. Using the information in the given diagram, describe the sets $A \cap B'$, $A' \cap C$, $B \cap C$, $(A \cap B)' \cap C$.

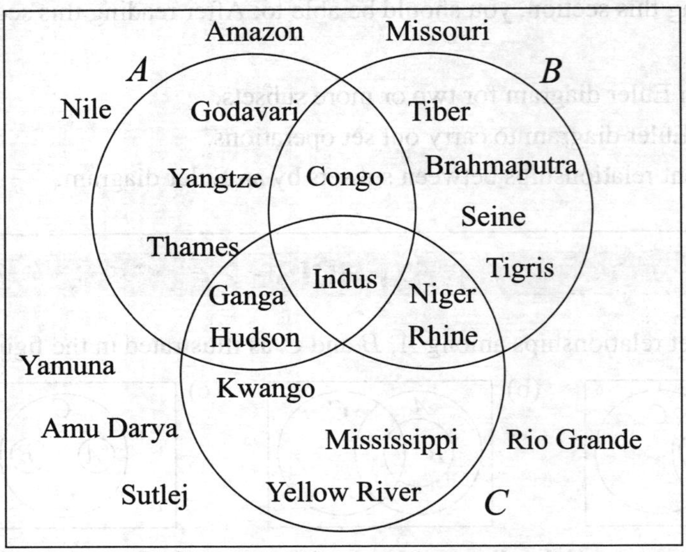

2.5 Venn Diagrams

In this section, we shall see how to use diagrams to demonstrate the various properties of set operations. Now, it is important to draw diagrams which are general rather than special, so we shall always draw the subsets in general configuration. Let us recall that for two or three subsets, the general configurations are as shown below:

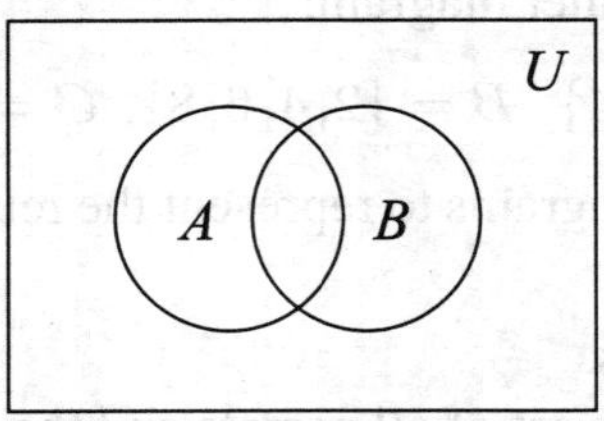

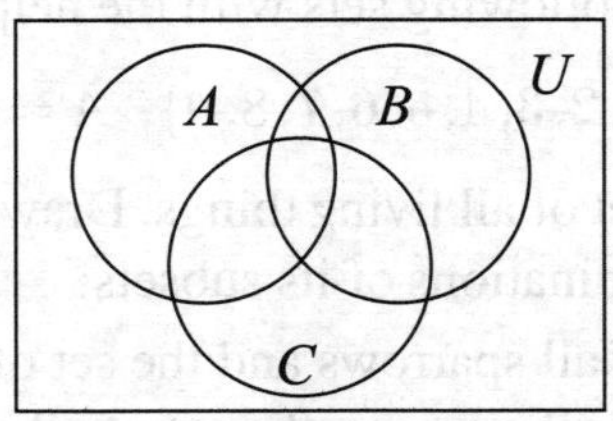

The methodical use of diagrams for this purpose was pioneered by John Venn[8] around 1880, and so they are also called **Venn diagrams.** A Venn diagram just shows the various ways in which sets *may* overlap. It can cover all cases via the simple convention that any particular region is allowed to be empty. Thus, the Venn diagram for two subsets includes the case when they are disjoint by allowing the overlap 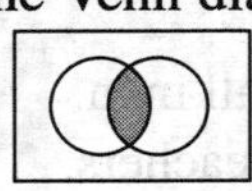 to be empty.

[8] You can learn more about John Venn's life and work in [W2].

Example 2.5.1 Suppose we wish to represent $(A \cup B) \cap (A \cup C)$ through a Venn diagram. $A \cup B$ and $A \cup C$ are represented by the  and  shaded portions, respectively, in the following diagrams:

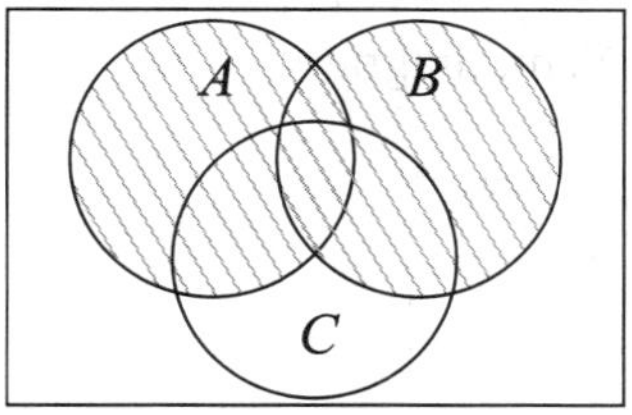

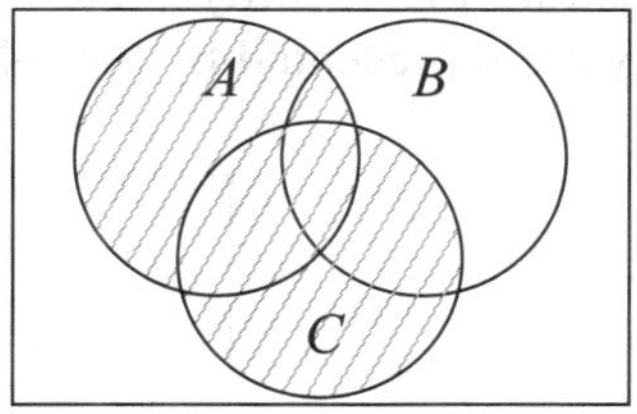

Now, $(A \cup B) \cap (A \cup C)$ consists of all the regions that are shaded in the first diagram *and* in the second diagram. Therefore, it is the  shaded region below:

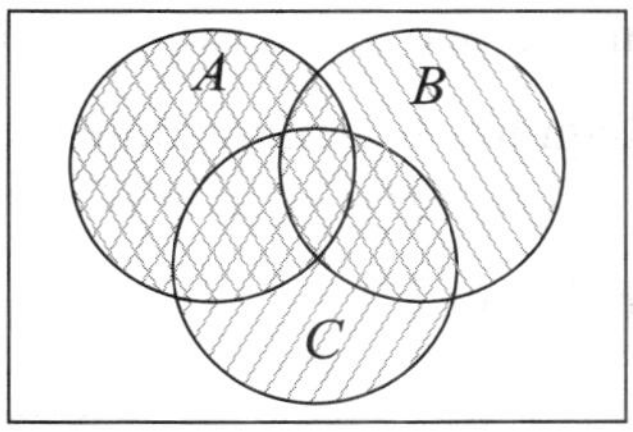

□

Now, we take up one of the distributive laws: $(A \cup B) \cap (A \cup C) = A \cup (B \cap C)$.

Example 2.5.2 We first show A and $B \cap C$ as the shaded portions of a Venn diagram:

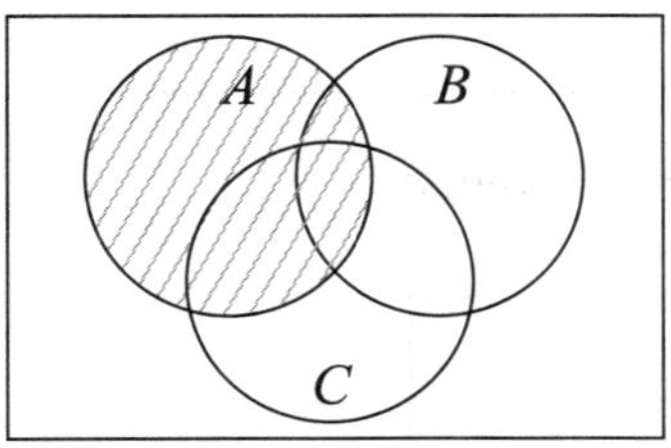

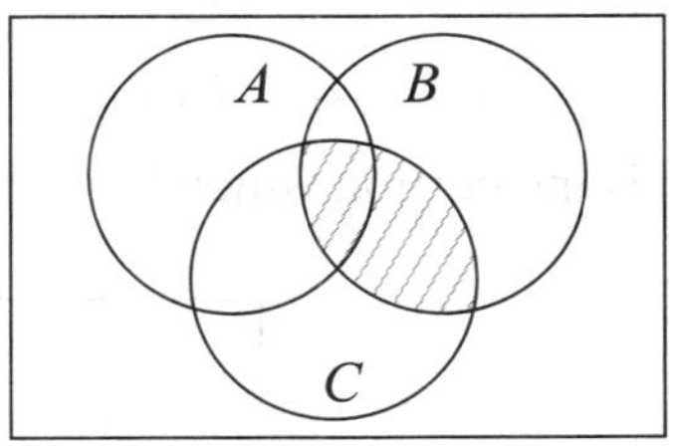

We form the union $A \cup (B \cap C)$ by pooling the two shaded regions into one:

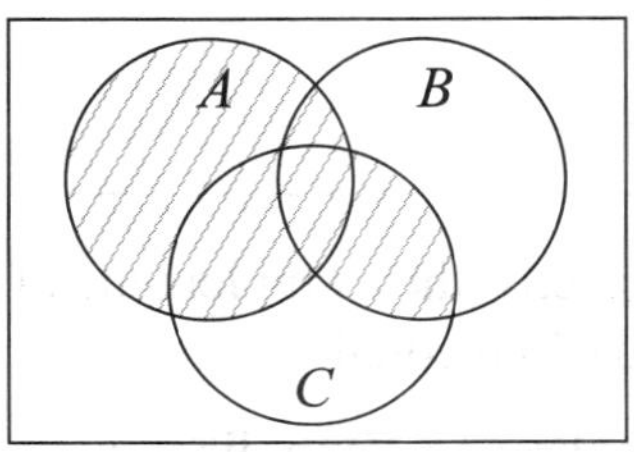

Compare this with the region obtained in the previous example. We see that they are exactly the same! This gives the equality

$$(A \cup B) \cap (A \cup C) = A \cup (B \cap C) \qquad \square$$

Task 2.5.1 Demonstrate the identity $(A \cap B) \cup (A \cap C) = A \cap (B \cup C)$ by showing that both sides of the identity are represented by the following Venn diagram:

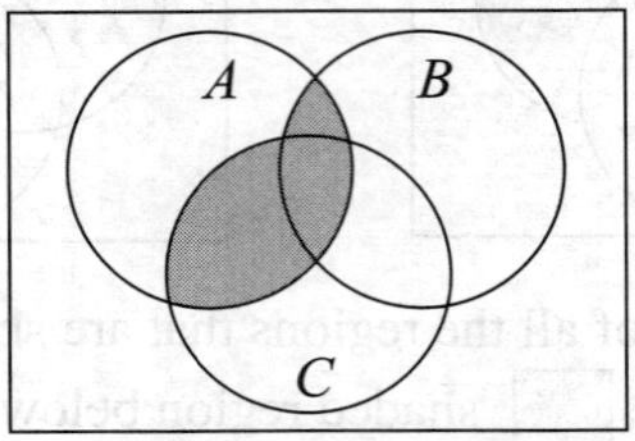

We round off this section by taking up one of De Morgan's laws.

Example 2.5.3 We shall use Venn diagrams to demonstrate the identity

$$(A \cup B)' = A' \cap B'$$

We first draw the Venn diagrams for A' and B':

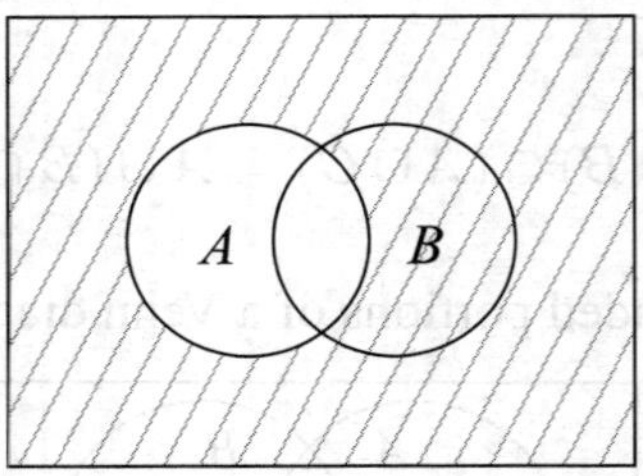

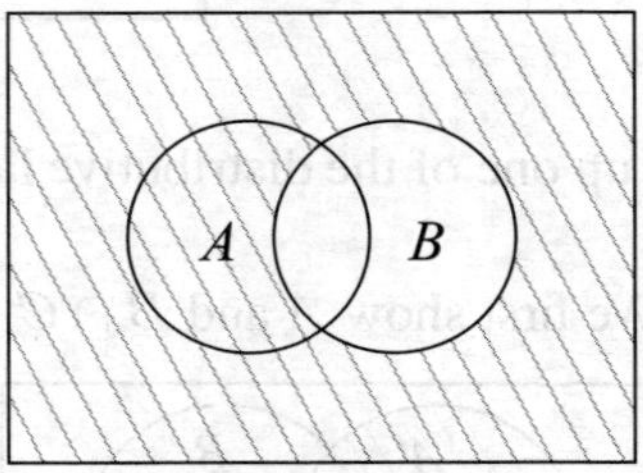

The intersection is the portion which is shaded in both diagrams:

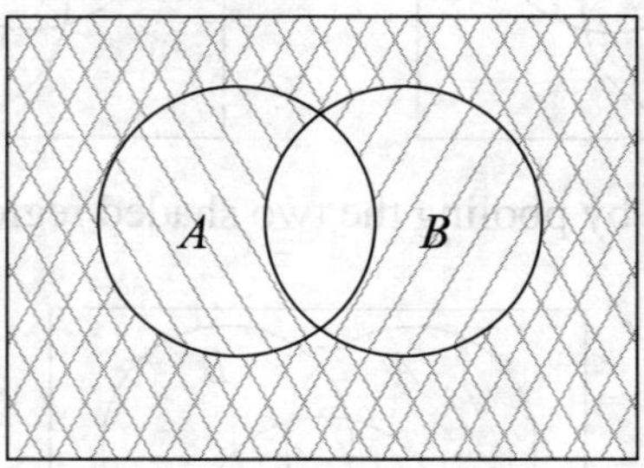

We see that the  portion is exactly the complement of $A \cup B$ (we have drawn that Venn diagram earlier) and this gives the desired equality. $\square$

This approach of comparing sets through Venn diagrams can also be used to show one set is a subset of another.

Example 2.5.4 We have already worked out the following diagrams:

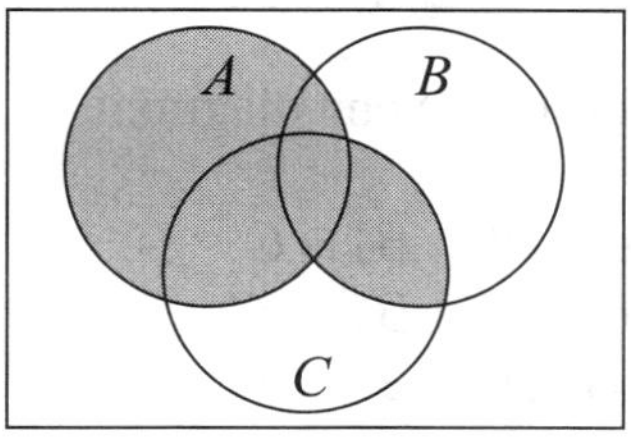

$A \cup (B \cap C)$

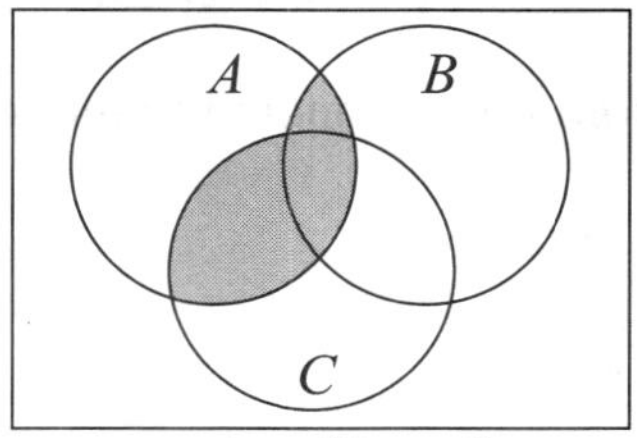

$A \cap (B \cup C)$

The shaded region in the first diagram completely contains the one in the second diagram. This means that the first set always contains the second set: $A \cap (B \cup C) \subseteq A \cup (B \cap C)$. □

Finally, we can use Venn diagrams to deduce that two combinations of sets may not be equal, and then identify conditions under which they would be equal.

Example 2.5.5 Consider the combinations $X = A \cap (B \cup C)$ and $Y = B \cap (A \cup C)$. Their Venn diagrams are:

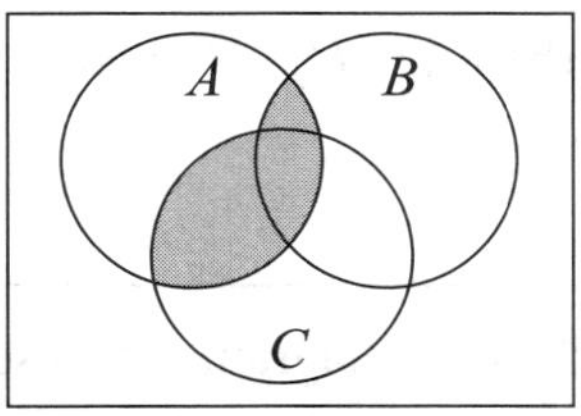

$A \cap (B \cup C)$

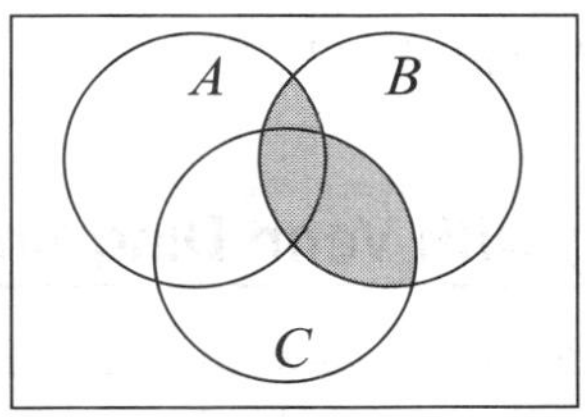

$B \cap (A \cup C)$

It is evident that neither shaded region is contained in the other. So, *in general*, neither of $A \cap (B \cup C)$ and $B \cap (A \cup C)$ is contained in the other. But *sometimes* one may be contained in the other. For example, if $B \cap C \cap A' = \emptyset$, then $B \cap (A \cup C)$ actually only occupies 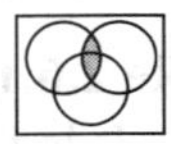 and in that case $B \cap (A \cup C) \subseteq A \cap (B \cup C)$.

Similarly, if $B \cap C \cap A' = A \cap B' \cap C = \emptyset$, then we even have equality: $B \cap (A \cup C) = A \cap (B \cup C)$. □

Quick Review

After reading this section, you should be able to:

1. Represent any combination of two or three sets by shading appropriate regions of a Venn diagram.
2. Use Venn diagrams to demonstrate set identities.
3. Use Venn diagrams to demonstrate that one set is a subset of another.
4. Use Venn diagrams to show that two combinations of sets may not be equal and identify conditions under which equality does hold.

EXERCISE 2.5

1. Shade the regions corresponding to the following sets on a Venn diagram:

(a) $(A \cup B') \cap C$
(b) $(A' \cap B') \cap C$
(c) $(A \cap B) \cap C'$
(d) $(A' \cap B) \cap C'$
(e) $(A' \cup B)' \cap C$
(f) $(A' \cup B') \cap C$

2. Verify using Venn diagrams:

(a) $(A \cup B) \cup (A \cup C) = A \cup (B \cup C)$
(b) $(A \cap B) \cap (A \cap C) = A \cap (B \cap C)$
(c) $(A \cap B)' = A' \cup B'$

3. With the help of Venn diagrams, prove the following are always true:
(a) $A \cap (B \cup C) \subseteq (A \cap B) \cup C$
(b) $(A \cap B) \cup (B \cap C) \subseteq (C \cap A^c) \cup (A \cap B)$

4. Use Venn diagrams to show that the following are not always equal:

(a) $(A \cap B) \cup B$ and $A \cap B$
(b) $(A \cap B) \cup C$ and $A \cap (B \cup C)$

2.6 Counting with Venn Diagrams

In the last two sections, we saw that the Euler and Venn diagrams represent various relations among sets in a pictorial way. On the one hand, we learnt to draw these diagrams given certain information about sets. On the other, we learnt to read and interpret given diagrams. In this section, we shall learn another application of Venn diagrams—counting objects with various combinations of properties.

Example 2.6.1 Suppose a survey of 100 people is conducted regarding the type of coffee they like to drink. It is found that 36 people like espresso, 55 like cappuccino and 20 like both. First, let us draw a Venn diagram to represent the information given to us. The large set in which the action is taking place is the set U consisting of the 100 people surveyed. In U, there are two subsets: one consisting of espresso lovers and the other consisting of cappuccino lovers. Let us denote them by sets E and C, respectively. We, therefore, have the Venn diagram as given below.

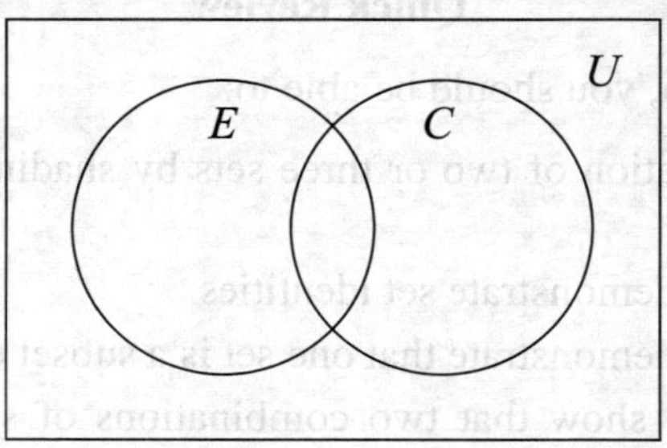

The two circles cut the rectangle into four regions, each region corresponding to a particular combination of preferences. To see the distribution of these preferences, we write in each region the number of people with corresponding tastes. First, since 20 people like both espresso and cappuccino, we place 20 in the intersection of E and C. Thirty-six people like espresso, out of which 20 are already placed. Therefore, we are left with 16 people who only like espresso. Similarly, 35 (= 55 − 20) people only like cappuccino. This totals up to 16 + 20 + 35 = 71. But we had 100 people. Therefore, 100 − 71 = 29 people like neither espresso nor cappuccino. So we place 29 outside the circles E and C.

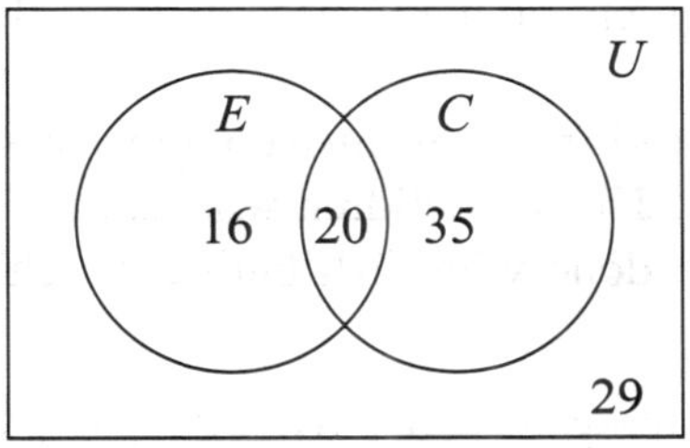

□

Example 2.6.2 In a survey of children who saw three different shows at Disney World, the following information was gathered. Thirty-nine children liked *The Little Mermaid*, 43 children liked *101 Dalmatians*, 56 children liked *Mickey Mouse*, 7 children liked *The Little Mermaid* and *101 Dalmatians*, 10 children liked *The Little Mermaid* and *Mickey Mouse*, 16 children liked *101 Dalmatians* and *Mickey Mouse*, 4 children liked all the three and 6 children did not like any of the shows. We wish to answer the following questions:

(i) How many students were surveyed?
(ii) How many liked only *The Little Mermaid*?
(iii) How many liked only *101 Dalmatians*?
(iv) How many liked only *Mickey Mouse*?

Let U be the set of all children surveyed. We have to consider three subsets of U:

(a) Let L be the set of children who liked *The Little Mermaid*.
(b) Let M be the set of children who liked *Mickey Mouse*.
(c) Let D be the set of children who liked *101 Dalmatians*.

The following Venn diagram represents the data:

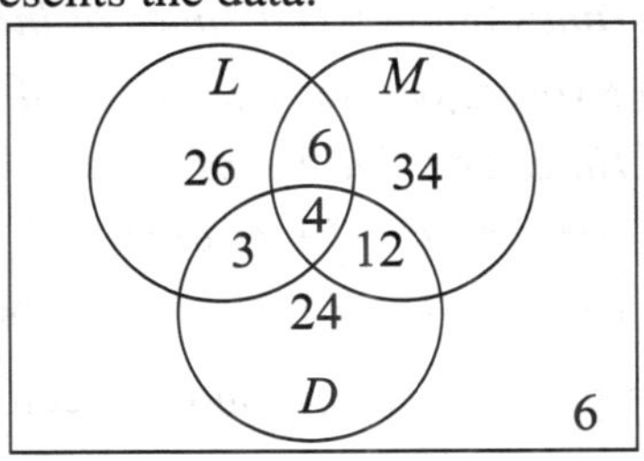

Since 4 children liked all the three shows, we have 4 in the region representing $L \cap M \cap D$.

There are 7 children who liked *The Little Mermaid* and *101 Dalmatians*, out of these 4 are already placed. Hence, we have 3 in the region representing $L \cap D$ excluding the region $L \cap M \cap D$.

Now, 10 children liked *The Little Mermaid* and *Mickey Mouse* and 16 children liked *101 Dalmatians* and *Mickey Mouse*; in each case, 4 of these are already placed. So we have 6 in the region of $L \cap M$, excluding the region $L \cap M \cap D$ and 12 in the region $M \cap D$, excluding the region $L \cap M \cap D$.

Further, 39 children liked *The Little Mermaid*. Of these, $6 + 4 + 3 = 13$ children have already been placed in the circle of M. So we are left with 26 children.

Mickey Mouse was liked by 56 children. Of these, $6 + 4 + 12 = 22$ are already placed. So we are left with 34 children.

Similarly, 24 children are placed in the exclusive region of set D, representing those children who only liked *101 Dalmatians*. The six children who did not like any show are placed outside these three circles. Once we are done with distributing the children in various regions, we can answer the above questions.

(i) Total number of children surveyed $= 26 + 34 + 24 + 6 + 3 + 12 + 4 + 6 = 115$.

(ii) Number of children who only liked *The Little Mermaid* is 26.

(iii) Number of children who only liked *101 Dalmatians* is 24.

(iv) Number of children who only liked *Mickey Mouse* is 34. □

Example 2.6.3 The Venn diagram below represents the readership of *Hindustan Times* (H), *The Times of India* (T) and *The Indian Express* (I) in a locality of Delhi.

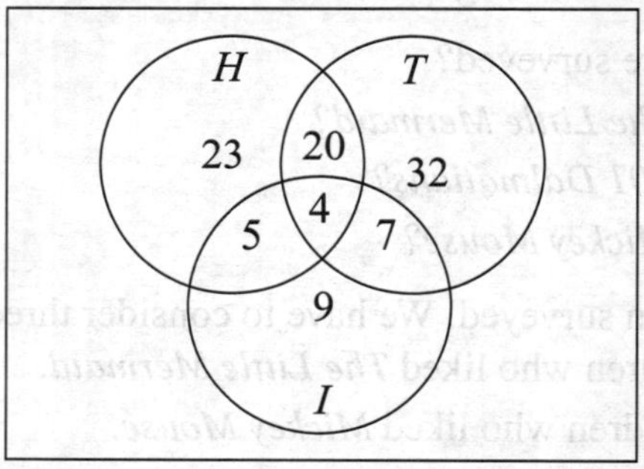

(i) How many read *Hindustan Times*?

(ii) How many read *Hindustan Times* but not *The Times of India*?

(iii) How many read *The Indian Express* but not *The Times of India*?

(iv) How many read *Hindustan Times* and *The Times of India*?

(v) How many read *Hindustan Times* or *The Times of India*?

(vi) How many read *Hindustan Times* or *The Times of India* but not *The Indian Express*?

As shown below, Figures (i)–(vi) give the corresponding region of the Venn diagram associated with parts (i)–(vi) above.

Figure (i)

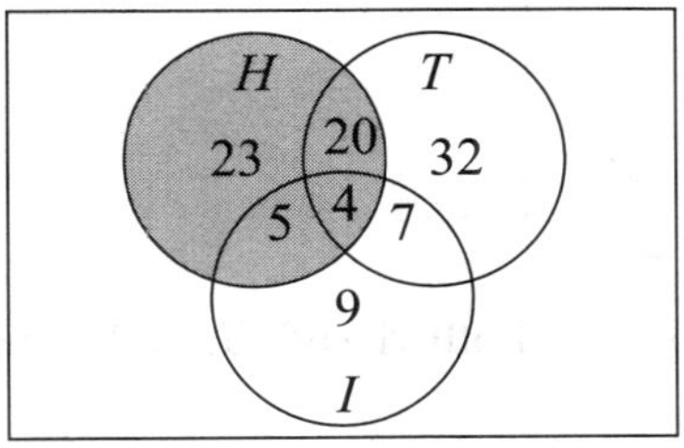

Figure (ii)

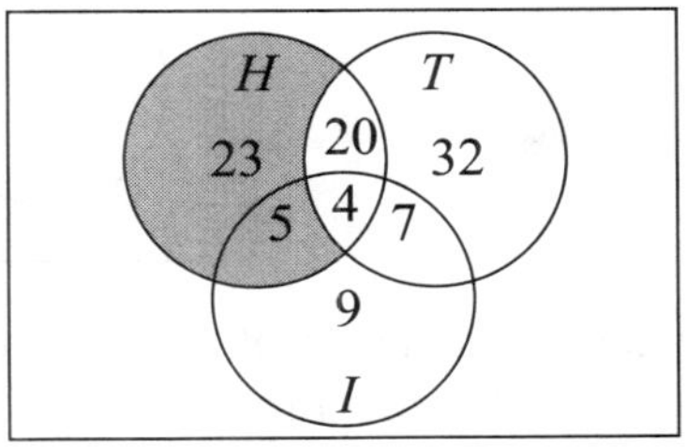

Figure (iii)

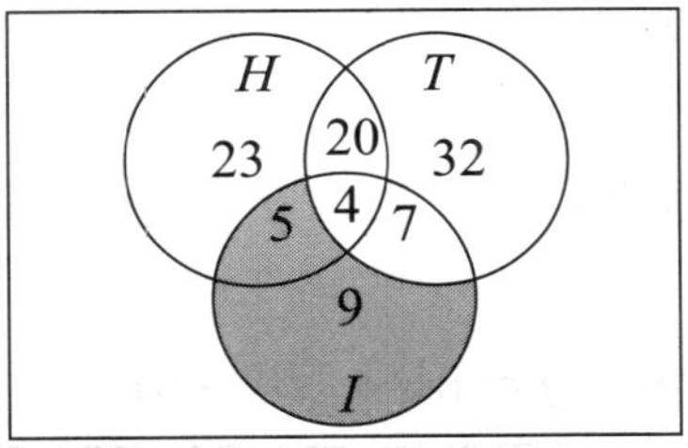

Figure (iv)

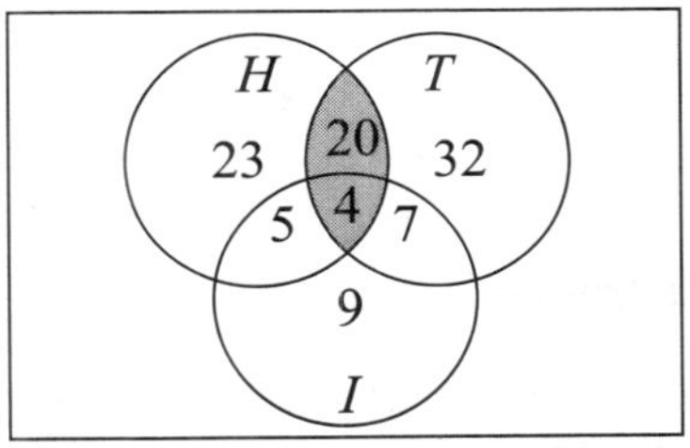

Figure (v)

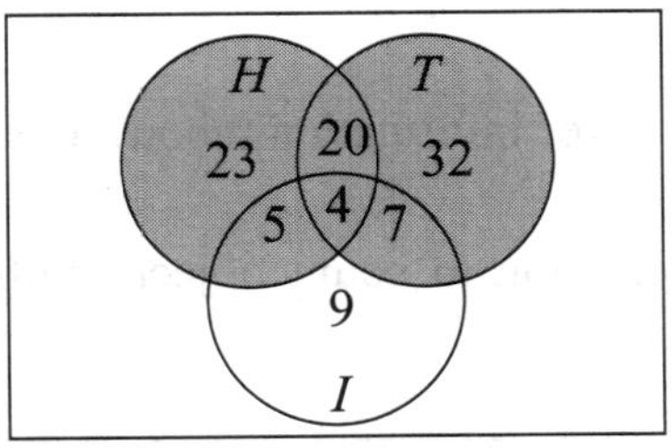

Figure (vi)

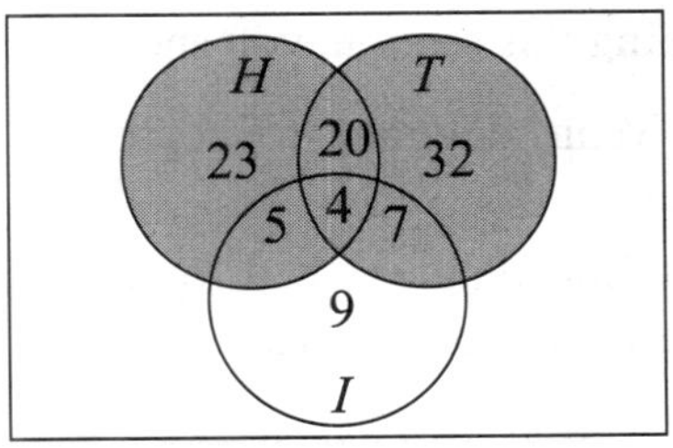

Therefore,

(i) The number of people who read *Hindustan Times* is $23 + 20 + 4 + 5 = 52$.

(ii) The number of people who read *Hindustan Times* but not *The Times of India* is $23 + 5 = 28$.

(iii) The number of people who read *The Indian Express* but not *The Times of India* is $9 + 5 = 14$.

(iv) The number of people who read *Hindustan Times* and *The Times of India* is $20 + 4 = 24$.

(v) The number of people who read *Hindustan Times* or *The Times of India* is $23 + 5 + 20 + 4 + 32 + 7 = 91$.

(vi) The number of people who read *Hindustan Times* or *The Times of India* but not *The Indian Express* is $23 + 20 + 32 = 75$. □

An interesting application of Venn diagrams is to create formulas connecting numbers of elements in various sets. Consider two subsets A and B of U. The following diagram represents the number of elements in various intersections:

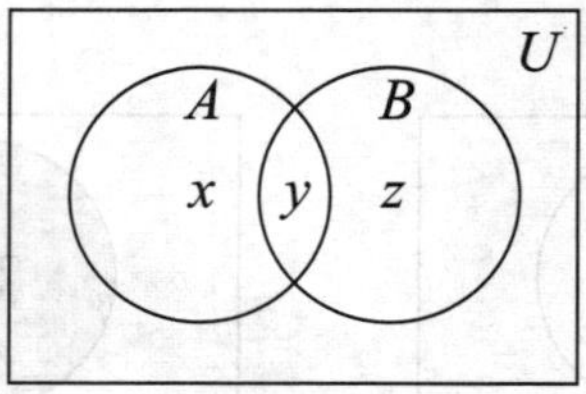

Specifically, we have x members of A alone, y members of both A and B, and z members of B alone. We can see from the diagram that

$$|A| = x + y$$

$$|B| = z + y$$

$$|A \cap B| = y$$

$$|A \cup B| = x + y + z$$

We can conclude from this that

$$|A \cup B| = x + y + z = (x + y) + (z + y) - y = |A| + |B| - |A \cap B|$$

Quick Review

After reading this section, you should be able to:

1. Use a Venn diagram to represent quantitative information about overlapping categories.
2. Read a Venn diagram with quantitative information to count objects with certain combinations of properties.

EXERCISE 2.6

1. In a class, 12 students study mathematics, 10 study economics, 5 study both and 6 study neither.
   (a) How many students are there in the class?
   (b) How many students study only mathematics or economics?
2. In a group of 1,000 people, 700 can speak Hindi and 500 can speak English. If all the people speak at least one of the two languages, find
   (a) The number of people who can speak both the languages.
   (b) The number of people who speak only one language.
3. In an exam, 60% of the candidates passed in mathematics, 70% of the candidates passed in English and 10% candidates failed in both the subjects. If 300 candidates passed in both the subjects, find the total number of candidates that appeared in the exam assuming that each candidate took the test in only these two subjects.

4. A group of 90 children went for a school picnic. The teacher noticed the following:

   3 had a sandwich, a cold drink and an ice-cream
   24 had a sandwich
   5 had a sandwich and a cold drink
   30 had a cold drink
   10 had a cold drink and an ice-cream
   38 had an ice-cream
   8 had a sandwich and an ice-cream

   How many children
   (a) Had only a sandwich?
   (b) Had either a cold drink or an ice-cream?
   (c) Had only a sandwich and an ice-cream?
   (d) Had at most two items?

5. A total of 65 students staying in a hostel were asked if they would like to keep a washing machine (W) or a refrigerator (R) or a microwave (M) in their rooms. The Venn diagram below represents the response given by them.

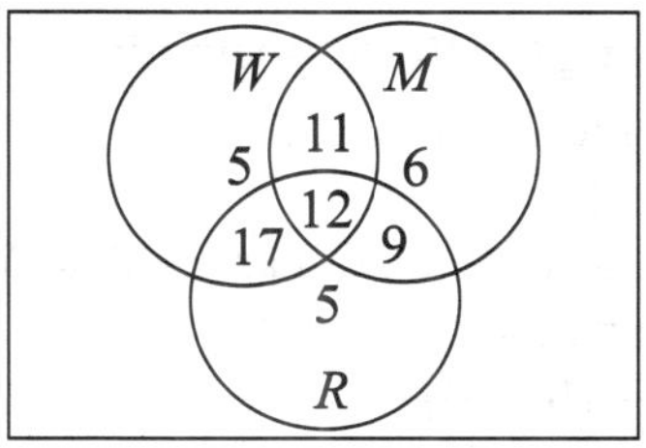

   How many students
   (a) Wanted exactly one of the appliances in the room?
   (b) Wanted at least two of the appliances in the room?
   (c) Wanted a refrigerator and a microwave but not a washing machine?
   (d) Wanted either a washing machine or a refrigerator?

6. Let A, B and C be finite subsets of U. Use a Venn diagram to show that

$$|A \cup B \cup C| = |A| + |B| + |C| - |A \cap B| - |A \cap C| - |B \cap C| + |A \cap B \cap C|$$

Review Exercises

1. Express the following sets in set-builder notation (there may be more than one way):
   (a) $A = \{$Helium, Neon, Argon, Krypton, Xenon, Radon$\}$
   (b) $B = \{$A, E, L, P, S, T$\}$
   (c) $C = \{$January, March, May, July, August, October, December$\}$
   (d) $D = \{$Mowgli, Baloo, Bagheera, Shere Khan, Kaa, $\ldots\}$
   (e) $E = \{1, \frac{1}{2}, \frac{1}{3}, \frac{1}{4}, \ldots, \frac{1}{100}\}$
   (f) $F = \{1, 8, 27, 64, \ldots\}$
   (g) $G = \{$Gita, Quran, Bible, Guru Granth Sahib, Avesta, $\ldots\}$

(h) $H = \{2, 3, 5\}$
(i) $I = \{2, 4, 8, 16, 32, \ldots\}$
(j) $J = \{16, 25, 34, 43, 52, 61, 70\}$

2. Which of the following describes a set?
   (a) The collection of tall boys in your class.
   (b) The collection of all boys in your class.
   (c) The collection of all good cricketers in the Indian team.
   (d) The collection of interesting subjects being taught to you.
   (e) The collection of subjects being taught to you.
   (f) The collection of all months starting with the letter A.
   (g) The collection of all novels written by Shakespeare.
   (h) The collection of all multiples of 4.
   (i) The collection of good teachers in your college.
   (j) The collection of all integers less than 5.

3. Describe the following sets using the Roster method:
   (a) $K = \{x \mid x \text{ is a letter of the word PETALS}\}$
   (b) $L = \{x \in \mathbb{R} \mid x^2 - 7x + 12 = 0\}$
   (c) $M = \{x \mid x \text{ is a movie that won the Oscar for Best Movie in 2010}\}$
   (d) $N = \{x \mid x \text{ is an odd prime factor of 1024}\}$
   (e) $O = \{x \in \mathbb{N} \mid x^2 - 2x - 15 = 0\}$
   (f) $P = \{x \in \mathbb{N} \mid x \text{ is a divisor of 70}\}$
   (g) $Q = \{x \mid x = 2^n - 3, n \in \mathbb{N}\}$
   (h) $R = \{x \in \mathbb{Q} \mid x^2 - 3 = 0\}$
   (i) $S = \{x \in \mathbb{N} \mid 7 < x < 8\}$
   (j) $T = \{x \mid x \text{ is a letter of the word AGGREGATE}\}$

4. Refer to the sets listed in 1 and 3 above. List the sets that are
   (a) Finite (b) Infinite (c) Singleton (d) Empty

5. Which of the following statements are true? Justify your answer. Also write a correct form of each incorrect statement.
   (a) $\mathrm{a} \subseteq \{\mathrm{a}, \mathrm{b}, \mathrm{c}\}$
   (b) $1 \in \{x \mid x \text{ is a factor of 25}\}$
   (c) $\emptyset \in \{x \mid x \text{ is an even factor of 35}\}$
   (d) Sets $A = \{\{a\}\}$ and $B = \{a\}$ are equal.
   (e) $\{1, 2, 3\} \in \{\{1\}, \{1, 2, 3\}, 4, 5\}$
   (f) The sets $A = \{x \mid x \text{ satisfies the equation } x^2 - 8x + 12 = 0\}$ and $B = \{x \mid x \text{ is an even factor of 6}\}$ are equal.
   (g) $\mathcal{P}(A) \cup \mathcal{P}(B) = \mathcal{P}(A \cup B)$
   (h) The number of subsets of $A = \{1, 2, 3, \ldots, 8\}$ is 16.
   (i) $\{\phi\}$ is a subset of every set.
   (j) The set $\{x \in \mathbb{R} \mid x + 5 = 5\}$ is an empty set.

6. Let $U = \{x \mid x \in \mathbb{N}, x \leq 30\}$, $A = \{1, 2, 3, \ldots, 14, 15\}$, $B = \{2, 4, 6, \ldots, 20\}$, $C = \{\,3, 6, 9, \ldots, 30\}$ and $D = \{x \mid x$ is a prime natural number less than $30\}$. Find

   (a) $A \cup (B \cap D)$
   (b) $C \cap (B \cap A)$
   (c) $B \cup (A \cap D)$
   (d) $A' \cap B'$
   (e) $(A \cup B)'$
   (f) $C' \cup D'$
   (g) $(C \cap D)'$
   (h) $(A \cup B') \cap (B \cup D')$
   (i) $A \cup (B \cap C)'$
   (j) $B \cap (A \cap D)'$

7. For three subsets A, B and C of U, draw the most general Euler diagram that will represent the following relations among them:

   (a) $A \subseteq B \subseteq C$
   (b) $A \subseteq C$ and $B \subseteq C$
   (c) $A \subseteq C$, $B \subseteq C$ and $A \cap B = \emptyset$
   (d) $A \cap B = \emptyset$ and $B \cap C = \emptyset$

8. Shade the following regions on appropriate Venn diagrams.

   (a) $A' \cap B$
   (b) $A \cup B'$
   (c) $(A \cap B') \cup (A' \cap B)$
   (d) $(A \cup C') \cap B'$
   (e) $A' \cup (B \cup C)$
   (f) $A \cap (B' \cup C')$

9. Let U be the set of all humans, M be the set of males, S be the set of college students and H be the set of people whose height is above 5 feet. Draw a suitable Venn diagram for these. Shade the regions that depict the following:
   (a) College students having height more than 5 feet.
   (b) People who are not male and have a height less than 5 feet.
   (c) All the people who are male less than 5 feet but are not college students.
   (d) All the people who are either college students or above 5 feet but not male.

10. With the help of Venn diagrams, prove the following are always true:
    (a) $A \cap B' = (A' \cup B)'$
    (b) $(A \cap B') \cup (A \cap C) = A \cap (B' \cup C)$
    (c) $(A \cap B) \cup (A \cap C) \cup (B \cap C) = (A \cup B) \cap (A \cup C) \cap (B \cup C)$
    (d) $(A \cup B) \cap C \subseteq A \cup (B \cap C)$
    (e) $(A \cap B') \cup (B \cap A') \subseteq A \cup B$
    (f) $(C \cap A') \cup (A \cap B) \subseteq C \cup (A \cap B)$

11. Use Venn diagrams to show that the following are not always equal:

    (a) $(A \cup B) \cap B$ and $A \cup B$
    (b) $(A \cup B) \cap C$ and $A \cup (B \cap C)$

12. In a class of 50 students, each student travels to college by riding the metro or a bus or both. If 32 ride the metro and out of these 17 do not ride a bus, find the number of students that use

    (a) both the modes of transport
    (b) bus but not metro
    (c) at most one mode of transport

13. A market survey was conducted of 1,000 consumers for two soap brands X and Y. It was reported that 860 liked brand X, 725 liked brand Y and 600 liked both. Find the number of consumers who

(a) liked only brand X
(b) liked only brand Y
(c) did not like any of the two brands

14. A survey reported that out of 2,000 citizens, 1,000 can speak Hindi, 800 can speak English and 500 can speak neither English nor Hindi. Find the number of citizens who can speak
    (a) both English and Hindi
    (b) only English
    (c) only Hindi

15. Of the members of three societies, 21 are in the film society, 26 in the dramatic society and 29 in the debating society. Fourteen are in the film and dramatic societies, 15 in dramatic and debating societies and 12 in the film and debating societies. If eight members are in all the three societies, find
    (a) the total number of members.
    (b) the number of members who are in exactly two societies.
    (c) the number of members who are in at least two societies.
    (d) the number of members who are in only one society.

16. The modes of transport of the employees working in a certain office are given in the diagram below.

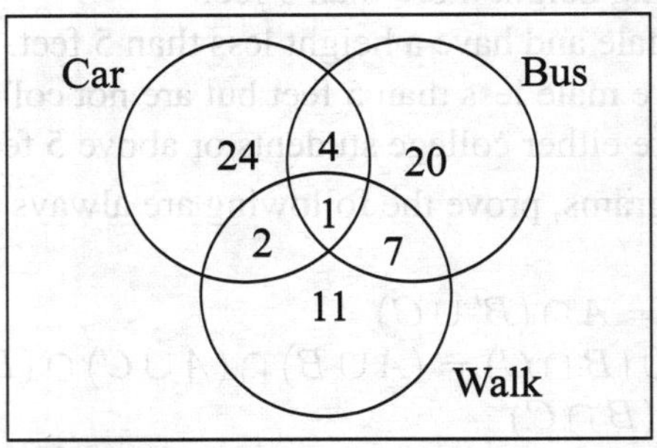

Answer the following questions:
    (a) How many walk to office?
    (b) How many take a bus or car but do not walk?
    (c) How many use only a car?
    (d) How many were surveyed?
    (e) How many take a car and walk but do not take a bus?

17. Hundred people were asked who were the following characters in Mahabharata: Ghatotkach, Abhimanyu and Subhadra. The following data were collected: 25 people did not know any of these, 3 people knew all the three, 48 people knew about Abhimanyu or Subhadra but not Ghatotkach, 40 people knew about Abhimanyu, 21 people knew who at least two of these characters were, 7 people knew about Ghatotkach and Abhimanyu but not about Subhadra, 8 people knew who Ghatotkach and Subhadra were. How many people knew about Subhadra?

Bibliography

Books and Articles

[1] W P Berlinghoff and K E Grant. 1992. *A Mathematics Sampler: Topics for the Liberal Arts*. 3rd edition. New York: Rowman & Littlefield Publishers.

[2] R F Blitzer. 2015. *Thinking Mathematically*. 6th edition. Boston, MA: Prentice Hall, 2015.

[3] H B Enderton. 1977. *Elements of Set Theory*. Cambridge, MA: Academic Press.

[4] Richard Hammack. 2013. *Book of Proof*. Richmond, VA: Richard Hammack. `http://www.people.vu.edu/~rhammak/BookOfProof/` (accessed: 27 January 2017).

[5] Donald Herrick. 1970. *Mathematics for Liberal Arts Students*. Boston, MA: Prindle,Weber & Schmidt.

[6] Ronald P Morash. 1987. *Bridge to Abstract Mathematics*. New York, NY: Random House, 1987.

[7] Michael L O'Leary. 2015. *A First Course in Mathematical Logic and Set Theory*. Hoboken, NJ Wiley-Blackwell, 2015.

Websites

[W1] J J O' Connor and E F Robertson. *Cantor*. `www-history.mcs.st-and.ac.uk/Biographies/Cantor.html`

[W2] J J O' Connor and E F Robertson. *Venn*. `history.mcs.st-and.ac.uk/Biographies/Venn.html`.

[W3] Rod Pierce. *Set Theory Index*, Math is Fun. `http://www.mathsisfun.com/sets/` (accessed 27 January 2017).

3

CHAPTER

Logic and Reasoning

> I argue very well. Ask any of my remaining friends. I can win an argument on any topic, against any opponent. People know this, and steer clear of me at parties. Often, as a sign of their great respect, they don't even invite me.
>
> —*Dave Barry, an American writer and humorist.*

While one would not want to suffer the fate presented in the quote above, one can still ask: What is an argument? How can one win arguments? Will it help if one can be like the Vulcan Spock, in Star Trek, for whom logic is the only basis for presenting arguments. In this chapter, we will explore the basic principles of logic that are used to analyse statements, the reasoning behind valid arguments and the inferences that we can draw from reasoning with logic.

The word *logic* has its roots in a Greek word 'logos' which when loosely translated means 'reason'. The ancient Greek philosophers used this to mean 'reasoned speech'. Formal logic is studied in the disciplines of philosophy as well as mathematics. Modern-day formal logic has its origins in the works of **George Boole.**[1]

3.1 Statements

> Let us start at the very beginning, for it is a very good place to start.
>
> (From the song *Do-Re-Mi* of the movie *The Sound of Music*.)

For us, the beginning will be to figure out what we mean formally by the term *statement*. By a **statement** we mean a sentence that is declarative, that is, it declares or states something. From now on, when we use the term statement, we mean a sentence as described above. A basic assumption that we will make is that all *statements* are either *true* or *false* (but not both). The two possible choices 'True' or 'False' are called the **truth value** of the given statement.

[1] You can find out about the life and works of George Boole in [W1].

Example 3.1.1 The following are examples of **statements.**

(a) Today is a Sunday.
(b) Every rose is red.
(c) Some people like music.
(d) No human being can live forever.
(e) The number 2 is less than the number 3.
(f) The set $\{x \in \mathbb{N} \mid x < -1\}$ is empty.
(g) I like Hindi films and English songs.
(h) If it rains, then I will not be able to go outside. □

Is there a difference that you notice between the first six statements and the last two? Statements (g) and (h) are built out of two statements using connectives like *and* and *if, then*, whereas the statements (a)–(f) cannot be broken up further into statements. Statements that cannot be broken up further are called **simple statements,** whereas statements that are built up out of other statements using connectives are called **compound statements.** The statements (a)–(f) are examples of simple statements; statements (g) and (h) are examples of compound statements.

Example 3.1.2 The examples listed below are **not statements.**

(a) What is your name?
(b) How many marks did you get?
(c) When did you buy this dress?
(d) 'Stop arguing!'
(e) Bring me that book.
(f) Please come to my party.
(g) Please do not cry. □

Questions as in Example 3.1.2 (a)–(c), *commands* as in Example 3.1.2 (d) and (e) and *requests* as in Example 3.1.2 (f) and (g) are **not statements.**

Example 3.1.3 Some more examples of statements are listed below.

(a) Today is a Monday.
(b) I live in Delhi.
(c) She wears skirts.
(d) Some cats are cute.
(e) All women have long hair.
(f) No dog is dirty. □

If we consider the statements in Example 3.1.3, we notice that there is a difference between the statements (a)–(c) and the statements (d)–(f). The latter have the words 'all', 'no' and 'some' in them. The next section will make precise what these differences are when we analyse statements.

Quick Review

After reading this section, you should be able to:

- Decide whether a given sentence is a statement.
- Give examples of sentences that are not statements.
- Differentiate between simple statements and compound statements.

EXERCISE 3.1

1. Which of the following sentences are **statements?**
   (a) The sun rises in the East.
   (b) Akbar was a great king.
   (c) Stop that at once.
   (d) If it does not rain, then I can go out.
   (e) What is a statement?
2. Label the following statements as **simple** and **compound.**
   (a) Today is not a weekday.
   (b) All roses are red.
   (c) Nisha can sing and dance.
   (d) If a dog barks, then it seldom bites.
   (e) You can eat or drink.

3.2 Analysing Statements

While analysing statements we wish to identify key ingredients from the point of view of logic. The first step is to identify whether the given statement is a simple or a compound statement.

A working definition of a simple statement is that there are no connectives like 'and', 'or', 'if, then'. Formally (from a grammatical point of view), a simple statement has a single subject and a predicate. A simple statement basically describes just one idea and has only one verb. If the given statement is a compound statement, then we have to identify the simple statements from which the compound statement has been constructed. The analysis is then carried out on the building blocks, which are the simple statements. Inferences are then drawn about the compound statement.

Let us concentrate on simple statements now. There are several points of interest when we study or analyse a simple statement. We should find out whether it is a statement with quantifiers or without quantifiers. Given our real-life experience, we should be able to decide whether the statement is true or false. In order to do this, we have to understand the meaning of quantifiers rather well. For example, let us consider the following question.

Decide using real-life experience whether the following statements are true or false.

1. All roses are red.
2. Some roses are red.
3. No rose is red.
4. Today is a Saturday.

Before answering the above question, you need to be sure as to what 'all', 'some' and 'no' mean in their respective statements. On the other hand, it is fairly easy to make a decision about the fourth part. If the day on which you are answering the question is a Saturday, then your answer will be true; if it is not a Saturday, then the answer will be false. It will be worthwhile to spend some time on statements with quantifiers.

Simple Statements with Quantifiers

Words such as 'all', 'some' or 'none' are examples of quantifiers. In a statement, quantifiers basically indicate that the statement is dealing with 'quantities' or 'numbers'. Statements with quantifiers are called **quantified statements.** Some examples are given below.

- All girls have long hair.
- Some girls have long hair.
- None of the girls have long hair.

All stands for **each and every.** So a statement like 'All roses are red' will be false from our real-life experience. This is because there is at least one rose which is not red in colour. Note that we just need the fact that there is one non-red rose to make the statement 'All roses are red' false. Similarly, the sentence 'All girls have long hair' is also false from our real-life experience as there is at least one girl with short hair. The statement 'All green plants have chlorophyll' is a true statement. This is because the presence of chlorophyll in the plants causes the green colour. 'All humans are mortal' is also a true statement as every human being eventually dies. The word 'every' is sometimes used instead of 'all'.

We can also interpret a statement of the type 'All dogs are poodles' from a set theoretic point of view. For example, if we let A be the set of all dogs and B be the set of all poodles, then the statement 'All dogs are poodles' is a statement of the form 'All A are B'. Note that 'All dogs are poodles' will be true if and only if each and every dog is a poodle, that is, if and only if $A \subseteq B$. Of course, in the specific example that we are considering, there are dogs that are not poodles (see Figure 3.2.1). So $A \nsubseteq B$. Therefore, the statement 'All dogs are poodles' is false.

Some stands for **at least one.** So a statement like 'Some roses are red' will be true from our real-life experience. This is because there is a rose which is red in colour. There may be more than one rose that is red in colour but what concerns us is whether there is at least one. Again 'Some girls have long hair' is true as there are girls (and so at least one girl) with long hair. On the other hand the statement 'Some snakes have legs' is clearly false as no snake has legs, that is, there is not even one snake which has legs. A statement like, 'some deer are carnivorous' is false as there is no deer which is predatory or which feeds on other animals.

Figure 3.2.1 Dogs: Labrador and Poodle. [C2] and [C3]

None or **no** stand for **not even one.** A statement like 'No rose is red' can be true only if not even one rose is red in colour. But as we have seen earlier there are roses that are red in colour and therefore the statement 'No rose is red' is false from our real-life experience. The statement 'None of the girls have long hair' is false from our experience as there are girls with long hair. The statements 'No cat can fly' and 'No new born human baby can walk' are true statements.

Quick Review

After reading this section, you should be able to:

- **Differentiate between quantified statements and statements without quantifiers.**
- **Decide the conditions under which a given quantified statement will be true.**
- **Decide the conditions under which a given quantified statement will be false.**

EXERCISE 3.2

1. Use your everyday experience to decide the truth value of the following statements. Please give reasons for your answers.
   (a) All women in India wear bindis.
   (b) None of the boys in this class are shorter than 3 feet.
   (c) Some deer are herbivores.
   (d) Some deer are not herbivores.
   (e) The set $\{x \in \mathbb{N} \mid 2 < x < 3\}$ is an empty set.
2. Using a set theoretic interpretation, answer the following questions.
   (a) If the statement 'All A are B' is true, then which of the following statements will also be true?
      (i) Some A are B.
      (ii) Some A are not B.
      (iii) None of the A are B.
   (b) If the statement 'Some A are B' is true, then which of the following statements will also be true?

(i) Some B are A.
(ii) All A are B.
(iii) None of the B are A.

(c) If the statement 'None of the A are B' is true, then which of the following statements will also be true?
(i) Some A are not B.
(ii) Some B are A.
(iii) Some B are not A.

3. Write down using the notation of sets, a relation between sets A, B that represents as accurately as possible the following statements.
(a) All A are B.
(b) Some A are B.
(c) No A is a B.
(d) Some A are not B.

4. Read the passage given below.
Usha is a young girl studying in Lovely Public School in Delhi. She travels by bus to her school which is 13 km from her house. She studies in eighth standard. Her class has 45 students, of these 25 are girls and 20 are boys. Her favourite subject is English, but she dislikes mathematics. She is usually given some homework every weekday and has tests every Wednesday. Now, answer the following questions.
(a) State whether true or false based **only on the information in the passage above.** In case the data given in the passage is insufficient to decide whether the given statements are true or false, then state 'Data is insufficient'. Give reasons for your answers.
(i) Some students in Delhi have to travel by bus to school.
(ii) No student of Lovely Public School stays beyond 5 km of the school.
(iii) Some students in the eighth standard do not like mathematics.
(iv) Some students in the eighth standard like English.
(v) Some students in the eighth standard do not like English.
(vi) There are many more girls than boys in Usha's class.
(vii) All eighth standard classes have more girls than boys.
(viii) If Usha has a test, then it must be a Wednesday.
(ix) If it is a Wednesday, then Usha has a test.
(x) Usha has homework to do on Mondays.
(b) If Usha was not given any homework and she tells her mother that she has completed all her homework, is she lying?

3.3 Negating Simple Statements

In the previous section, we analysed the truth value of simple statements with quantifiers. This analysis was based on our real-life experience. In this section, we will move from the concrete to the abstract. Given any simple statement, we know that it is either true or false, but not both. Our task in this section is to understand the concept of 'negation' of a simple statement. We will use small letters such as p, q, r, s and so on to represent statements.

Indeed, given any statement p, we can define the **negation of** p to be a statement which has the opposite truth value to that of p. The negation of p is denoted by $\sim p$ and is also called **not** p.

Let us begin with some examples where we negate simple statements without quantifiers. Let p be the statement 'Today is a Friday'. Then $\sim p$ is the statement 'Today is not a Friday'. It is obvious that if p is true, that is, if today is indeed a Friday, then the statement 'Today is not a Friday' will be false. On the other hand, if p is false, that is, today is a day other than Friday, then the statement 'Today is not a Friday' will be true. Thus, we see that p and $\sim p$ have the opposite truth values in this example. Suppose q is the statement 'Zehra likes coffee. Then $\sim q$ is the statement 'Zehra does not like coffee'. Similarly, if r is the statement 'John knows to drive a car', then its negation would be $\sim r$, that is, 'John does not know to drive a car'.

So what exactly do we mean by saying 'not p' has the opposite truth value? By this we mean that, 'not p' or equivalently '$\sim p$' is false whenever p is true and $\sim p$ is true whenever p is false. Given any statement p, its negation $\sim p$ always exists. In the rest of this section, we will concentrate on the form and structure that $\sim p$ has when p is a simple statement.

Example 3.3.1 Given below are some more examples of simple statements without quantifiers and their negations (Figure 3.3.1).

(a) p: Today is not a Monday.
$\sim p$: Today is a Monday.

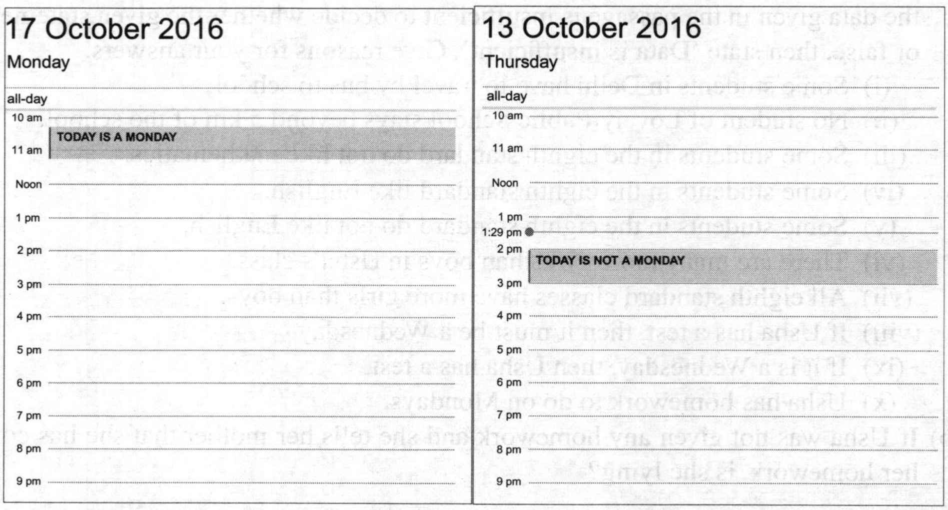

Figure 3.3.1 Negation: Example 1. [C4]

(b) q: I live in Delhi.
$\sim q$: I do not live in Delhi.

(c) r: She is wearing a skirt.

$\sim r$: She is not wearing a skirt.

(d) s: The number 2 is less than the number 3.
$\sim s$: The number 2 is not less than the number 3.

(e) t: The set $\{x \in \mathbb{N} \mid x < -1\}$ is empty.
$\sim t$: The set $\{x \in \mathbb{N} \mid x < -1\}$ is not empty. □

Sometimes there is a tendency to mistake 'opposites' for negations. For example, let p be the statement 'The number of mammals is lesser than the number of insects'. A common mistake made is to negate p as 'The number of mammals is greater than the number of insects'. In this instance, 'lesser' has been replaced by its opposite 'greater'. However, the negation of p is the statement $\sim p$ which is 'The number of mammals is not lesser than the number of insects' or equivalently 'The number of mammals is greater than or equal to the number of insects'. Similarly, if q were the statement 'Hari is shorter than Susan', then $\sim q$ will be 'Hari is not shorter than Susan', and the statement 'Hari is taller than Susan' **is not** the negation of q.

Note that when we negate a statement p, we are not bothered about its truth value with respect to our real-life experience, but instead we try to create a statement which will have the exact opposite truth value to that of p.

A simple fact to note is that the negation of $\sim p$ is p, that is, **not not** p is the same as p. The statement 'not not p' is called a **double negation** and we take $\sim (\sim p)$ to be p. For example, consider the statement q given by 'It is not true that Meera does not like to drink milk'. If we take p as the statement 'Meera likes to drink milk', then the given statement q is $\sim (\sim p)$. The statement q is actually the same as p, namely that, 'Meera likes to drink milk'.

Example 3.3.2 Given below are two examples of statements which are double negations.

(a) It is not the case that today is not a Sunday. Here, p is the statement 'Today is a Sunday'. The given statement is $\sim (\sim p)$ and is the same as p.

(b) It is not true that the Sahara Desert does not have a hot climate. The above statement is of the form $\sim (\sim p)$ and is the same as the statement p given by 'The Sahara Desert has a hot climate'. □

We turn our attention now to negating simple statements with quantifiers. Once again, we will not be concerned here about the truth value of the given statement p based on real life but rather on constructing a new statement $\sim p$ that will have the opposite truth value from p. So that if p were true, $\sim p$ would be false; and if p were false, $\sim p$ would be true.

Recollect that 'all' stands for 'each and every'. What would be the negation of the statement p where p stands for 'All birds fly'. To find the answer, let us first examine when p would be true and when it would be false.

'All birds fly' would be true only if each and every bird flies. It would be false if there is at least one bird that does not fly. We want a statement which would be false when p is true, and if p is false, the statement should be true. In the process of examining when p is true and when it is false, we have already made progress. We saw that p would be false if there is at least one bird that does not fly. Consider the statement 'Some birds do not fly'. If p is true, then each and every bird flies and so the statement 'Some birds do not fly' will be false (as each and every bird flies). On the other hand, If p is false, then it must be the case that there is at least one bird which does not fly; thus, the statement 'Some birds do not fly' becomes true. Consequently, if

p: All birds fly,

then

$\sim p$: Some birds do not fly.

Some more examples of statements with 'all' and their negations are given below. Is there a pattern that you can spot in negating a simple statement with the quantifier 'all' in it?

Example 3.3.3 Given below are examples of simple statements with the quantifier **all** and their negations.

(a) p: All girls have long hair.
$\sim p$: Some girls do not have long hair.

(b) q: All deer are herbivores.
$\sim q$: Some deer are not herbivores.

(c) r: All fish swim.
$\sim r$: Some fish do not swim.

(d) s: All men have moustaches.
$\sim s$: Some men do not have moustaches.

(e) t: All girls are tall.
$\sim t$: Some girls are not tall. □

There is an obvious pattern to the negations given above.

> The negation of
>
> **'All** $\cdots$ **are** $\cdots$'
>
> is
>
> **'Some** $\cdots$ **are not** $\cdots$'.

Let us now negate statements with the quantifier 'some'. Let the statement p be 'Some birds fly'. Now p will be true only if there is at least one bird that flies and p will be false if there is no bird that flies. Consider the statement 'No bird flies'. If p is true, then 'No bird flies' is false; and if p is false, then 'No bird flies' is true. Thus, $\sim p$ is ' No bird flies'. Some more examples are given below.

Example 3.3.4 Given below are examples of simple statements with the quantifier **some** and their negations.

(a) p: Some girls have long hair.
$\sim p$: No girl has long hair.

(b) q: Some deer are herbivores.
$\sim q$: No deer is a herbivore.

(c) r: Some fish swim.
$\sim r$: No fish swim.

(d) s: Some men have moustaches.
$\sim s$: No man has a moustache.

(e) t: Some girls are tall.
$\sim t$: No girl is tall. □

Again there is an obvious pattern to the negations given above.

> The negation of
>
> **'Some** $\cdots$ **are** $\cdots$'
>
> is
>
> **'None** $\cdots$ **are** $\cdots$'.

How do we negate statements of the type **'Some** $\cdots$ **are not** $\cdots$'? For example, what is the negation of 'Some birds do not fly'? We saw earlier that if p had the form **'All** $\cdots$ **are** $\cdots$', then $\sim p$ had the form **'Some** $\cdots$ **are not** $\cdots$'. Since $\sim (\sim p)$ is p, we see that

> The negation of
>
> **'Some** $\cdots$ **are not** $\cdots$'
>
> is
>
> **'All** $\cdots$ **are** $\cdots$'.

Thus, the negation of 'Some birds do not fly' is 'All birds fly'. Using the fact that $\sim (\sim p)$ is p, we can also note that

> The negation of
>
> **'No** $\cdots$ **is** $\cdots$'
>
> is
>
> **'Some** $\cdots$ **is** $\cdots$'

Thus, the negation of 'No deer is a carnivore' will be 'Some deer is a carnivore'. We end this section with some more examples.

Example 3.3.5 Given below are examples of simple statements with quantifiers and their negations.

(a) p: No girl has long hair.
$\sim p$: Some girl has long hair.

(b) q: All tigers are herbivores.
$\sim q$: Some tigers are not herbivores.

(c) r: Some humans swim.
$\sim r$: No human swims.

(d) s: Some men do not have long hair.
$\sim s$: All men have long hair.

(e) t: No new born baby can walk.
$\sim t$: Some new born babies can walk. □

Quick Review

After reading this section, you should be able to:

- Negate a simple statement without a quantifier.
- Identify double negations.
- Negate a statement of the type 'All ⋯ are ⋯'.
- Negate a statement of the type 'Some ⋯ are ⋯'.
- Negate a statement of the type 'Some ⋯ are not ⋯'.
- Negate a statement of the type 'None ⋯ are ⋯'.

EXERCISE 3.3

1. Negate the following statements.
   (a) All women in India wear bindis.
   (b) Some women in India wear bindis.
   (c) None of the women in India wear bindis.
   (d) Some women in India do not wear bindis.
   (e) The set $\{x \in \mathbb{N} \mid 2 < x < 4\}$ has one element.

2. Choose the statement that is the same as the given double negation.
   (a) q: 'It is not true that there is no water in the well' is the same as
      (i) There is water in the well.
      (ii) There is no water in the well.
      (iii) There may not be water in the well.
      (iv) There may be water in the well.
   (b) q: 'It is not the case that no Indian women wear bindis' is the same as
      (i) All Indian women wear bindis.
      (ii) Some Indian women wear bindis.

(iii) No Indian woman wears bindis.
(iv) Indian women do not wear bindis.

(c) q: 'It is not the case that some basketball players are not tall' is the same as
(i) Some basketball players are tall.
(ii) Some basketball players are short.
(iii) All basketball players are tall.
(iv) No basketball player is tall.

(d) q: 'It is not the case that no bird can swim' is the same as
(i) Some bird can swim.
(ii) No bird can swim.
(iii) All birds can swim.
(iv) Some bird cannot swim.

(e) q: 'It is not true that all flowers are red' is the same as
(i) All flowers are red.
(ii) No flower is red.
(iii) Some flower is red.
(iv) Some flower is not red.

3.4 Compound Statements: An Introduction

In the previous section, we analysed simple statements from the point of view of logic. As a first step, we learnt to use our real-life experience to decide whether a simple statement is true or false. The negation of a statement was defined and we negated simple statements. In this section, we will proceed further with the analysis of statements by turning our attention to compound statements. As mentioned previously, a compound statement is a statement constructed from two or more simple statements by using logical connectives such as 'and', 'or', 'but', 'if...then' and so on.

We will concern ourselves with the case where two simple statements have been linked by a connective to make a compound statement.

Compound statements constructed from two simple statements will be studied in three categories: conjunctions, disjunctions and implications.

Let p and q be simple statements. Then, the following compound statements define conjunction and disjunction. Implications will be studied in subsequent sections.

The compound statement 'p and q' is called a **conjunction.** It is denoted by $p \wedge q$ and is also called 'p **meet** q'. Some examples are as follows:

(a) 'I have a dog and I have a cat'. Here, p is the statement 'I have a dog' and q is 'I have a cat'. The compound statement can be written symbolically as $p \wedge q$.

(b) 'She can sing and dance'. Here, p is the statement 'She can sing' and q is 'She can dance'. The compound statement can be written symbolically as $p \wedge q$.

Sometimes 'but' or 'while' is used instead of 'and'. For example, 'Today is a Sunday but the shops are open', 'The boys are wearing blue socks, while the girls are wearing red socks'.

The compound statement 'p or q' is called a **disjunction.** It is denoted by $p \vee q$ and is also called 'p **join** q'. Some examples are as follows:

(a) 'I have a dog or I have a cat'. Here, p is the statement 'I have a dog' and q is 'I have a cat'. The compound statement can be written symbolically as $p \vee q$.
(b) 'She can sing or dance'. Here, p is the statement 'She can sing' and q is 'She can dance'. The compound statement can be written symbolically as $p \vee q$.

As with simple statements, we will first explore the truth values of conjunctions and disjunctions both via real-life experiences and in abstract. Every statement has a negation; we will also learn to negate conjunctions and disjunctions in subsequent sections.

We end this section with the concept of **equivalent statements.** A statement p is **equivalent** to the statement q if p and q always have the exact same truth value. That is, if p is true, then q must be true; and if p is false, then q must be false. We denote this by $p \equiv q$. If there is an instance when p has a different truth value from q, then we say that p is **not equivalent** to q and denote it as $p \not\equiv q$. Some examples are as follows:

(a) 'I have a dog and I have a cat' is equivalent to 'I have a cat and I have a dog'.
(b) 'She can sing or dance' is equivalent to 'She can dance or sing'.
(c) 'Some girls wear ribbons' is equivalent to 'There is a girl wearing a ribbon'.
(d) 'Some dogs are poodles' is not equivalent to 'Some dogs are not poodles'.

Try to give reasons for why the above statements are equivalent or not. Is the statement 'If it rains, then I will not be able to play' equivalent to 'It does not rain or I will not be able to play'? In a few pages hence we will equip ourselves with tools which will help us decide in a systematic way as to when two statements are equivalent.

Quick Review

After reading this section, you should be able to:

- Recognise a compound statement and identify the simple statements from which it has been constructed.
- Recognise and work with conjunctions and disjunctions.
- Decide when two statements are equivalent.

EXERCISE 3.4

1. Categorise the following compound statements into conjunctions and disjunctions. Identify the simple statements from which the compound statements have been constructed. Write the given compound statement symbolically.
   (a) Today is a Sunday and I can sleep late.
   (b) She can sketch, but she cannot paint.
   (c) He can walk or he can run.
   (d) Every tiger is a carnivore, but not all carnivores are tigers.
   (e) He can eat chocolates or candy bars.

2. Let p, q and r be the statements. Then show the following.
   (a) $p \equiv p$.
   (b) If $p \equiv q$, then $q \equiv p$.
   (c) If $p \equiv q$ and $q \equiv r$, then $p \equiv r$.
   (d) $(p \wedge q) \equiv (q \wedge p)$.
   (e) $(p \vee q) \equiv (q \vee p)$.

3.5 Conjunctions and Disjunctions

We saw a few examples of conjunctions and disjunctions in the previous section. We will now study conjunctions and disjunctions further by trying to determine their truth values on the basis of the truth values of the simple statements from which they have been constructed. A word of caution: mathematically the connective 'and' stands for 'both'. So when we say 'she can dance and she can sing', we mean that she can both dance and sing. Continuing in the mathematical vein, the connective 'or' is an 'inclusive or'. When we say 'she can sing or she can dance', we mean that either she can only sing or she can only dance or that she can do both. This is indeed the same underlying concept that we use when we talk about the union of two sets: if A and B are sets, then $A \cup B$ is the set whose elements belong to A **or** B. The 'or' here means that the element could either be in just A or just in B or in both. These concepts have to be used when we want to determine the truth values of conjunctions and disjunctions.

Example 3.5.1 Consider the following statements:
p: I have a cat.
q: I have a dog
$p \wedge q$: I have a cat and I have a dog.

The statement 'I have a cat and I have a dog' is true only when I have both a cat and a dog. If it turns out that I have only one of a cat or a dog or if I do not have either a cat or a dog, then the statement 'I have a cat and I have a dog' will be false. Thus, $p \wedge q$ will be true if and only if both p and q are true. □

Therefore, we have the following rule for the **truth value of a conjunction.**

> $p \wedge q$ **will be true if and only if both** p **and** q **are true.** $p \wedge q$ **will be false if either** p **is false or** q **is false.**

Below are some examples where the above rule is used.

Example 3.5.2 Using your everyday experience, decide on the truth value of the following statements.

(a) Every dog is an Alsatian, but every tiger is a cat (Figure 3.5.1).
Here, p is 'Every dog is an Alsatian' and q is 'Every tiger is a cat'. The above statement is the conjunction $p \wedge q$. By our real-life experience, p is false as there are dogs that are not Alsatians, whereas every tiger belongs to the cat family. So in this case, the given conjunction $p \wedge q$ will be false as p is false.

Figure 3.5.1 Tigers are cats. [C4]

(b) All Alsatians are dogs, while all dogs are mammals.
Here, p is 'All Alsatians are dogs' and q is 'All dogs are mammals'. The above statement is the conjunction $p \wedge q$. By our real-life experience, p is true as Alsatians or German Shepherds are a breed of dog. Further, q is also true as dogs belong to the class of mammals. So in this case, the given conjunction $p \wedge q$ will be true as both p and q are true.

(c) All humans swim and all birds can fly.
Here, p is 'All humans swim' and q is 'All birds can fly'. The above statement is the conjunction $p \wedge q$. By our real-life experience, p is false as there are human beings who cannot swim and q is also false as penguins are birds which cannot fly. So in this case, the given conjunction $p \wedge q$ will be false as both p and q are false. Note that if we knew that one of p or q is false, then $p \wedge q$ will be false.

(d) Some men do not have long hair and no woman has long hair.
Here p is 'Some men do not have long hair' and q is 'No woman has long hair'. The above statement is the conjunction $p \wedge q$. By our real-life experience, p is true as there is at least one man who has short hair and q is false as there is at least one woman with long hair. So in this case, the given conjunction $p \wedge q$ will be false as q is false. □

Now let us turn our attention to disjunctions.

Example 3.5.3 Consider the following statements.

p: I have a cat.

q: I have a dog

$p \vee q$: I have a cat or I have a dog.

The statement 'I have a cat or I have a dog' will be true in each of the following three circumstances: (a) If I only have a cat. (b) If I only have a dog. (c) If I have both a cat and a dog. (Recall that 'or' here is inclusive in its usage.) The statement 'I have a cat or I have a dog' will be false only when I neither have a cat nor do I have a dog. Thus, the only condition under which $p \vee q$ will be false is if both p and q are false; and $p \vee q$ will be true if either statement p is true or statement q true (or both). □

Thus, we have the following rule for the **truth value of a disjunction.**

> $p \vee q$ **will be false if and only if both** p **and** q **are false.** $p \vee q$ **will be true if either** p **is true or** q **is true.**

Below are some examples where the above rule is used.

Example 3.5.4 Using your everyday experience, decide on the truth value of the following statements.

(a) Every dog is an Alsatian or every tiger is a cat.
Here, p is 'Every dog is an Alsatian' and q is 'Every tiger is a cat'. The above statement is the disjunction $p \vee q$. By our real-life experience, p is false as there are dogs that are not Alsatians, whereas every tiger belongs to the cat family. So in this case, the given disjunction $p \vee q$ will be true as q is true.

(b) All Alsatians are dogs or dogs are mammals (Figure 3.5.2).

All Alsatians are dogs.

Figure 3.5.2 All Alsatians are dogs. [C5]

Here, p is 'All Alsatians are dogs' and q is 'All dogs are mammals'. The above statement is the disjunction $p \vee q$. Note that p is true as Alsatians or German Shepherds are a breed of dog. Further, q is also true as dogs belong to the class of mammals. So in this case, the given disjunction $p \vee q$ will be true as both p and q are both true. (Note we only need one to be true for $p \vee q$ to be true.)

(c) All humans swim or all birds can fly.
Here p is 'All humans swim' and q is 'All birds can fly'. The above statement is the disjunction $p \vee q$. Clearly, p is false as there are human beings who cannot swim and q is also false as penguins are birds which cannot fly. So in this case, the given disjunction $p \vee q$ will be false as both p and q are both false.

(d) Some men do not have long hair or no woman has long hair.
Here, p is 'Some men do not have long hair' and q is 'No woman has long hair'. The above statement is the disjunction $p \wedge q$. Here, p is true as there is at least one man who has short hair and q is false as there is at least one woman with long hair. So in this case the given disjunction $p \vee q$ will be true as q is true. □

An efficient way to represent the possible truth values of a disjunction or a conjunction is to use the concept of a truth table. The next section deals in detail with truth tables. We end this section with a Quick Review and Exercise.

Quick Review

After reading this section, you should be able to:

- Decide the conditions under which a conjunction will be true.
- Decide the conditions under which a conjunction will be false.
- Decide the conditions under which a disjunction will be true.
- Decide the conditions under which a disjunction will be false.

EXERCISE 3.5

1. Write the following conjunctions and disjunctions in a symbolic form by specifying the statements p and q. Use your everyday experience to decide the truth value of p, q and then that of the corresponding disjunction or conjunction. Give reasons for your answers.
   (a) I like watching Hindi movies and listening to rock music.
   (b) Every rose is a flower and every flower is red.
   (c) No boy in the Maths class is greater than 3 feet and no girl in Maths class is less than 3 feet.
   (d) Some deer are not herbivores or some tigers are not carnivores.
   (e) All Indian women wear sarees or all Indian men wear dhotis.

2. From your real-life experience, give examples of simple statements p and q satisfying the conditions given below. For each of the examples that you give, find the truth value of $p \wedge q$ and $p \vee q$.

(a) p is true, q is true.
(b) p is true, q is false.
(c) p is false, q is true.
(d) p is false, q is false.

3.6 Truth Tables

Let p and q be simple statements. Consider the following compound statement: $(\sim p) \vee q$.

Imagine the following game. You are told a truth value of p and that of q, and you have to calculate the truth value of $(\sim p) \vee q$. If you give the right answer, then you win ₹100, and if you get it wrong, you lose ₹100. You are again given a chance to answer the same question but with a different choice of truth values of p and of q. Now you can win ₹200 if you get the answer right, but if you answer wrong, you lose ₹200. Suppose, that this game continues with the amount you earn doubling each time you answer correctly and losing a similar amount if you answer wrongly. What is the maximum amount you can possibly earn or lose?

The maximum amount you can earn or lose will depend on how many different truth value inputs you can have for the simple statements p, q. Now, for both p and for q, we have two choices each for the input truth values, namely true and false. Thus, the total number of possibilities for the input truth values are $2 \times 2 = 4$. Thus, if for each of these four input scenarios, you get the output, namely the truth value of $(\sim p) \vee q$ is correct, then you stand to win a handsome sum of ₹100 + ₹200 + ₹400 + ₹800 = ₹1500. Also, you do not want to lose any money so you wish to use your understanding of negations, disjunctions to determine the truth value of $(\sim p) \vee q$ for each of the four input scenarios correctly. So let us make sure that you win the ₹1500 possible by learning more about finding truth values of compound statements from those of their simple components.

It is easiest to represent these input truth values and the corresponding output truth values in a table. In fact, the table can be constructed in such a way that the intermediate steps can also be reflected. Such a table is called a **truth table.** Before we construct the truth table for the two compound statements above, let us begun with easier truth tables, that of the negation, disjunction and conjunction.

Table 3.6.1 Truth Table for Negation

p	$\sim p$
T	F
F	T

T stands for 'True' and F stands for 'False'. In Table 3.6.1, the first column consisting of the two truth values for p is called the input column. p is the only input variable here. The second column with truth values of $\sim p$ is the output column. This table tells us that when p is true, $\sim p$ is false and when p is false, then $\sim p$ is true.

We have used colour schemes to help differentiate between the input columns, columns representing intermediate steps and the output columns. The input columns have been coloured

■, the columns representing intermediate steps, if any, are coloured ▒, and the output columns are mostly coloured □.

Table 3.6.2 Truth Table for Disjunction

p	q	$p \vee q$
T	T	T
T	F	T
F	T	T
F	F	F

In Table 3.6.2, as can be seen the by the colour code used, the first column consisting of truth values for p and the second column consisting of truth values for q are the input columns. The third column with truth values of $p \vee q$ is the output column. Note that there are five rows. The first row is the header row in which the columns are labelled by statements. The subsequent rows are determined by the possible choices we have of truth value combinations of the input variables. In this case, the input variables are p and q, each can take a value of true (T) or false (F). Thus, the possible combinations for truth values for p and q are TT, TF, FT and FF. These form the rest of the rows of the truth table.

Table 3.6.3 Truth Table for Conjunction

p	q	$p \wedge q$
T	T	T
T	F	F
F	T	F
F	F	F

In Table 3.6.3, the first and second columns are the input columns and the third column is the output column. In this case too, the input variables are p and q, and each can take a value of true (T) or false (F). Thus, the possible combinations for truth values for p and q are TT, TF, FT and FF. Each row of the truth table gives us information about the output truth value based on the input truth values. For example, the third row says that if p is false and q is true, then $p \wedge q$ must be false.

Now, we are in a position to create truth tables for compound statements, involving negations or disjunctions or conjunctions. Let us consider the first example we began with namely $(\sim p) \vee q$.

In Table 3.6.4, the first and second columns are the input columns. The fourth and final column is the output column, and it displays the truth values of $(\sim p) \vee q$. In this case too, the input variables are p and q. Note that there is one other column in between, namely, the third column labelled by $\sim p$. This column represents the output of an intermediate step and is colour coded accordingly. The truth values of $\sim p$ are opposite to the truth values of p given in the first column.

Table 3.6.4 Truth Table for $(\sim p) \vee q$

p	q	$\sim p$	$(\sim p) \vee q$
T	T	F	T
T	F	F	F
F	T	T	T
F	F	T	F

The truth values in the final column are calculated on the basis of the inputs from the third column and second columns. For example, consider the first row: the truth value of $\sim p$ is F and the truth value of q is T; hence, the truth value of $(\sim p) \vee q$ will be T. This can be seen from the third row of the disjunction table.

Let p, q and r be the simple statements and let us consider the compound statement $(p \vee q) \wedge r$. This is a compound statement that has been constructed out of more than two simple statements. Such statements can be studied by breaking down the analysis to compound statements involving only two simple statements at a time.

Table 3.6.5 Truth Table for $(p \vee q) \wedge r$: Stage I

p	q	r	$p \vee q$	$(p \vee q) \wedge r$
T	T	T	?	?
T	T	F	?	?
T	F	T	?	?
T	F	F	?	?
F	T	T	?	?
F	T	F	?	?
F	F	T	?	?
F	F	F	?	?

In Table 3.6.5, there are three input variables. These are p, q and r. Each of these '3' statements can take '2' values: true (T) and false (F). Thus, we will have $2 \times 2 \times 2 = 2^3 = 8$ input rows. These are TTT, TTF, TFT, TFF, FTT, FTF, FFT, FFF. Once these input truth values have been entered in the truth table, then the truth values of $p \vee q$ in the fourth column have to calculated. The values for $p \vee q$ will depend on the corresponding input truth values for p and q as entered in Columns 1 and 2. We continue with the next stage of the table where we will fill the fourth column. Since the statement in the fourth column is $p \vee q$, we can use Table 3.6.2 which describes the truth values of a disjunction.

Now, we are in a position to complete the truth table for $(p \vee q) \wedge r$. For completing the final column, our input Columns will be the intermediate Column 4 (which gives the values for $p \vee q$) and Column 3 which gives the values for r. Since the final statement is a conjunction, we will need information from Table 3.6.3 to complete the final column.

Table 3.6.6 Truth Table for $(p \vee q) \wedge r$: Stage II

p	q	r	$p \vee q$	$(p \vee q) \wedge r$
T	T	T	T	?
T	T	F	T	?
T	F	T	T	?
T	F	F	T	?
F	T	T	T	?
F	T	F	T	?
F	F	T	F	?
F	F	F	F	?

Table 3.6.7 Truth Table for $(p \vee q) \wedge r$: Stage III

p	q	r	$p \vee q$	$(p \vee q) \wedge r$
T	T	T	T	T
T	T	F	T	F
T	F	T	T	T
T	F	F	T	F
F	T	T	T	T
F	T	F	T	F
F	F	T	F	F
F	F	F	F	F

Although we have shown the intermediate stages of the truth table construction (Tables 3.6.5 and 3.6.6), normally all the stages are shown together as in Table 3.6.7.

Constructing Truth Tables

It is time to sum up the steps that are required to construct a truth table.

The **first step is to identify the input variables** in the given statement. For example, in the statement $(p \vee q) \wedge r$, there are three input variables, namely p, q and r. In the statement $\sim p$, there is only one input variable, namely p. The statement $[p \rightarrow (\sim q)] \wedge [r \vee s]$ has four input variables p, q, r and s.

The **number of input variables decides the number of input rows** in a table. If there are n input variables, the truth table will have 2^n input rows. There will also be a header row labelled by the statements. Thus, the truth table for $(p \vee q) \wedge r$ will have $2^3 = 8$ input rows, the truth table for

$\sim p$ will have $2^1 = 2$ input rows and the truth table for $[p \to (\sim q)] \wedge [r \vee s]$ will have $2^4 = 16$ input rows.

Broadly, the **number of connectives in the statement and the number of input variables decide the number of columns in a truth table.** For calculating the number of columns, $\sim$ will also be considered a connective. The number of columns will be the sum of the number of connectives and the number of input variables.

In the statement $(p \vee q) \wedge r$, we have two connectives, namely, $\vee$ and $\wedge$. We also have three input variables: p, q and r. The truth table for $(p \vee q) \wedge r$ will have $2 + 3 = 5$ columns.

For the statement $\sim p$, we will have $1 + 1 = 2$ columns as there is only one connective ($\sim$) and only one input variable (p).

The statement $[p \to (\sim q)] \wedge [r \vee s]$ has four connectives ($\to, \sim, \wedge, \vee$) and four input variables (p, q, r, s). Thus, its truth table will have $4 + 4 = 8$ columns.

Of course, we still have not learnt the truth table for $p \to q$. Once we know this, we can complete a truth table for $[p \to (\sim q)] \wedge [r \vee s]$. We present below the header row for $[p \to (\sim q)] \wedge [r \vee s]$ (Table 3.6.8).

Table 3.6.8 Header Row of the Truth Table for $[p \to (\sim q)] \wedge [r \vee s]$

p	q	r	s	$\sim q$	$p \to \sim q$	$r \vee s$	$[p \to (\sim q)] \wedge [r \vee s]$

Tautologies and Contradictions

A **tautology** is a statement which is always true. It cannot be false for any truth value of its input variables.

Consider the statement $p \vee \sim p$. If p is true, then we know that $\sim p$ is false and $p \vee \sim p$ will be true. If p is false, then we know that $\sim p$ is true, so $p \vee \sim p$ will be true. Thus, $p \vee \sim p$ is always true for any truth value of the variable p and is an example of a tautology. This establishes the fact that a compound statement of the type 'p or not p' will always be true.

On the other hand, consider the statement $p \wedge q$. If p is true and q is false, then $p \wedge q$ is false. Thus, there are truth values of the input variable for which the given statement $p \wedge q$ is false. Hence, it is not a tautology.

We can use a truth table to decide if a statement is a tautology. Recall that the final column in the truth table represents the truth value of the given statement. So a given statement is a tautology if and only if every entry in the final column is T. Let us consider some statements and create their truth tables to decide whether they are tautologies or not.

Example 3.6.1 Construct a truth table for the statement $(\sim p \wedge q) \vee (p \vee \sim q)$. Decide whether it is a tautology or not. Give reasons.

The truth table of Example 3.6.1 will have $5 + 2 = 7$ columns and 5 rows and is given in Table 3.6.9.

Note that the final column of Table 3.6.9 has all entries as T. Thus, the statement $(\sim p \wedge q) \vee (p \vee \sim q)$ is a tautology.

Table 3.6.9 Truth Table for $(\sim p \wedge q) \vee (p \vee \sim q)$

p	q	$\sim p$	$\sim q$	$\sim p \wedge q$	$p \vee \sim q$	$(\sim p \wedge q) \vee (p \vee \sim q)$
T	T	F	F	F	T	T
T	F	F	T	F	T	T
F	T	T	F	T	F	T
F	F	T	T	F	T	T

□

Example 3.6.2 Construct a truth table for the statement $(p \vee q) \wedge (\sim p \vee \sim q)$. Decide whether it is a tautology or not. Give reasons.

The truth table for $(p \vee q) \wedge (\sim p \vee \sim q)$ is given in Table 3.6.10.

Table 3.6.10 Truth Table for $(p \vee q) \wedge (\sim p \vee \sim q)$

p	q	$\sim p$	$\sim q$	$p \vee q$	$\sim p \vee \sim q$	$(p \vee q) \wedge (\sim p \vee \sim q)$
T	T	F	F	T	F	F
T	F	F	T	T	T	T
F	T	T	F	T	T	T
F	F	T	T	F	T	F

As the last row of the final column of Table 3.6.10 has F as an entry, the statement $(p \vee q) \wedge (\sim p \vee \sim q)$ is not a tautology. □

Supposing that r is a statement that is a tautology, what can we say about $\sim r$? It is obvious that $\sim r$ would take the truth value F for all input truth values. In other words, $\sim r$ will always take truth value 'false'.

A **contradiction** is a statement which is always false. It cannot be true for any truth value of its input variables. Clearly, the negation of a tautology is a contradiction and vice versa.

Consider the statement $p \wedge \sim p$. If p is true, then we know that $\sim p$ is false, and, therefore, $p \wedge \sim p$ will be false. If p is false, then we know that $\sim p$ is true, so $p \wedge \sim p$ will be false. Thus, $p \wedge \sim p$ is always false for any truth value of the variable p and is an example of a contradiction. This shows that a compound statement of the type 'p and not p' will always be false and hence a contradiction.

On the other hand, consider the statement $p \vee q$. If p is true and q is false, then $p \vee q$ is true. Thus, there are truth values of the input variable for which the given statement $p \vee q$ is true. Hence, it is not a contradiction.

Truth tables can be used in this case too to decide if a statement is a contradiction. It is obvious that a given statement will be a contradiction if and only if every entry in the output or the final column of its truth table is F. Let us consider some statements and create their truth tables to decide whether they are contradictions or not.

Example 3.6.3 Construct a truth table for the statement $(\sim p \wedge q) \wedge (p \vee \sim q)$. Decide whether it is a contradiction or not. Give reasons.

Table 3.6.11 Truth Table for $(\sim p \wedge q) \wedge (p \vee \sim q)$

p	q	$\sim p$	$\sim q$	$\sim p \wedge q$	$p \vee \sim q$	$(\sim p \wedge q) \wedge (p \vee \sim q)$
T	T	F	F	F	T	F
T	F	F	T	F	T	F
F	T	T	F	T	F	F
F	F	T	T	F	T	F

Note that the final column of Table 3.6.11 has all entries as F. Thus, the statement $(\sim p \wedge q) \wedge (p \vee \sim q)$ is a contradiction. □

Now, consider the statement $(p \vee q) \wedge (\sim p \vee \sim q)$ of Example 3.6.2. Its truth table is given in Table 3.6.10. The final column entry of the second row is T; thus, the statement $(p \vee q) \wedge (\sim p \vee \sim q)$ is not a contradiction.

Establishing Logical Equivalence Using Truth Tables

In Section 3.4, we defined equivalent statements. Recall that two statements r and s are equivalent or 'logically equivalent' if they always have the same truth value. We denote this as $r \equiv s$. Truth tables can be used to determine whether two statements are logically equivalent or not.

Let r and s be the compound statements built up from the same input variables. Then, we can construct truth tables for both r and s, with the exact same input rows. Then, $r \equiv s$ if and only if the output column for r and s are identical. If both the output columns for r and s are the same, then it is obvious that for each set of input truth values, r and s have the exact same truth value and hence are logically equivalent. Now, let us consider some examples.

Example 3.6.4 Construct truth tables for the pairs of statements given below. Decide whether they are logically equivalent or not. Give reasons.

(i) $\sim p \wedge q, \sim (p \vee \sim q)$.
(ii) $\sim (p \vee \sim q), \sim p \vee \sim q$.

Instead of constructing two truth tables with the same input rows for the pair of statements given in Example 3.6.4 (i), we create a single truth table in which the two final columns represent the statements $\sim p \wedge q$ and $\sim (p \vee \sim q)$ (Table 3.6.12).

Table 3.6.12 Truth Table for $\sim p \wedge q$ and $\sim (p \vee \sim q)$

p	q	$\sim p$	$\sim q$	$p \vee \sim q$	$\sim p \wedge q$	$\sim (p \vee \sim q)$
T	T	F	F	T	F	F
T	F	F	T	T	F	F
F	T	T	F	F	T	T
F	F	T	T	T	F	F

As the two final columns representing the truth values of $\sim p \wedge q$ and $\sim (p \vee \sim q)$ are identical, we can state that $(\sim p \wedge q) \equiv \sim (p \vee \sim q)$.

Now, we consider the joint truth table for $\sim (p \vee \sim q)$ and $\sim p \vee \sim q$ as shown in Table 3.6.13. We note that the two final columns are not identical. In fact, the entries in the second input row and fourth input row for the two final columns differ. This shows that there are the same input truth values for which the two given statements have different truth values.

Table 3.6.13 Truth Table for $\sim (p \vee \sim q)$ and $\sim p \vee \sim q$

p	q	$\sim p$	$\sim q$	$p \vee \sim q$	$\sim (p \vee \sim q)$	$\sim p \vee \sim q$
T	T	F	F	T	F	F
T	F	F	T	T	F	T
F	T	T	F	F	T	T
F	F	T	T	T	F	T

For example, using the second input row, we see that when p is true and q is false, the statement $\sim (p \vee \sim q)$ is false, whereas the statement $\sim p \vee \sim q$ is true. Thus, $\sim (p \vee \sim q)$ is not logically equivalent to $\sim p \vee \sim q$. We denote this by $\sim (p \vee q) \not\equiv \sim p \vee \sim q$. □

Quick Review

After reading this section, you should be able to:

- Construct truth tables for compound statements involving disjunctions and conjunctions.
- Define a tautology and a contradiction.
- Decide whether the given statement is a tautology or a contradiction on the basis of the truth values in the final column of a truth table.
- Decide when two compound statements created out of the same input variables are logically equivalent on the basis of the two final column entries of the truth table.

EXERCISE 3.6

1. Which of the following are tautologies and which are contradictions? Give reasons.
   (a) $(p \vee r) \vee \sim (p \wedge q)$.
   (b) $(p \vee q) \wedge \sim (\sim q \wedge r)$.
   (c) $\sim [(p \wedge r) \wedge (\sim p \wedge \sim q)]$.
   (d) $(\sim p \wedge r) \vee (\sim p \vee r)$.
   (e) $\sim p \wedge (\sim q \wedge p)$.

2. Which of the following pairs are logically equivalent? Give reasons.
   (a) $(\sim p \wedge q) \vee (\sim p \vee q)$, $(\sim p \vee q)$.
   (b) $(p \vee q) \wedge r$, $(p \wedge r) \vee (q \wedge r)$.

(c) $(p \wedge q) \vee r$, $(p \vee r) \wedge (q \vee r)$.
(d) $(\sim p \vee q) \wedge \sim (\sim q \wedge r)$, $(p \wedge q) \vee r$.
(e) $(\sim p \wedge r) \vee (\sim p \vee r)$, $\sim [(p \vee r) \wedge (\sim r)]$.

3. Write down the truth table for the following statements. Using the truth table, decide if any pairs are logically equivalent. Are any of the statements tautologies or contradictions? Give reasons for all your answers.
   (a) $\sim p \vee (\sim q \wedge \sim r)$.
   (b) $\sim [\sim p \wedge (q \vee r)]$.
   (c) $\sim [p \wedge (q \vee r)]$.
   (d) $(\sim p \wedge (r \vee q)) \wedge (p \wedge (r \vee q))$.

3.7 De Morgan's Laws

The negation of a given statement was defined in Section 3.3. Given any statement r, the negation of r, denoted as $\sim r$, is the statement which always has a truth value that is, opposite of the truth value of r. In Section 3.3, we learnt to negate simple statements with and without quantifiers. Since then, we have become acquainted with some compound statements, namely disjunctions and conjunctions. If a given statement r is a conjunction or a disjunction, we should be able to negate r. **Augustus De Morgan**[2] discovered the 'rules' or 'laws' that allow us to negate disjunctions and conjunctions. Before we describe these laws, let us consider the truth table of $\sim (p \vee q)$ (Table 3.7.1).

Table 3.7.1 Truth Table for $\sim (p \vee q)$

p	q	$p \vee q$	$\sim (p \vee q)$
T	T	T	F
T	F	T	F
F	T	T	F
F	F	F	T

Note that in the final column, we have three rows with F and one row with T. Notice the similarity with the table for 'and' (see Table 3.6.3). So $\sim (p \vee q)$ is a statement that takes a true value if and only if p takes the value false, q takes the value false, which happens if and only if $\sim p$ and $\sim q$ are true.

So it is reasonable to conjecture that the negation of $p \vee q$ is $\sim p \wedge \sim q$. We prove precisely this by showing that $\sim (p \vee q)$ is logically equivalent to $\sim p \wedge \sim q$. Similarly, we also show that $\sim (p \wedge q)$ is logically equivalent to $\sim p \vee \sim q$.

[2] You can find out about the life and works of Augustus de Morgan in [W2].

As in the previous section, we construct joint truth tables and use them to show the required logical equivalences (Table 3.7.2).

Table 3.7.2 Truth Table for $\sim (p \vee q)$ and $\sim p \wedge \sim q$

p	q	$\sim p$	$\sim q$	$p \vee q$	$\sim (p \vee q)$	$\sim p \wedge \sim q$
T	T	F	F	T	F	F
T	F	F	T	T	F	F
F	T	T	F	T	F	F
F	F	T	T	F	T	T

As the two final columns representing the truth values of $\sim (p \vee q)$ and $\sim p \wedge \sim q$ are identical, we can state that $\sim (p \vee q) \equiv \sim p \wedge \sim q$.

Similarly, we note that the two final columns of Table 3.7.3 representing the truth values of $\sim (p \wedge q)$ and $\sim p \vee \sim q$ are identical, so we have $\sim (p \wedge q) \equiv \sim p \vee \sim q$.

Table 3.7.3 Truth Table for $\sim (p \wedge q)$ and $\sim p \vee \sim q$

p	q	$\sim p$	$\sim q$	$p \wedge q$	$\sim (p \wedge q)$	$\sim p \vee \sim q$
T	T	F	F	T	F	F
T	F	F	T	F	T	T
F	T	T	F	F	T	T
F	F	T	T	F	T	T

We are now in a position to negate disjunctions and conjunctions. The two laws stated below give us the precise rule.

> **De Morgan's Laws**
>
> Let p and q be statements. Then
>
> $$\sim (p \vee q) \equiv \sim p \wedge \sim q$$
>
> $$\sim (p \wedge q) \equiv \sim p \vee \sim q\,.$$

Let us see how to use the laws given above to negate conjunctions and disjunctions.

Example 3.7.1 Write the following conjunctions and disjunctions in a symbolic form by specifying the statements p and q. Negate the statements p and q in each case and then negate each of the given statements.

(a) Today is a Sunday and the market is closed.

(b) She likes dark chocolate or she likes white chocolate.

(c) Some students like mathematics, while some students do not like history.

(d) All birds fly or no animal flies.

In Example 3.7.1, part (a), the given statement is of the form $p \wedge q$, where

p : Today is a Sunday.

q : The market is closed.

By De Morgan's laws, $\sim (p \wedge q) \equiv \sim p \vee \sim q$. Thus, we need to find the negation of p as well as q. Clearly,

$\sim p$: Today is not a Sunday.

$\sim q$: The market is not closed.

Therefore, the negation of the given statement, $\sim (p \wedge q)$, will be given by

'Today is not a Sunday or the market is not closed'.

Let us analyse the other parts similarly. In Example 3.7.1, part (b), the given statement is of the form $p \vee q$, where

p : She likes dark chocolate.

q : She likes white chocolate.

By De Morgan's laws, $\sim (p \vee q) \equiv \sim p \wedge \sim q$. Clearly,

$\sim p$: She does not like dark chocolate.

$\sim q$: She does not like white chocolate.

Therefore, the negation of the given statement, $\sim (p \vee q)$, is

'She does not like dark chocolate and she does not like white chocolate'.

In the above examples, the simple the statements were the statements without quantifiers, the last two parts of Example 3.7.1, however, involve simple statements with quantifiers. The method of analysis is still the same. In part (c), the statement is of the form $p \wedge q$, where

p : Some students like mathematics.

q : Some students do not like history.

By De Morgan's laws, $\sim (p \wedge q) \equiv \sim p \vee \sim q$. Now,

$\sim p$: No student likes mathematics.

$\sim q$: All students like history.

Therefore, the negation of the given statement, $\sim (p \wedge q)$, will be given by

'No student likes mathematics or all students like history'.

In part (d) of Example 3.7.1, the statement is of the form $p \vee q$, where

$$p : \text{All birds fly.}$$

$$q : \text{No animal flies.}$$

By De Morgan's laws, $\sim (p \vee q) \equiv \sim p \wedge \sim q$. Now,

$$\sim p : \text{Some birds do not fly.}$$

$$\sim q : \text{Some animals fly.}$$

Therefore, the negation of the given statement, $\sim (p \vee q)$, will be given by

'Some birds do not fly and some animals fly'. □

Another way in which we can use De Morgan's laws is to negate more complex compound statements involving conjunctions and disjunctions. For example,

$$\begin{aligned} \sim [p \vee (\sim q \wedge r)] &\equiv \sim p \wedge \sim (\sim q \wedge r) && \text{since } \sim (s \vee t) \equiv \sim s \wedge \sim t \\ &\equiv \sim p \wedge [\sim (\sim q) \vee \sim r] && \text{since } \sim (s \wedge t) \equiv \sim s \vee \sim t \\ &\equiv \sim p \wedge (q \vee \sim r) && \text{since } \sim (\sim s) \equiv s \end{aligned}$$

There are several other laws that govern conjunctions and disjunctions. Some of these are listed below. These laws can be established by proving the logical equivalences using truth tables. Some of these might already have been listed as exercises or otherwise in earlier sections.

Laws Governing Disjunctions and Conjunctions

Let p, q and r be the statements. Then,

1. Law of idempotence

   $p \vee p \equiv p$, $\quad p \wedge p \equiv p$.

2. Law of absorption

   $p \vee (p \wedge q) \equiv p$, $\quad p \wedge (p \vee q) \equiv p$.

3. Commutative law

   $p \vee q \equiv q \vee p$, $\quad p \wedge q \equiv q \wedge p$.

4. Associative law

   $p \vee (q \vee r) \equiv (p \vee q) \vee r)$, $\quad p \wedge (q \wedge r) \equiv (p \wedge q) \wedge r)$

5. Distributive law

   $p \vee (q \wedge r) \equiv (p \vee q) \wedge (p \vee r), p \wedge (q \vee r) \equiv (p \wedge q) \vee (p \wedge r)$

Do notice the similarity of the rules described above for 'or', 'and' with the rules followed by 'union', 'intersection' of sets.

Quick Review

After reading this section, you should be able to:

- Prove the equivalences that give rise to De Morgan's laws.
- Be able to analyse and negate disjunctions and conjunctions.
- Be able to negate complex compound statements involving disjunctions and conjunctions.

EXERCISE 3.7

1. Write the following conjunctions and disjunctions in a symbolic form by specifying the statements p, q. Negate the statements p, q in each case and then negate each of the following statements.
   (a) No scoundrel is a ruffian, while some thieves are not scoundrels.
   (b) None of us is out of breath or some of us are not fat.
   (c) You wear a tie to the interview or you do not get hired.
   (d) She is beautiful but modest.
   (e) All rainy days are muggy and some summer days are not hot.
2. Negate the following statements using De Morgan's laws. Write all the steps involved with reasons for every step.
   (a) $(p \wedge q) \vee r$
   (b) $(p \wedge r) \vee (q \wedge r)$
   (c) $(\sim p \vee r) \wedge (\sim p \wedge r)$
   (d) $(p \vee r) \wedge (\sim r)$
   (e) $(\sim p \wedge q) \vee (\sim p \vee q)$

3.8 Conditional or Implicative Statements

In the previous sections, we studied disjunctions and conjunctions, the truth tables associated with them as well as their negations. Another important compound statement is the 'implicative statement' or 'implication' which we will define below.

The compound statement, 'If p, then q' is called an **implication.** It is denoted by $p \rightarrow q$ and is also called 'p **implies** q. Some examples are as follows:

(a) 'If I have a dog, then I have a cat'. Here, p is the statement 'I have a dog' and q is 'I have a cat'. The compound statement can be written symbolically as $p \rightarrow q$.

(b) 'If she can sing, then she can dance'. Here, p is the statement 'She can sing' and q is 'She can dance'. The compound statement can be written symbolically as $p \rightarrow q$.

In this section, we study 'implications' in detail. We will consider truth values and the truth table associated with implications, and learn how to negate implications. Other topics that will help us to understand implications better will also be considered.

Let us recall the definitions and examples that were introduced in Section 3.4. Let p and q be the statements. Then the compound statement 'If p, then q' is defined to be an implicative statement or simply put 'an implication'. It is sometimes also written as 'p implies q' and symbolically represented by $p \rightarrow q$. The statement p in $p \rightarrow q$ is called the 'antecedent' and q is called the 'consequent'. For example, let p be the statement 'I study hard' and let q be the statement 'I will do well in my logic examination'. Then the implication denoted by $p \rightarrow q$ will be the statement 'If I study hard, then I will do well in my logic examination'.

So when will $p \rightarrow q$ be false and when will it be true? Maybe the way to proceed is to think of when the statement 'If I study hard, then I will do well in my logic examination' be false. Clearly, if one does well in the logic examination, then the implication cannot be false. However, if one did study well but did not do well in the logic examination, then it is obvious that the statement 'If I study hard, then I will do well in my logic examination' will be false. Thus, $p \rightarrow q$ is false if p is true but q is false. Indeed, this is the only scenario in which $p \rightarrow q$ can be false; in all other circumstances, the statement $p \rightarrow q$ will be true. We can now state the following truth value for an implication.

The implication $p \rightarrow q$ will be false if and only if p is true and q is false.

The statement $p \rightarrow q$ will be true in all other circumstances, that is, if p is false; if p and q are both true. The above statement is enough to construct the truth table for an implication (see Table 3.8.1).

Table 3.8.1 Truth Table for Implication

p	q	$p \rightarrow q$
T	T	T
T	F	F
F	T	T
F	F	T

Now that the truth table for an implication is available, the truth table whose header row was given in Table 3.6.8 can be completed. Note that since there are four input variables, this truth table will have 16 input rows other than the header. This is left as an exercise for the reader.

The circumstance of the implication $p \rightarrow q$ being true when p is false seems strange. An implication $p \rightarrow q$ that is true because p is false is sometimes called a 'vacuously true' statement. Let us investigate this a little more. Consider the statement 'All people in this room are talking'.

This statement is false only if there is at least one person in the room who is not talking. Thus, the statement will be true if there are actually no people in the room.

Note that the statement 'All people in this room are talking' can be re-written equivalently in a implicative form, namely, 'If there are people in the room, then they will talk'. So the antecedent p here is 'There are people in this room' and the consequent q is 'They will talk'. So if there are no people in the room, that is, if p is false, then the statement '$p \rightarrow q$' will be true or equivalently 'All people in this room are talking' will be true.

This can be used by children to logically state the truth by saying 'I have done all my homework' when they were not actually given any homework. (Note that the equivalent implicative version is 'If I am given homework, then I do it.')

We have seen a 'logical equivalence' between statements of the form 'All A are B' and the implicative statement 'If A, then B'.

Another equivalence is that the statement 'No A are B' is equivalent to 'If A, then not B. Thus,

'All p are q' is logically equivalent to **'If p, then q'.**
'No p are q' is logically equivalent to **'If p, then not q'.**

Example 3.8.1 In what follows some more examples of implicative statements are given.

(a) If it rains, then I will stay home. This is of the form $p \rightarrow q$, where p is 'It rains' and q is 'I will stay home'.

(b) If I do not study, then I will not get good grades. This can be regarded to be of the form $\sim p \rightarrow \sim q$, where p is 'I study' and q is 'I get good grades'.

(c) If you do not walk regularly, then you will fall ill. Here, p is 'You walk regularly', q is 'You will fall ill' and the statement is $\sim p \rightarrow q$.

(d) If you watch too much television, then your eyesight will not be good. Here, p is 'You watch too much television', q is 'Your eyesight will be good' and the statement is $p \rightarrow \sim q$.

(e) All poodles are dogs. Here, p is 'It is a poodle' and q is 'It is a dog' and the statement is 'If it is a poodle, then it is a dog', namely $p \rightarrow q$.

(f) All ants live in communities. Here, p is 'It is an ant' and q is 'It lives in a community'. The statement is 'If it is an ant, then it lives in a community', namely $p \rightarrow q$.

(g) None of the deer eat meat. Here, p is 'It is a deer' and q is 'It eats meat'. The statement is 'If it is a deer, then it does not eat meat', namely $p \rightarrow \sim q$.

(h) No girl has long hair. Here, p is 'She is a girl' and q is 'She has long hair'. The statement will be 'If she is a girl, then she does not have long hair', namely $p \rightarrow \sim q$. □

Example 3.8.2 Let p be the statement 'The tea is too hot' and q be the statement 'I will burn my lips'. In this example, we shall write each of the statements below symbolically using p and q.

(a) If the tea is too hot, then I will burn my lips. This is of the form $p \rightarrow q$.

(b) If I burn my lips, then the tea is too hot. This is of the form $q \rightarrow p$.

(c) If the tea is not too hot, then I do not burn my lips. This is of the form $\sim p \rightarrow \sim q$

(d) If I don't burn my lips, then the tea is not too hot. This is of the form $\sim q \rightarrow \sim p$. □

The statements in above example are all related to $p \rightarrow q$ and are important enough to be given special names.

Suppose statements p, q and the implicative statement $p \rightarrow q$ are given. We call the statement $p \rightarrow q$ the **direct statement.** The statement $q \rightarrow p$ is called the **converse statement.** The statement $\sim p \rightarrow \sim q$ is called the **inverse statement** and the statement $\sim q \rightarrow \sim p$ is called the **contrapositive statement.** Please note that converse, inverse and contrapositive statements can only be defined after we have designated the direct statement.

Example 3.8.3 Let p be the statement 'I study' and q be the statement 'I do well in my exams'. Let the direct statement be $p \rightarrow q$, namely 'If I study, then I do well in my exams'. Then

(a) The converse statement $q \rightarrow p$ will be 'If I do well in my exams, then I have studied'.

(b) The inverse statement $\sim p \rightarrow \sim q$ will be 'If I do not study, then I will not do well in my exams'.

(c) The contrapositive statement $\sim q \rightarrow \sim p$ will be 'if I did not do well in my exams, then I have not studied'. □

Consider the direct statement and the contrapositive statement in Example 3.8.3. We know that the direct statement 'If I study then I do well in my exams' will be false if and only if I study but do not do well in the examination. Similarly, the contrapositive statement 'if I did not do well in my exams, then I have not studied' will be false if and only if I did not do well in my exams but have studied.

Recall that 'but' is the same as the connective 'and'. Also, recall that the statement 'A and B' is logically equivalent to 'B and A'. Using these two facts we see that both the direct statement and the contrapositive statement are false if and only if I have studied and have not done well in the examinations. Thus, we have informally established that the direct and contrapositive statements have the same truth values in all circumstances. In other words, the direct and the contra-positive statements are logically equivalent.

Similarly, one can establish that the converse and inverse statements are also logically equivalent. Let us now examine this formally through truth tables.

The input variables p and q are common to all the four statements. Consequently, we create a single truth table with four output columns representing the direct, contrapositive, converse and inverse, respectively.

Table 3.8.2 shows that the direct statement $p \rightarrow q$ and the contrapositive statement $\sim q \rightarrow \sim p$ are logically equivalent. Similarly, the converse statement $q \rightarrow p$ and the inverse statement $\sim p \rightarrow \sim q$ are logically equivalent.

Table 3.8.2 Truth Table Showing Logical Equivalences for Implications

p	q	$\sim p$	$\sim q$	$p \to q$	$\sim q \to \sim p$	$q \to p$	$\sim p \to \sim q$
T	T	F	F	T	T	T	T
T	F	F	T	F	F	T	T
F	T	T	F	T	T	F	F
F	F	T	T	T	T	T	T

Let $p \to q$ be the direct statement. Then

> The **direct statement** $p \to q$ is logically equivalent to the **contrapositive statement** $\sim q \to \sim p$.
>
> The **converse statement** $q \to p$ is logically equivalent to the **inverse statement** $\sim p \to \sim q$.

Are there any logical equivalences that can be established between disjunctions and implications, between conjunctions and implications? Let us explore this by considering an example. Let p be the statement 'You study' and let q be the statement 'You will fail'. The disjunction '$p \vee q$' will be the statement 'You study or you will fail'. Can we think of an implicative form for this statement? The implicative statement that conveys the same meaning seems to be 'If you don't study, then you will fail', namely $\sim p \to q$. We can explore this 'equivalence' informally through this example and can also formally check for logical equivalence through truth tables.

A disjunction $p \vee q$ is false if and only if both p and q are false. Thus, the statement 'You study or you will fail' will be false if and only if you don't study and you do not fail. On the other hand, 'If you don't study then you will fail' will be false if and only if you don't study and you don't fail. Thus, we see that the truth values of $p \vee q$ and $\sim p \to q$ are the same in all cases (Table 3.8.3).

Table 3.8.3 Truth Table Showing $(p \vee q) \equiv (\sim p \to q)$

p	q	$\sim p$	$p \vee q$	$\sim p \to q$
T	T	F	T	T
T	F	F	T	T
F	T	T	T	T
F	F	T	F	F

Question 2 parts (a)–(c) in Exercise 3.4 show certain properties that logical equivalence between statements satisfies. We can use these, DeMorgan's laws and the above equivalence between disjunctions and implications to establish a logical equivalence between conjunctions and implications. Let p and q be the statements. Then

$$\begin{aligned} p \wedge q &\equiv \sim [\sim (p \wedge q)] \\ &\equiv \sim [\sim p \vee \sim q] \\ &\equiv \sim [\sim (\sim p) \rightarrow \sim q] \\ &\equiv \sim [p \rightarrow \sim q]\,. \end{aligned}$$

This discussion shows that the statement 'p and q' is logically equivalent to 'It is not the case that if p, then not q'. Symbolically, we have $p \wedge q \equiv \sim (p \rightarrow \sim q)$. Let us illustrate this with an example. The statement 'You study and you will not fail' is logically equivalent to 'It is not the case that if you study, then you will fail'.

Let p and q be the statements. Then

The statement $p \vee q$ is **logically equivalent** to $\sim p \rightarrow q$.
The statement $p \wedge q$ is **logically equivalent** to $\sim (p \rightarrow \sim q)$.

This also provides us a way to find the negation of an implication. Again using properties of logical equivalences and already established equivalences, we note that

$$\begin{aligned} \sim (p \rightarrow q) &\equiv \sim [p \rightarrow \sim (\sim q)] \\ &\equiv p \wedge \sim q\,. \end{aligned}$$

Thus, the negation of the implication 'If p, then q' is the conjunction 'p and not q'. We also note that $p \rightarrow q$ is false if and only if p is true and q is false. The conjunction $p \wedge \sim q$ is true if and only if p is true and $\sim q$ is true. Thus we see that the conjunction $p \wedge \sim q$ is true if and only if p is true and q is false. This establishes once again that $p \rightarrow q$ and $p \wedge \sim q$ have the opposite truth values in all cases. So they are negations of each other.

Let p and q be statements. Then

The **negation** of **'If** p, **then** q**'** is the statement **'p and not q'**.   Symbolically, $\sim (p \rightarrow q) \equiv p \wedge \sim q$.

Example 3.8.4 In the following examples, we shall write down equivalent disjunctions for implications or vice versa.

(a) Consider the implication 'If you do not carry an umbrella, then you will get soaked'. It is of the form $\sim p \rightarrow q$, where p is the statement 'You carry an umbrella' and q is the statement 'You will get soaked'. Since $\sim p \rightarrow q \equiv p \vee q$, the given statement will be equivalent to 'You carry an umbrella or you will get soaked'.

(b) 'If you carry a heavy bag, then you will get a back ache' will be logically equivalent to 'You do not carry a heavy bag or you will get a back ache'. Note that here p is 'You carry a heavy bag' and q is 'You will get a back ache' and the statement $p \rightarrow q \equiv \sim p \vee q$.

(c) 'If you do not run fast, then you will not fall' will be logically equivalent to 'You run fast or you will not fall'.

(d) 'You should drink coffee or tea' is logically equivalent to 'If you do not drink coffee, then you should drink tea' (Figure 3.8.1).

You should drink coffee or tea.

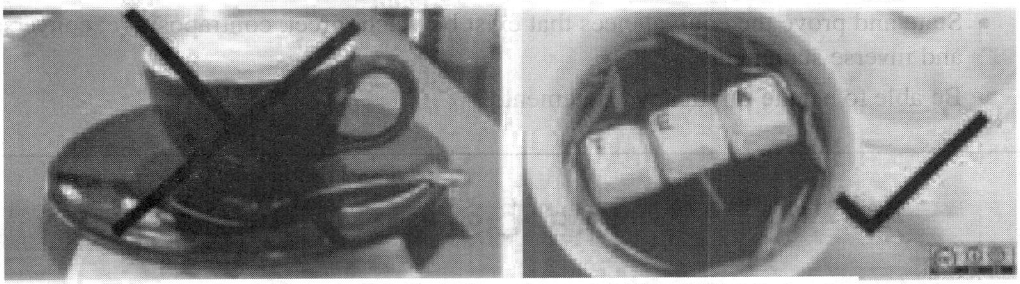

If you do not drink coffee, then you should drink tea.

Figure 3.8.1 Coffee or tea. [C1] and [C6]

(e) 'You should not play music loudly or you will wake the neighbours' is logically equivalent to 'If you play music loudly, then you will wake the neighbours'.

(f) 'You should not eat road side food or you will not feel well' is logically equivalent to 'If you eat road side food, then you will not feel well'. □

Example 3.8.5 In the following examples, we shall write down the negations for the implications.

(a) Consider the implication 'If you carry an umbrella, then you will not get soaked'. It is of the form $p \rightarrow q$, where p is the statement 'You carry an umbrella' and q is the statement 'You will not get soaked'. Since $\sim (p \rightarrow q) \equiv p \wedge \sim q$, the negation of the given statement will be equivalent to 'You carry an umbrella and you get soaked'.

(b) The negation of 'If you carry a heavy bag, then you will get a back ache' will be 'You carry a heavy bag and you do not get a back ache'. Note that here p is 'You carry a heavy bag' and q is 'You will get a back ache'.

(c) The negation of 'If you do not run fast, then you will not fall' will be 'You do not run fast and you fall'.

(d) 'You do not drink coffee and you do not drink tea' is the negation of 'If you do not drink coffee, then you should drink tea'.

(e) 'You play music loudly and you do not wake the neighbours' is the negation of 'If you play music loudly, then you will wake the neighbours'.

(f) 'You eat road side food and you feel well' is the negation of 'If you eat road side food, then you will not feel well'. □

Quick Review

After reading this section, you should be able to:

- Construct truth tables for compound statements involving implications.
- Write down converse, inverse and contrapositive statements of a direct implicative statement.
- State and prove the equivalences that exist between direct, contrapositive; converse and inverse statements.
- Be able to negate implicative statements.

EXERCISE 3.8

1. For the given p and q, write down the statements $\sim (p \rightarrow q)$, $\sim p \rightarrow q$, $p \rightarrow \sim q$ and $\sim p \rightarrow \sim q$.
   (a) p: you eat carrots; q: you have good eyesight.
   (b) p: you don't commit a traffic violation; q: your insurance premium does not go up.
   (c) p: you don't carry an umbrella; q: it will rain.
   (d) p: you exercise rarely; q: you do not stay healthy.
2. For each of the following conditional statements, write down the converse, inverse and contrapositive statements.
   (a) If we get a salary increase, then we will be happy.
   (b) If a dog wags its tail, then it won't bite.
   (c) All tigers are predators.
   (d) No beggars are choosers.
   (e) If all scoundrels are ruffians, then some pirates are scoundrels.
3. Negate each of the conditional statements in Question 2.
4. For each of the conditional statements in Question 2, write down a disjunction that it is logically equivalent to.
5. Find the truth value of the given statements under the conditions specified.
   (a) $(\sim s \wedge p) \rightarrow (\sim q \rightarrow s)$, when p and s are true and q is false.
   (b) $(p \rightarrow q) \vee (r \wedge \sim s)$ when p is false, but q, r and s are true.

(c) $\sim p \rightarrow \{\sim [(q \vee \sim r) \wedge (w \wedge \sim q) \vee (u \vee \sim w)]\}$, when p, s and w are true and q, r and u are false. (Hint: Think carefully, you don't have to actually work out the truth value of each part.)

6. Make a truth table for each of the statements given below. Are any of the statements Tautologies? Are any two of them logically equivalent?
   (a) $\sim p \rightarrow q$.
   (b) $(p \wedge q) \rightarrow (\sim p \rightarrow q)$.
   (c) $(\sim p \wedge \sim q) \rightarrow q$.

3.9 Analysis of Arguments

We often find that if we wish to demand something, then we have to justify the demand by giving a good reason. In the academic realm, we have to carefully argue our case to support statements that we believe in. In law, evidence and sound arguments are often the basis on which one can win a case. What makes a good argument? When we say that an argument is sound or that the reasoning is clear or justified what do we mean? In this section, we analyse the statements or parts that make an argument. We then use the analysis to decide if the argument is valid or not.

Given below are two arguments.

Example 3.9.1 The local bank was robbed. The robber was seen getting away in a Maruti 800 car. Suhel is known to have a record of theft and he owns a Maruti 800. Therefore, Suhel is the person who robbed the local bank. □

Example 3.9.2 The local bank was robbed. The person who robbed the local bank left the only unauthorised finger print on the safe. Suhel's fingerprint matches the unauthorised one left on the safe. Therefore, Suhel is the person who robbed the local bank. □

In any argument, there are two parts. The first part consists of **evidence** or **premises.** The second part is the **conclusion.** If the premises being true force the conclusion to be true, then we say that the argument is valid. Otherwise, the argument is considered invalid.

Let us analyse Example 3.9.1.

Premises:

(a) The local bank was robbed.
(b) The robber was seen getting away in a Maruti 800 car.
(c) Suhel is known to have a record of theft and he owns a Maruti 800.

Conclusion: Suhel is the person who robbed the local bank.

We see here that even if all the premises are true, it is possible for someone else other than Suhel to have robbed the local bank. This is because the robber was not identified, only the type of getaway car was. So even if every premise were true, it would not force the conclusion to be true. Thus, this is an invalid argument.

Now let us consider Example 3.9.2.

Premises:

(a) The local bank was robbed.

(b) The person who robbed the local bank left the only unauthorised finger print on the safe.

(c) The unauthorised fingerprint left on the safe matches Suhel's fingerprint.

Conclusion: Suhel is the person who robbed the local bank.

We see in the above case that if all the premises are true, then no one other than Suhel could have robbed the local bank. The truth of the premises force the conclusion to be true. Hence this is a valid argument. Of course, in reality it would be an extremely foolish bank robber who would leave his or her fingerprint on the safe. So, in a sense Premise (b) is weak. However, if Premises (a), (b) and (c) are true, then the conclusion has to be true.

Analysing Arguments

The aim in this part is to learn how to use logic formally to analyse arguments. The following steps give the details.

1. Symbolise (consistently) all the premises and the conclusion. This means that we identify the basic statements or input variables, and then express the premises and conclusions in terms of the basic statements. Being consistent means not giving two different symbols to the same or connected basic statements or using the same symbol for two different statements.
2. Create a truth table for the argument having a column for each input variable or basic statement appearing in the argument. Further a column each for the premises and a final column for the conclusion. The truth values of the premise and conclusion columns will be calculated on the basis of the truth value of the input variables.
3. If there is a row in the truth table where every entry for the premises is true but the corresponding entry for the the conclusion is false, then such a row is called a **bad row.**
4. If the truth table for the argument has a bad row, then it is an invalid argument; otherwise, it is a valid argument.

Let us analyse the arguments given below using the steps described earlier. Both arguments involve 'seeing Movie A', but are they both valid or not?

Example 3.9.3 If I see Movie A, then I will be able to discuss Movie A in class. I discussed Movie A in class. Therefore, I saw Movie A. □

Example 3.9.4 I have to see either Movie A or Movie B. I cannot see Movie B. Therefore, I have to see Movie A. □

Let us symbolise the argument of Example 3.9.3.

Basic statements:

p: I see Movie A.

q: I discuss Movie A in class.

Premises:
P_1: $p \rightarrow q$ (If I see Movie A, then I will be able to discuss Movie A in class.)
P_2: q (I discussed Movie A in class.)
Conclusion:
C: p (I saw Movie A.)
Table 3.9.1 gives the truth table for the argument of Example 3.9.3.

Table 3.9.1 Truth Table for Argument of Example 3.9.3

Input	Variables	P_1	P_2	C
p	q	$p \rightarrow q$	q	p
T	T	T	T	T
T	F	F	F	T
F	T	*T*	*T*	*F*
F	F	T	F	F

In the second last row of Table 3.9.1, we note that the entries for P_1 and P_2 are T but the entry for C is F. This is an example of a **bad row,** that is, even though the premises are true, the conclusion is false. Thus, this argument is an invalid argument.

Now, let us analyse the argument of Example 3.9.4 (Table 3.9.2).

Basic statements:
p: I have to see Movie A.
q: I have to see Movie B.
Premises:
P_1: $p \vee q$
P_2: $\sim q$
Conclusion:
C: p

Table 3.9.2 Truth Table for Argument of Example 3.9.4

Input	Variables	P_1	P_2	C
p	q	$p \vee q$	$\sim q$	p
T	T	T	F	T
T	F	T	T	T
F	T	T	F	F
F	F	F	T	F

The only row in which both the premises P_1 and P_2 take value T is the fourth row. However, here C also takes the value T. So this truth table does not have a bad row. Hence, this is a valid argument.

From the above analysis, we can note certain patterns. If any argument can be symbolised in the same pattern as in Example 3.9.3, namely

$$\begin{array}{l} P_1 : p \to q \\ P_2 : \quad q \\ \hline C : \quad p \end{array}$$

then the argument will have the same truth table as given in Table 3.9.1 and hence will have a bad row. Hence, we can conclude that any argument which can be symbolised in the above pattern is an invalid argument. This pattern of invalid reasoning is called **fallacy of the converse.**

Similarly, we find that if an argument can be symbolised in the same pattern as Example 3.9.4, namely

$$\begin{array}{l} P_1 : p \vee q \\ P_2 : \sim q \\ \hline C : \quad p \end{array}$$

then the argument will have the same truth table as given in Table 3.9.2 and so will not have a bad row. Hence, we can conclude that any argument which can be symbolised in the above pattern is an valid argument. This pattern of valid reasoning is called **disjunctive syllogism.**

In the next section, we consider some common patterns of reasoning and examples.

Patterns in Reasoning

As seen towards the end of the last section, there are certain 'common patterns' of reasoning. In this section, we will list a few common patterns of both valid and invalid reasonings. If an argument has the same symbolic pattern as any of the common patterns, then we can conclude immediately whether the argument is valid or not. We begin with listing some valid common patterns of reasoning and examples of arguments that follow that pattern.

Valid Patterns of Reasoning

I. *Disjunctive Syllogism*

Any argument that can be symbolised to have the following pattern is a valid argument, and this valid pattern of reasoning is called disjunctive syllogism:

$$\begin{array}{l} P_1 : p \vee q \\ P_2 : \sim q \\ \hline C : \quad p \end{array}$$

Similarly, any argument that can be symbolised to have the following pattern will also be valid by disjunctive syllogism:

$$\begin{array}{l} P_1 : p \vee q \\ P_2 : \sim p \\ \hline C : \quad q \end{array}$$

Example 3.9.5 I will drink coffee or I will eat a chocolate. I could not eat a chocolate. Therefore, I will drink coffee. □

In the above example, let p be the statement 'I will drink coffee' and q be the statement 'I will eat a chocolate'. Then, the argument has the first of the patterns given above and hence is a **valid** argument by **disjunctive syllogism.** All the following examples are valid by disjunctive syllogism.

Example 3.9.6 Arif is in Agra or Arif is in Delhi. Arif is not in Delhi. Therefore, Arif is in Agra. □

Example 3.9.7 Sybil will take her dog for a walk or Sybil will feed her pet cat. Sybil could not feed her pet cat. Therefore, Sybil will take her dog for a walk. □

Example 3.9.8 The Vice-Chancellor will either declare a holiday on Monday or she will hold the convocation on Monday. The Vice-Chancellor could not declare a holiday on Monday. Hence, she will hold the convocation on Monday. □

II. *Contrapositive Reasoning*

Any argument that can be symbolised to have the following pattern is a valid argument, and this valid pattern of reasoning is called **contrapositive reasoning:**

$$\begin{array}{l} P_1 : p \rightarrow q \\ P_2 : \ \sim q \\ \hline C : \ \sim p \end{array}$$

One way of showing validity of argument is by creating a truth table for the pattern and showing that the truth table has no bad row. Recall that a bad row occurs if the premises are true but the conclusion is false. Note that the conclusion $\sim p$ will be false if and only if p is true. Also, P_2 will be true if and only if $\sim q$ is true, that is, if and only if q takes the false value. But when p is true and q is false, then P_1, namely, $p \rightarrow q$ is forced to be false. Thus, we can see that we will never have input truth values for p and q which makes C false and P_2, P_1 true. So, there can be no bad row in a truth table for this pattern of argument.

Example 3.9.9 All tigers are carnivores. Tom is not a carnivore. Therefore, Tom is not a tiger. □

Recall that 'All tigers are carnivores' is the same as 'If it is a tiger, then it is a carnivore'. In the above example, let p be the statement 'It is a tiger' and q be the statement 'It is a carnivore'. Then, the argument has the pattern given above and hence is a **valid** argument by **contrapositive reasoning.** All the following examples are valid by contrapositive reasoning.

Example 3.9.10 If I study, then I will do well in my logic and reasoning test. I did not do well in my logic and reasoning test. Hence, I did not study (Figure 3.9.1). □

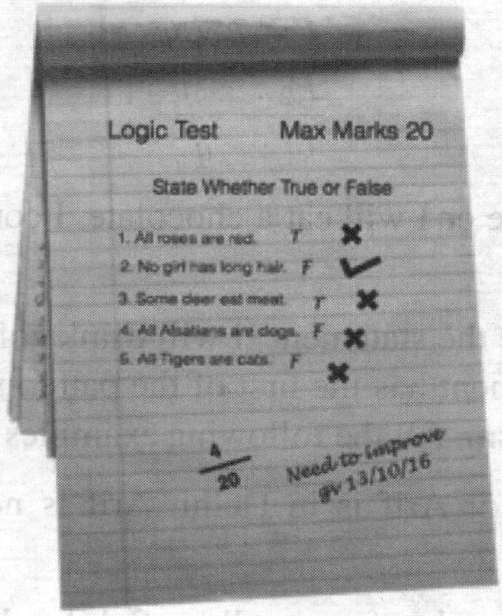

Figure 3.9.1 I did not study. [C4]

Example 3.9.11 If I drink coffee at night, then I will not be able to sleep properly. I slept properly. Therefore, I did not drink coffee at night. □

Example 3.9.12 If Lisa does not memorise the passage, then she will not be able to recite it. Lisa was able to recite the passage. Therefore, Lisa did memorise the passage. □

III. *Direct Reasoning*

The following pattern is a valid argument. If an argument in a symbolic form has this pattern, then the argument will be valid by **direct reasoning:**

$$\begin{array}{l} P_1 : p \to q \\ \underline{P_2 : \quad p} \\ C : \quad q \end{array}$$

If C is false and P_2 is true, then p must be true and q must be false. But when p is true and q is false, $p \to q$, namely P_1 will be false. Thus, the truth table for this pattern will never have a bad row and hence will be a valid argument.

Example 3.9.13 All cats can climb trees. Kitty is a cat. Therefore, Kitty can climb a tree (Figure 3.9.2). □

Note that, 'All cats can climb trees' is the same as the statement 'If it is a cat, then it can climb trees'. In the above example, let p be the statement 'If it is a cat' and q be the statement 'It can climb trees'. Then, the argument has the pattern given above and hence is a **valid** argument by **direct reasoning.**

Figure 3.9.2 Kitty, the cat, climbing a tree. [C7]

All the following examples listed below are valid by direct reasoning.

Example 3.9.14 If I practise running everyday, then I will be able to complete the marathon. I practised running everyday. Therefore, I completed the marathon. □

Example 3.9.15 If I watch television at night, then I will not be able to fall asleep. I watched television at night. Therefore, I was not able to fall asleep. □

Example 3.9.16 If Maria learns her multiplication tables, then she will be able to answer the mental mathematics problems. Maria learnt her multiplication tables. Therefore, Maria was able to answer the mental mathematics problems. □

IV. *Transitive Reasoning*

The following valid pattern of reasoning is called **transitive reasoning.** If an argument can be symbolised to have the same pattern as follows, then we say that it is a valid argument using transitive reasoning:

$$\begin{array}{c} P_1 : p \to q \\ \underline{P_2 : q \to r} \\ C : p \to r \end{array}$$

It is easy to see that C can be false if and only if p is true, but r is false. Let us assume that C is false. Then, r is false, so $q \to r$, namely P_2 will be true only if q is also false. But when p is true and q is false, $p \to q$, namely P_1 will be false. Thus, the truth table for this pattern will never have a bad row and hence will be a valid argument.

Example 3.9.17 All monkeys are mammals. All mammals look after their young. Therefore, all monkeys look after their young (Figure 3.9.3). □

Recall that 'All monkeys are mammals' is the same as 'If it is a monkey, then it is a mammal'. In the above example, let p be the statement 'It is a monkey', q be the statement 'It is a mammal' and let r be the statement 'It looks after its young'. Then, the argument has the pattern given above and hence is a **valid** argument by **transitive reasoning.**

Figure 3.9.3 Langur monkey with baby. [C4]

All the following examples are valid by transitive reasoning.

Example 3.9.18 If I study, then I will do well in my test. If I do well in my test, I will get a scholarship. So if I study, then I will get a scholarship. □

Example 3.9.19 If I wake up on time, then I will be able to catch the bus. If I catch the bus then I will be able to attend the first lecture. Thus, if I wake up on time, then I will be able to attend the first lecture. □

Example 3.9.20 If Mohan does not submit his assignment on time, then he will get a low grade. If Mohan gets a low grade, then he will fail the course. Therefore, if Mohan does not submit his assignment on time, then he will fail the course. □

Invalid Patterns of Reasoning

I. *Fallacy of the Converse*

As seen in Example 3.9.3, any argument that can be symbolised to have the pattern as follows is an invalid argument, and this invalid pattern of reasoning is called **fallacy of the converse:**

$$\begin{array}{l} P_1 : p \rightarrow q \\ P_2 : \quad q \\ \hline C : \quad p \end{array}$$

Example 3.9.21 All birds can fly. A bat can fly. Therefore, a bat is a bird (Figure 3.9.4). □

Recall that, the statement 'All birds can fly' is the same as the statement 'If it is a bird, then it can fly'. In the above example, let p be the statement 'It is a bird' and q be the statement 'It can fly'. Then, the argument has the pattern above and hence is an **invalid** argument by **fallacy of the converse.**

Figure 3.9.4 Bat, a bird? [C8]

All the following examples are invalid by fallacy of the converse.

Example 3.9.22 If I attend all my classes, then I will do well in my courses. I do well in my courses. Therefore, I have attended all my classes. □

Example 3.9.23 If I eat too many chocolates, then I will suffer from indigestion. I suffered from indigestion. So I ate too many chocolates. □

Example 3.9.24 If today is a holiday, then I will be able to read my book. I read my book. Therefore, today is a holiday. □

II. *Fallacy of the Inverse*

The following pattern is an invalid argument, and this invalid pattern of reasoning is called **fallacy of the inverse.** Any argument with this pattern symbolically will be invalid by fallacy of the inverse:

$$\begin{array}{l} P_1 : p \to q \\ P_2 : \ \sim p \\ \hline C : \ \sim q \end{array}$$

To show that the pattern above constitutes an invalid argument we have to show that the truth table for the argument has a bad row. Alternatively we can show that there are input truth values of p and q for which P_1 and P_2 are true but C is false.

The conclusion C will be false if and only if $\sim q$ is false or if q has the input value true. The premise P_2 will be true if and only if $\sim p$ is true or equivalently if p takes the input value false. When p is false and q is true, $p \to q$, namely P_1 will be true. Thus, for p false and q true, P_1 and P_2 are true and C is false. Hence, the argument is invalid.

Example 3.9.25 All policemen are brave. Joe is not a policeman. Therefore, Joe is not brave. □

The statement 'All policemen are brave' is the same as 'If you are a policeman, then you are brave'. Let p be the statement 'You are a policeman' and q be the statement 'You are brave'. Then, the argument has the pattern above and hence is an **invalid** argument by **fallacy of the inverse.** All the following examples are invalid by fallacy of the inverse.

Example 3.9.26 If I do my job well, then I will not get transferred. I do not do my job well. Therefore, I get transferred. □

The above argument seems very reasonable and at first glance one might be tempted to assume that the argument is sound. But the problem lies in the fact that even if all the premises are true, it is possible for the conclusion to be false. Imagine a scenario where I have not done my job well but have not got transferred. In this scenario, all the premises are true but the conclusion is false. The premise is actually silent about what happens if the job is not done well, and no conclusion can be reached in that case.

Example 3.9.27 If you do not eat potato chips, then you are not happy. You eat potato chips. Therefore, you are happy. □

Here too one can imagine a scenario where you have eaten potato chips, but you are also sad. Then, all the premises would be true but the conclusion will be false.

Example 3.9.28 If you sing well, then you will be able to appreciate music. You do not sing well. Therefore, you cannot appreciate music. □

Think of a person who cannot sing well but can appreciate music.

III. *False Chain*

Any argument that can be symbolised to have the following pattern is an invalid argument, and this invalid pattern of reasoning is called the **false chain:**

$$\begin{array}{l} P_1 : p \to q \\ P_2 : p \to r \\ \hline C : q \to r \end{array}$$

To show that the pattern below constitutes an invalid argument we have to show that the truth table for the argument has a bad row. Alternatively, we can show that there are input truth values of p, q and r, for which P_1 and P_2 are true but C is false. The conclusion C will be false if and only if q is true and r is false. Given that r is false, the premise P_2 will be true if and only if p is false. When p is false and q is true, $p \to q$, namely P_1 will also be true. Thus, for p false and q true and r false, P_1 and P_2 are true and C is false. Hence, the argument is invalid.

Any argument with the following pattern can also be shown to be invalid. This invalid pattern of reasoning is also called the **false chain:**

$$\begin{array}{l} P_1 : p \to q \\ P_2 : r \to q \\ \hline C : p \to r \end{array}$$

Example 3.9.29 If today is a government holiday, then the university is closed. If today is a Sunday, then the university is closed. Therefore, if today is a government holiday, then today is a Sunday. □

Let p be the statement 'Today is a government holiday', q be the statement 'The university is closed' and r be the statement 'Today is a Sunday'. Then, the argument has the pattern above and hence is an **invalid** argument by the **false chain.** All the following examples are invalid by the false chain.

Example 3.9.30 All cats are mammals. All cats are predators. Therefore, all mammals are predators (Figure 3.9.5). □

Example 3.9.31 If you like to eat chocolates, then you have sweet teeth. If you like to eat desserts, then you have sweet teeth. So if you like to eat chocolates, then you like to eat desserts. □

Figure 3.9.5 Cute Pika PREDATOR. [C4]

Example 3.9.32 All deer are herbivores. All deer can run fast. So all herbivores can run fast. □

IV. *Disjunctive Fallacy*

The following pattern is an invalid argument. Any argument that can be symbolised to have the same pattern will be invalid by **disjunctive fallacy:**

$$\begin{array}{l} P_1 : p \vee q \\ \underline{P_2 : \quad p} \\ C : \ \sim q \end{array}$$

We can show that there are input truth values of p and q for which P_1 and P_2 are true but C is false. The conclusion C will be false if and only if $\sim q$ is false or if q has the input value true. The premise P_2 will be true if and only if p takes the input value true. When p is true and q is true, $p \vee q$, namely P_1 will be true. Thus, for p true and q true, P_1 and P_2 are true and C is false. Hence, the argument is invalid.

Any argument with the following pattern can also be shown to be invalid. This invalid pattern of reasoning is also called **disjunctive fallacy:**

$$\begin{array}{l} P_1 : p \vee q \\ \underline{P_2 : \quad q} \\ C : \ \sim p \end{array}$$

Example 3.9.33 Reema likes to dance or she likes to sing. Reema likes to dance. Therefore, she does not like to sing. □

Let p be the statement 'Reema likes to dance' and q be the statement 'Reema likes to sing'. Then, the argument has the pattern above and hence is an **invalid** argument by **disjunctive fallacy.** All the following examples are invalid by disjunctive fallacy.

Example 3.9.34 He drinks coffee or he drinks tea. He drinks tea. So he does not drink coffee. □

Example 3.9.35 You read detective fiction or you read comics. You read detective fiction. Hence, you do not read comics. □

Example 3.9.36 Today is not a Sunday or I stay home. I stay home. So today is not a Sunday. □

We have now explored four patterns that fit a valid argument, namely disjunctive syllogism, contrapositive reasoning, direct reasoning and transitive reasoning. Similarly, we have also considered four patterns that give rise to an invalid argument, namely fallacy of the converse, fallacy of the inverse, false chain and disjunctive fallacy. However, it should be emphasised that many arguments when symbolised may not fit any of these eight patterns. In such cases, the arguments will have to be analysed for bad rows either using a truth table or directly. We end with an example of this kind.

Example 3.9.37 I got a scholarship and I got an A grade in logic and reasoning. I am good at logic or I get an A grade in logic and reasoning. Therefore, I am good at logic or I do not get a scholarship. □

In the above argument, let p be the statement 'I got a scholarship', q be the statement 'I got an A grade in logic and reasoning' and r be the statement 'I am good at logic'. Then, the argument can be symbolised as follows:

$$\begin{array}{l} P_1 : \; p \wedge q \\ \underline{P_2 : \; r \vee q} \\ C : q \vee \sim p \end{array}$$

We can analyse the above argument directly to see if we can locate a 'bad row', namely input truth values for p, q and r such that P_1 and P_2 are true but C is false. If we find even one 'bad row', then the argument will be invalid. If, on the other hand, we cannot find any 'bad row', then the argument will be valid.

Note that C is false if and only if q is false and $\sim p$ is false, that is, if and only if p is true and q is false. Now, if p is true and q is false, P_1, namely $p \wedge q$, will be false. Thus, we can never have input truth values such that C is false but P_1 is true. Hence, we will not have a 'bad row' for this argument. Consequently, this argument is a valid argument.

Quick Review

After reading this section, you should be able to:

- Symbolise an argument and make the corresponding truth table for the argument.
- Analyse the truth table of an argument for bad rows.
- Be able to decide if an argument is valid or invalid.
- Identify valid and invalid patterns of argument.

EXERCISE 3.9

1. Symbolise the following arguments and test their validity using a truth table.
   (a) If I want to be a lawyer, then I want to study logic. If I don't want to be a lawyer, then I don't like to argue. Therefore, If I like to argue, then I want to study logic.
   (b) If I buy cheap petrol, then my car runs badly. If I don't change the oil, then my car runs badly. Therefore, If I buy cheap petrol, then I don't change the oil.
   (c) If you are a clown, then you work in the circus. If you do not like cotton candy, then you do not work in the circus. Therefore, if you are a clown, then you like cotton candy.
   (d) If I do not pay my income taxes, then I file for an extension or I am a felon. I am not a felon and I did not file for an extension. Therefore, I paid my income taxes.
   (e) All Koala bears are cuddly. No cat is a Koala bear. Thus, no cat is cuddly.
2. Test the validity of the following arguments.
   (a) If I work in the garden, then I will get dirt under my nails. I did not get dirt under my nails. Therefore, I did not work in the garden.
   (b) If I do not change the engine oil in my car regularly, then the engine of my car will stop working properly. The engine of my car has stopped working properly. Thus, I did not change the engine oil in my car regularly.
   (c) All bats are mammals. All bats fly. Therefore, all mammals fly.
   (d) You will meet a beautiful stranger or you will stay home and clean your room. You did not meet a beautiful stranger. Therefore, you stayed home and cleaned your room.
   (e) If I do not tie my shoe laces, then I trip. I did not tie my shoe laces. Hence, I tripped.

Review Exercises

1. Which of the following sentences are **statements**?
   (a) Today is a Sunday and Rehana will watch a movie.
   (b) Please do not run across the road.
   (c) Do not jump red lights.

(d) Shyam can sing or he can dance.
(e) The Taj Mahal is in Agra.

2. Label the following statements as **simple** and **compound.**
   (a) Some men like sweets and chocolates.
   (b) Some flowers smell sweet and some flowers are brightly coloured.
   (c) The sun sets in the West.
   (d) Few children like mathematics.
   (e) Tom likes English and science.

3. Answer the following.
   (a) If the statement 'Some A are not B' is true, then will the statement 'Some B are not A' be true?
   (b) If the statement 'No A is B' is true, then will the statement 'No B is A' be true?
   (c) If the statement 'Some A are B' is false, then will the statement 'Some B are A' be false?
   (d) If the statement 'Some A are B' is false, then will the statement 'No A is B' be true?
   (e) If the statement 'All A are B' is false, then will the statement 'Some A are not B' be true?

4. Use your everyday experience to decide the truth value of the following statements. Please give reasons for your answers.
   (a) Every month has exactly 30 days.
   (b) Some months do not have 30 days.
   (c) Some months have 30 days.
   (d) No month has 30 days.
   (e) Today is not a Saturday.

5. Read the passage given below. State whether true or false based **only on the information in the passage above.** In case the data given in the passage is insufficient to decide whether the given statements are true or false, then state 'Data is insufficient'. Give reasons for your answers.

   Reena teaches mathematics at Ambedkar University Delhi. She has four classes a week. She does not teach on Wednesdays and on the weekend. She teaches both BA Honours first year students and BA Honours final year students. She has two classes a week for BA Honours first year students and two classes a week for the BA Honours final year students. Every Wednesday she holds office hours at the University from 10 am to 1 pm. She is available to students for consultations during this period. It has happened that on a few Saturdays she has had to come to the University to complete pending work. But on most Saturdays, Veena is able to stay home.

   (a) Reena teaches a class on all weekdays except Wednesdays.
   (b) Every Wednesday Reena is in the University and on some Saturdays also Reena comes to the University.
   (c) Reena teaches BA Honours first year students every week or on some Saturdays she does not have to teach BA Honours first year students.
   (d) Some Wednesdays Reena stays home.
   (e) Some Saturdays Reena does not come to the University.

6. Negate the following statements.
   (a) Every month has 30 days.
   (b) Some months do not have 30 days.
   (c) Some months have 30 days.
   (d) No month has 30 days.
   (e) Today is not a Saturday.
   (f) All boys wear shoes.
   (g) Some boys wear shoes
   (h) None of the boys wear shoes.
   (i) Some boys do not wear shoes.
   (j) Every boy in this class is shorter than 3 feet.
   (k) Some boys in this class is shorter than 3 feet.
   (l) None of the boys in this class are shorter than 3 feet.
   (m) Some boys in this class are not shorter than 3 feet.
   (n) There is a movie show on the television today.
7. Write down the statement equivalent to the given double negation.
   (a) It is not true that no deer is a herbivore.
   (b) It is not the case that some men do not have long hair.
   (c) It is not the case that no month has 30 days.
   (d) It is not true that today is not a Saturday.
   (e) It is not the case that no dog is a poodle.
8. Categorise the following compound statements into conjunctions and disjunctions. Identify the simple statements from which the compound statements have been constructed. Write the given compound statement symbolically.
   (a) He reads while he eats.
   (b) She likes to sing or dance.
   (c) A bat is a mammal but it can fly.
   (d) Rome burns while Nero fiddles.
   (e) He can try but he is bound to fail.
9. Let p and q be the statements. Show the following.
   (a) $p \wedge p \equiv p$.
   (b) $p \vee p \equiv p$.
   (c) if $p \equiv q$ then $(\sim p) \equiv (\sim q)$.
   (d) if $p \equiv (\sim q)$ then $\sim p \equiv q$.
   (e) $\sim (\sim p) \equiv p$.
10. Please write the following conjunctions and disjunctions in a symbolic form by specifying the statements p and q. Use your everyday experience to decide the truth value of p, q and then of the corresponding disjunction or conjunction. Please give reasons for your answers.
    (a) Every month has at least 30 days and some months have at most 31 days.
    (b) No man has long hair and some women have long hair.

(c) I like watching Hindi movies or listening to Rock music.
(d) Every rose is a flower or every flower is red.
(e) No boy in the Maths class is greater than 3 feet or no girl in the Maths class is less than 3 feet.
(f) Some deers are herbivores and some tigers are carnivores.
(g) All Indian women wear sarees and all Indian men wear dhotis.
(h) Some months have at least 30 days and some months have at most 31 days.
(i) No man has long hair or some women have long hair.

11. From your real-life experience, give examples of simple statements p and q satisfying the following conditions. For each of the examples that you give, find the truth value of $\sim p \wedge q$ and $\sim p \vee q$.

(a) p is true, q is true.
(b) p is true, q is false.
(c) p is false, q is true.
(d) p is false, q is false.

12. Are any of the statements tautologies or contradictions? Without using truth tables, see if you can give reasons. Write down the truth table for the following statements. Using the truth table, decide if any pairs are logically equivalent. Give reasons for all your answers.

(a) $\sim [(p \wedge \sim q) \wedge q]$.
(b) $p \vee (\sim q \wedge r)$.
(c) $(p \vee q) \wedge (p \vee \sim r)$.
(d) $\sim p \vee (q \vee \sim q)$.

13. Negate the following statements using De Morgan's laws. Write all the steps involved with reasons for every step.

(a) $(\sim p \vee q) \wedge \sim (\sim q \wedge r)$.
(b) $\sim p \vee \sim q$.
(c) $(p \vee q) \wedge r$.
(d) $(p \vee r) \wedge (p \vee q)$.
(e) $(p \wedge q) \vee (\sim p \wedge q)$.

14. Write the following conjunctions and disjunctions in a symbolic form by specifying the statements p, q. Negate the statements p, q in each case and then negate each of the given statements.

(a) Some roses are not red or all violets are blue.
(b) All baby animals are cute, while all cute animals are not babies.
(c) No tiger is a herbivore or all lions are carnivores.
(d) Some peacocks have beautiful tail feathers but all peahens are drab.
(e) Some women wear saris or some men wear trousers.

15. For each of the following conditional statements, write down the converse, inverse and contrapositive.

(a) If it is hot, then I will sweat.
(b) If there is no coffee, then I will drink tea.
(c) All Alsatians are dogs.
(d) No bat is a bird.
(e) If there is no rainfall, then there will be a drought.

16. For each of the conditional statements above, write a disjunction that it is logically equivalent to. Also, negate each of the above statements.
17. Test the validity of the following arguments.
    (a) If you are not polite, then you will not be treated with respect. You are not treated with respect. Therefore, you are not polite.
    (b) If you are kind to a puppy, then she will be your friend. You were not kind to that puppy. Therefore, she is not your friend.
    (c) All sneaks are devious. All swindlers are sneaks. Therefore, all swindlers are devious.
    (d) If I am literate, then I can read and write. I can read but I cannot write. Thus, I am not literate.
    (e) I will be a candidate for class representative elections or I will keep quiet. I stood for class representative elections. Therefore, I did not keep quiet.
18. Symbolise the following arguments and test their validity using a truth table.
    (a) If it rains or snows, then my roof leaks. My roof is leaking. Thus, it is raining or snowing.
    (b) If an animal is a squid, then it has tentacles. If an animal is an octopus, then it has tentacles. Therefore, if an animal is a squid, then it is an octopus.
    (c) I wash the dishes or I do not eat. I eat. Thus, I wash the dishes.
    (d) If you want to be a salesperson, then you have to dress well. You do not want to be a salesperson. Therefore, you do not dress well.
    (e) All Scorpio owners are used to trekking. Nisha is not a Scorpio owner. Therefore, Nisha is not used to trekking.

Bibliography

Books and Articles

[1] A G Hamilton. 2000. *Logic for Mathematicians*. Cambridge, UK: Cambridge University Press.

[2] Richard Hammack. 2013. *Book of Proof*. Richmond, VA: Richard Hammack. `http://www.people.vcu.edu/~rhammack/BookOfProof/` (accessed 27 January 2017).

Websites

[W1] J J O'Connor and E F Robertson. *George Boole*. `http://www-groups.dcs.st-and.ac.uk/history/Biographies/Boole.html` (accessed 27 Jan. 2017).

[W2] J J O'Connor and E F Robertson. *Augustus De Morgan*. `http://www-history.mcs.st-and.ac.uk/Biographies/De_Morgan.html` (accessed 27 January 2017).

[W3] James Wooland. *Liberal Arts Mathematics Homepage*. `http://www.math.fsu.edu/~wooland/` (assessed 27 January 2017).

Image Credits

[C1] Chumsdock Cheng. *Cup of Tea*. CC BY-SA 2.0 license. https://flic.kr/p/4QACSUflic.kr/p/4QACSU (accessed 27 January 2017).

[C2] Labrador photo has been taken from Wikimedia Commons, licensed under the Creative Commons Attribution-Share Alike 3.0 Unported (https://creativecommons.org/licenses/by-sa/3.0/deed.en) license, https://en.wikipedia.org/wiki/Labrador_Retriever\#/media/File:YellowLabradorLooking_new.jpg.

[C3] Poodle Photo has been taken from Wikimedia Commons, licensed under the Creative Commons Attribution 2.0 Generic (https://creativecommons.org/licenses/by/2.0/deed.en) license, https://en.wikipedia.org/wiki/Poodle#/media/File:Full_attention_(8067543690).jpg . Photographer: Tim Wilson (https://www.flickr.com/people/70816538@N00) from Blaine, MN, USA

[C4] Photo by Geetha Venkataraman.

[C5] https://en.wikipedia.org/wiki/German_Shepherd#/media/File:DSHwiki.jpg [photograph taken by Brigitte Mardorf]. Alsatian photo has been taken from Wikimedia Commons, licensed under the Creative Commons Attribution-Share Alike 3.0 Unported (https://creativecommons.org/licenses/bysa/3.0/deed.en) license.

[C6] Coffee Cup photo has been taken from Wikimedia Commons, licensed under the Creative Commons Attribution-Share Alike 3.0 Unported (https://creativecommons.org/licenses/by-sa/3.0/deed.en) license, photographer: Coffeecupgals. Tea Cup photo has been taken from the website Hack your cup of tea |Opensource.com, https://en.wikipedia.org/wiki/Coffee_cup#/media/File:Traditionalcappuccino.jpg. Photographer: chumsdock cheng (https://www.flickr.com/photos/chumsdock/2519019976/). Modified by Rikki Endsley. CC BY-SA 2.0 (https://creativecommons.org/licenses/by-sa/2.0/)

[C7] Source: https://commons.wikimedia.org/wiki/File:Cat_climbing_tree,_Uchimaki_Park.jpg. Photographer: sumomojam from Saitama, Japan. [CC BY-SA 2.1 jp (http://creativecommons.org/licenses/by-sa/2.1/jp/deed.en)], via Wikimedia Commons

[C8] Geetha Venkataraman. Original photo of Bat taken from Wikimedia Commons, Original photo: Oren Peles, Derivative work: User: MathKnight, licensed under the Creative Commons Attribution 2.5 Generic (https://creativecommons.org/licenses/by/2.5/deed.en) license. https://en.wikipedia.org/wiki/Bat#/media/File:PikiWiki_Israel_11327_Wildlife_and_Plants_of_Israel-Bat-003.jpg

4

CHAPTER

Data Analysis and Modelling

> There are wavelengths of light we cannot see and flavors we cannot taste. Why then, given our brains are wired the way they are, does the remark 'Perhaps there are thoughts we cannot think,' surprise you? Evolution, so far, may possibly have blocked us from being able to think in some directions; there could be unthinkable thoughts.
>
> —*R. W. Hamming (1915–98), a mathematician whose work is fundamental to modern digital communication techniques.*

In the earlier chapters of this book, you have been exposed to mathematics as a language. The focus was first on developing vocabulary (sets and numbers) and then on grammar (the rules of logic). One of the striking features of this study was the precision with which everything fitted together. This chapter, on the other hand, is all about how this language helps us deal with the real world. This is indeed the great mystery and charm of mathematics, that it is at the same time elegant and effective.

One may imagine that mathematics, over the course of history, has been subject to a process of evolution in which elegance of form and effectiveness in explaining phenomena have acted as twin criteria for natural selection. Ways of doing mathematics which were unduly narrow or ungainly have fallen by the wayside and others have replaced them. Accordingly, we should not view the mathematics of today as a finished product but as an ongoing human activity subject to change and even revolution. Such change could be driven by two related forces. First, the prevailing rules of mathematics may no longer suffice for us to express our thoughts and intuition. Second, there may be pressure from real-world problems for mathematics to be reshaped to explain and solve them.

In this chapter, we will begin to gain familiarity with how mathematics can be used to tackle information—especially numerical information—and convert data into insights. In order that you should become equipped to handle genuine real-world problems, not just artificial ones, we will also explore in some detail the use of computers in processing and analysing data. Specifically, we will cover the basic features and uses of spreadsheet programs.

4.1 Interacting with Data

Example 4.1.1 A piece of 20 cm long string has to be arranged into a rectangle. If we want the rectangle to have as much area as possible, what lengths should we choose for its sides?

We start by translating this problem into the symbolic language of mathematics. (It helps that it is about mathematical objects to begin with!) So we set names x and y for the lengths of the sides of the rectangle.

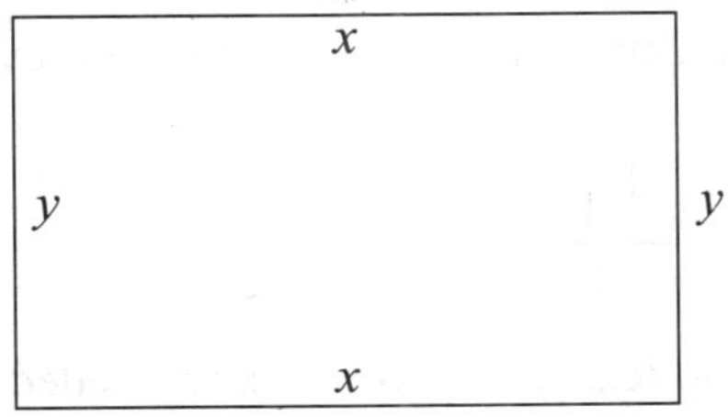

We are given that the perimeter of the rectangle is 20 cm. Since the perimeter is the sum of all four sides, we have the equation

$$2x + 2y = 20 \quad \text{or} \quad x + y = 10 \quad \text{or} \quad y = 10 - x$$

Now, as we vary the side x, the other side automatically takes the value $10 - x$, and so the area A of the rectangle is

$$A = xy = x(10 - x) = 10x - x^2$$

It is time to collect data. We take a few values of x and compute the corresponding value of A. For example, $x = 2$ gives $A = 2 \times (10 - 2) = 16$. We tabulate some initial results as follows:

x	1	2	4	6	8	9
A	9	16	24	24	16	9

These data suggest that the peak value is attained between $x = 4$ and $x = 6$. The symmetry of the data strongly suggests that the peak is actually attained at $x = 5$. Let us collect more evidence as follows:

x	4	4.5	5	5.5	6
A	24	24.75	25	24.75	24

At this point, the evidence that the maximum possible area is 25 cm^2 is strong enough for us to attempt a proof. Since we want to show that the area is at most 25, we consider the difference $25 - A$:

$$25 - A = 25 - 10x + x^2$$

We recognise that the last quantity is of the form $a^2 - 2ab + b^2 = (a - b)^2$ and so we obtain

$$25 - A = 25 - 10x + x^2 = (5 - x)^2$$

The quantity $(5-x)^2$ is always non-negative and so is $25-A$. This means in turn that A never exceeds 25. Thus, the maximum possible value of the area is in fact $A = 25\text{ cm}^2$, obtained when $x = y = 5$ cm and the rectangle is a square. □

This example serves to illustrate that the task before us can be split into two parts:

1. Collect appropriate data and present it in a form which highlights its key features.
2. Analyse the data and extract results, or at least hints towards possible results.

Consider Example 4.1.1 again. Our initial insight came from reviewing the data in the following table:

x	1	2	4	6	8	9
A	9	16	24	24	16	9

Had the situation been more complicated, we would have needed much more data before we felt safe in drawing conclusions. But as the amount of data grows, it becomes harder to absorb it, let alone extract information from it. We then need to present it in a way that is more friendly to the human eye. For example, the data in the table above could also have been presented graphically as follows:

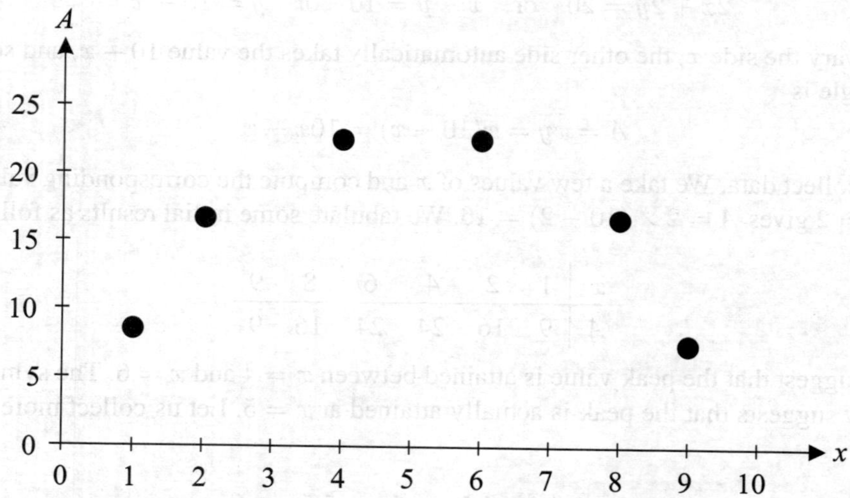

When we view the data graphically in this manner, we obtain an overview of the situation without even needing to take in the exact numbers involved. We immediately see the symmetry of the values and the location of the peak.

Ratio and Percentage

Numbers can be used in two fundamental ways: either to express an amount, or to express a relationship between two amounts. When a child says 'My friend has a hundred toys', she is using number in the first sense. However, if she says 'My friend has a hundred times as many toys as I do', she is using number in the second sense. Mathematically, such a comparison is expressed through a **ratio** and a mathematician making a similar comparison might say 'The ratio of my friend's theorems to mine is 100 : 1'.

Similarly, suppose we are told that the ratio of quantity A to quantity B is $4:2$. This means that there are 4 units of A to every 2 units of B. The relationship could also be expressed as there being 2 units of A for every unit of B. In other words, the ratios $4:2$ and $2:1$ express the same relationship and so they can be taken to be the same.

In general, if we say that the ratio of one quantity to another is $M:N$, we mean that there are M units of the first quantity for every N units of the second.

Ratios can be combined with each other. Let the ratio of A to B be $2:3$ and the ratio of B to C be $5:2$. We can work out the ratio of A to C. Start with 6 units of C. Corresponding to these, we have $5 \times 3 = 15$ units of B. Corresponding to these 15 units of B, we have $2 \times 5 = 10$ units of A. Thus, the ratio of A to C is $10:6$, which can also be expressed as $5:3$.

Task 4.1.1 The ratio of quantity A to quantity B is $M:N$, and the ratio of quantity B to quantity C is $P:Q$. What is the ratio of A to C? (Hint: Consider $Q \cdot N$ units of C.)

The ratio $M:N$ can also be expressed as the fraction M/N. The rule for multiplying fractions then corresponds to the rule for combining ratios (worked out in the previous task). For example, a ratio of $3:5$ can be expressed as the fraction $3/5$ and this can further be put in a decimal form: 0.6.

Another way of expressing a ratio or fraction is as a **percentage.** Here, one quantity is compared to 100 units of another. For example 75% (read as '75 percent') stands for the ratio $75:100$ and thus corresponds to the fraction $75/100 = 0.75$.[1] Percentages are normally used only when both quantities are being measured in the same units.

Task 4.1.2 Express the ratio $9:6$ as a fraction, a decimal and a percentage.

Percentages are commonly used to describe the increase or decrease in a quantity. A rise of 50% means that the amount of increase is 50% of the original. Thus, if an item was priced at ₹300 and the price is said to have gone up by 25% since, then the increase is ₹75 (25% of 300) and the final price is ₹375. If the same initial price is said to have decreased by 25%, then the final price is ₹225 ($= 300 - 75$). In the last situation, we may also say that the price has *changed* by -25%. The word 'change' is neutral and allows both increase and decrease, but the negative sign tells us that there was actually a decrease.

Example 4.1.2 In 2001, the average price of crude oil in the United States was \$23 per barrel. Ten years later, in 2011, it was \$87! This increase can be described using percentages in two ways. First, we convert the fraction $87/23$ into the decimal 3.78. This means that the final price is 378% of the starting price. Alternately, we focus on the increase in price (\$64) and calculate its ratio to the initial price: $64/23 = 2.78$. So we can also say that the price has increased by 278%. □

Raw and Grouped Data

In the remainder of this section, we take up the problem of processing numerical data (a list of numbers arising out of some measurements) and giving it a form which makes it easier to judge its key attributes.

[1] *Percent* can be read as *per cent* or *per hundred*, *centum* being the Latin word for hundred.

Consider the following collection of scores by a certain batsman in his first 60 innings of test cricket (Table 4.1.1)[2]:

Table 4.1.1 Runs Scored in the First 60 Innings Played in Test Cricket by a Certain Batsman

15	59	8	41	35	57	0	24	88	5	10	27	68	119	21
11	16	7	15	40	148	6	17	114	5	0	11	111	1	6
0	73	50	9	165	78	62	28	104	71	142	96	6	43	11
34	85	179	54	40	10	4	0	52	2	24	122	31	177	74

Scanning these scores, we see a good number of centuries and sense that this is an excellent player, especially when we further learn that he was unbeaten in 7 of these 60 innings. One way to give structure to these data is to arrange the scores in increasing order (Table 4.1.2):

Table 4.1.2 Scores Arranged in Increasing Order

0	0	0	0	1	2	4	5	5	6	6	6	7	8	9
10	10	11	11	11	15	15	16	17	21	24	24	27	28	31
34	35	40	40	41	43	50	52	54	57	59	62	68	71	73
74	78	85	88	96	104	111	114	119	122	142	148	165	177	179

This new arrangement of the data makes it easy to see certain features. We may be disappointed on seeing that a quarter of the scores are below 10, and then amazed to see that an equal number are over 70.

We can go further and put the scores into certain categories to bring out patterns. For example, we may decide to call scores of 24 or less 'low', scores of 25–74 'medium' and scores of 75 and above 'high'. Counting the number of scores in each range or category gives us a summary of the data which are easier to absorb and work with:

Table 4.1.3 Categorising the 60 Scores in Table 4.1.1 into Low, Medium and High

	Low (< 25)	Medium (25–74)	High (≥ 75)
Number of scores	27	19	14

In such a situation, the original unsorted data (the list of 60 scores) is called the **raw** or **primary data.** When we sort that into different categories, we create **grouped** or **processed data.** The individual categories into which we sort the data are called **classes** or **bins.** The number of data points in a class is its **frequency.** The **frequency table** of the grouped data lists the classes and their frequencies.

[2] All our cricket data have been obtained from the website HowSTAT! [W6].

Thus, Table 4.1.3 is a frequency table for grouped data coming from the raw data in Table 4.1.1. The first row describes the classes. Below each class we give its frequency.

The raw data may be quite complicated. For example, consider a situation where patients are being treated in a hospital for a certain disease. Various measurements would have been taken for each patient, such as body temperature, absence or presence of rash, blood cell counts, gender and age. Based on these measurements, which constitute the raw data, the concerned doctors would have classified the patient's state as, for example, 'critical', 'stable' or 'recovering'. Thus, the grouped data would consist of these three classes and the number of patients in each class.

When we classify raw data in this manner, there is a strong element of subjectivity in the choice of classes. The subjectivity encompasses both the number of classes and the precise rules used to allocate the data to classes. In our cricket example, we might well use 100 runs as a cut-off for 'high', or want another class for truly high scores, say over 200. The choice of classes depends not so much on the data itself but on the questions we wish to ask of the data. For example, we might want to know how often he has scored over 100 simply because this would enable easy comparison with databases which record the number of centuries scored by various batsmen. Then this would influence our cut-offs for the classes.

Task 4.1.3 Identify the batsman whose scores we are studying.

Quick Review

After reading this section, you should be able to:

- Express ratios as fractions, decimals or percentages.
- Use percentages to express the relative change in a quantity.
- Sort raw data into classes and draw up the frequency table.
- Draw conclusions from data in simple situations.

EXERCISE 4.1

1. Express the overall price change as a percentage:
   (a) It first increases by 50% and then falls by 50%.
   (b) It first falls by 50% and then rises by 50%.
2. A stone is thrown vertically upwards. The height h of the stone at t seconds after its launch is given by

$$h = 10t - 4.9t^2$$

where distance is measured in metres. Calculate the height of the stone for various values of t and estimate how high it goes before falling.

3. We wish to make a rectangular tray whose base has area 900 cm^2 and there is a rim of height 1 cm. Gather numerical evidence for the dimensions we should choose so that the amount of material used is the least possible. Present your evidence graphically.

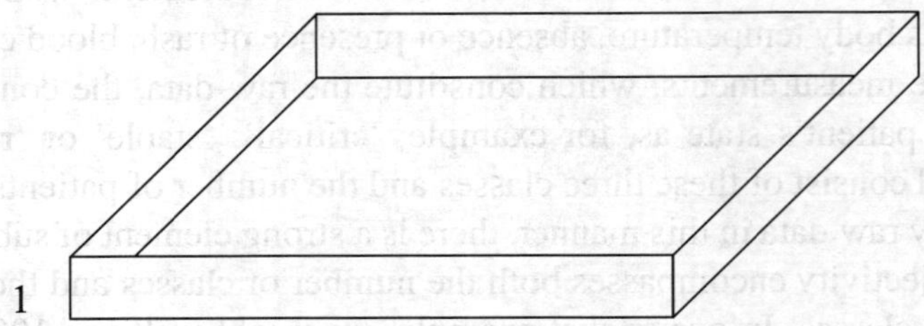

4. Classify the scores in Table 4.1.2 into the following two categories: bad (below 25) and good (25 or above). Draw the frequency table for this grouping.
5. Find a whole number so that half the scores in Table 4.1.1 are above it and half are below.

4.2 Introduction to Spreadsheet Programs

A spreadsheet program essentially provides a page on which we can arrange our calculations, in much the same way that we would work on a sheet of paper. It is easier to arrange our work neatly if the sheet of paper is divided into rows and columns, and so a spreadsheet program offers us the same format.

The most popular spreadsheet program at present is Excel, which is part of the Microsoft Office set of programs.[3] An alternative, available free, is LibreOffice Calc.[4] You could use either program to implement the techniques we are going to discuss in this chapter, since the basic commands are identical in both. We will give our descriptions in terms of Excel, but we shall try to point out all the occasions when Calc uses a different syntax.[5] At this point, you should make sure that one of these is available to you. Preferably, you should be sitting in front of a computer with your selected spreadsheet program running on it.

Basic Arithmetic in Spreadsheets

The simplest use of a spreadsheet program is as a desk calculator: we can use it to perform simple operations such as addition, subtraction, multiplication and division. We select a cell and type in the expression to be calculated, making sure to begin the expression with an equal (=) sign. For example, to compute $5 + 7$, we type $= 5 + 7$ in the selected cell and then press the Enter key (Figures 4.2.1 and 4.2.2).

---

[3] See Microsoft website [W8].

[4] See LibreOffice website [W7].

[5] We have worked out the examples using Excel 2013 and LibreOffice 5 on a computer with a Windows operating system. Certain steps may differ on other operating systems or with other versions of the programs.

	A	B	C	D	E
1					
2					
3					
4					
5					
6					
7					

Figure 4.2.1 A spreadsheet program offers you a space divided into rows and columns, into which you can enter your numbers and calculations. The columns are labelled by alphabets (A, B, C, . . .) and the rows by numerals (1, 2, 3, . . .). Each box or *cell* is labelled by its column and row. In this figure, the top left cell which has the thick border is labelled A1 because it is in Column A and Row 1.

	A	B	C
1	= 5 + 7		
2			

$\longrightarrow$

	A	B	C
1	12		
2			

Figure 4.2.2 We type = 5 + 7 in cell A1 and press Enter to get 12.

The following table shows how to perform all the basic operations:

Operation	Example	What you should type	Result
Addition	$4 + 2$	$= 4 + 2$	6
Subtraction	$4 - 2$	$= 4 - 2$	2
Multiplication	4×2	$= 4 * 2$	8
Division	$4 \div 2$	$= 4/2$	2
Exponentiation	4^2	$= 4\hat{}2$	16

More involved calculations can be carried out with the help of brackets. For example, to calculate $(2^{2^2} - 1) \div 3$, we type $= (2\hat{}(2\hat{}2) - 1)/3$.

Task 4.2.1 Carry out the following calculations in a spreadsheet:

1. $4 + 5 \times (4 - 2)$
2. $4 + 5 \times 4 - 2$
3. $\dfrac{1}{2^3} \div \dfrac{1}{3^2}$
4. 2^{40}

The results of the first three calculations in this list of tasks should hold no surprises for you, provided you are careful in your use of brackets (especially in the third problem). One of their purposes is to verify that spreadsheets follow the usual rules regarding a sequence of arithmetic operations—multiplication and division are given priority over addition and subtraction. The result of the last calculation, on the other hand, is a strange looking entity: 1.09951E+12. This is to be read as 1.09951×10^{12} and is obtained by rounding off the actual value of 1099511627776.

A number calculated in one cell can be used in a new calculation by simply referring to the label of its cell. For example, suppose the two numbers calculated in Task 4.2.1 were placed in cells A1 and B1. Then, we could calculate their sum in cell C1 by typing

= A1+B1

Here the program recognises the cell references A1 and B1 as locations of numbers and uses the corresponding numbers in the calculation. Thus, the cell references work like variables.

Spreadsheet Functions

Suppose you had the numbers $1, 3, 5, 7, 9, 11, 13, 15$ and 17 entered in sequence in cells A1–A9, and you wanted to calculate their sum in cell B1. Then you could type the following

= A1+A2+A3+A4+A5+A6+A7+A8+A9

in the B1 cell. There is also a shortcut, using the inbuilt command or function Sum:

= Sum(A1:A9)

The program recognises A1:A9 as a list consisting of the cells starting with A1 and ending at A9. Similarly, one could find the product of all these numbers by typing

= Product(A1:A9)

Spreadsheet programs have a host of such inbuilt functions which can shorten our work. We will soon take up various such functions that allow us to summarise the main features of given data. (All the inbuilt functions in Excel can be accessed from the Formulas tab.)

Grouping Data with Spreadsheets

Excel provides a Frequency function which calculates frequencies of classes from the raw data. We will show how to sort the data in Table 4.1.1 into the classes '0–50', '50–100', '100–150' and '150–200'. In each class, the lower limit is included and the upper limit is excluded. Thus, 50 belongs to the class '50–100' and not to '0–50'.

1. We open an Excel worksheet and enter the 60 scores in Column A. Thus, they occupy the cells A1:A60.
2. In Column B, we enter the limits 50, 100 and 150 in cells B2:B4.
3. We have four classes so we select four cells in which the frequencies will eventually be entered. We select cells C2:C5 so that the frequencies will be entered next to the class limits.

4. While keeping C2:C5 selected, type =Frequency(A1:A60, B2:B4).
5. Press the Ctrl and Shift keys simultaneously and, keeping them pressed, hit Enter (Figure 4.2.3).

Note Step 5 in particular. Frequency is an **array function** as its output is an array or list of numbers rather than a single number. All array functions have to be executed with this Ctrl–Shift–Enter combination.

	A	B	C	D
1	15	Bins	Frequency	
2	59	50	=FREQUENCY(A1:A60,B2:B4)	
3	8	100		
4	41	150		
5	35			
6	57			
7	0			
8	24			
9	88			
10	5			

→

	A	B	C	D
1	15	Bins	Frequency	
2	59	50	37	
3	8	100	13	
4	41	150	7	
5	35		3	
6	57			
7	0			
8	24			
9	88			
10	5			

Figure 4.2.3 Executing the Frequency function in Excel. Excel takes in the three limits of 50, 100 and 150, and creates four classes: below 50, 50–100, 100–150 and above 150. The respective frequencies have been calculated to be 37, 13, 7 and 3.

The same steps also work in LibreOffice Calc.

Copying from One Cell to Another

We can copy text or numbers from one spreadsheet cell to another in the usual way:

1. Left click on the cell you want to copy, and then press Ctrl-C.
2. Left click on the cell where you want to paste this material, and then press Ctrl-V.

Alternately, right click on the selected material and left click on Copy, then right click on the target cell and click on an appropriate Paste Option. We can also simultaneously copy and paste the contents of several adjacent cells as illustrated in Figure 4.2.4.

Let us try this procedure on a cell containing a formula.

	A	B	C
1	25	=A1+A2	
2	50		
3	75		

→

	A	B	C
1	25	75	
2	50		
3	75		

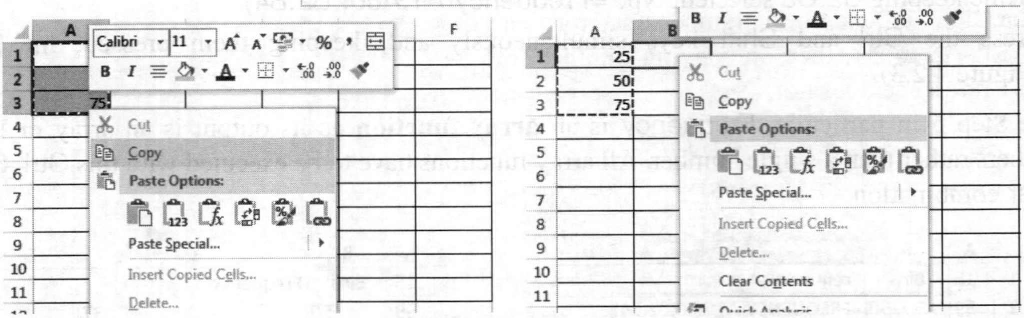

Figure 4.2.4 Copying the contents of cells A1:A3 into cells B1:B3.

Now, we copy cell B1 and paste in cell B2. We get the following:

	A	B	C
1	25	75	
2	50	125	
3	75		

Surprise! We copied 75 and got 125. Evidently, the formula changed when we copied the cell. Let's look at the formulas behind the numbers:

	A	B	C
1	25	=A1+A2	
2	50	=A2+A3	
3	75		

The cell references have changed. To see the pattern of how they change, let's paste the contents of B1 in a few more cells. The resulting formulas and output are shown below:

	A	B	C
1	25	=A1+A2	=B1+B2
2	50	=A2+A3	=B2+B3
3	75	=A3+A4	=B3+B4

→

	A	B	C
1	25	75	200
2	50	125	200
3	75	75	75

We see that the references are **dynamic;** they change according to the relative position of the cell where we paste the formula. For example, suppose we paste the content of B1 in C3. This is one position to the right and two positions below the original. We find that every cell reference changes by the same amount: the column gets shifted to the right by one and the row gets shifted down by two. Thus, A1+A2 becomes B3+B4.

As you gain experience with spreadsheets, you will find that this twist actually makes life easier for us, as it enables patterns and relationships to be easily replicated in other places.

Sometimes we want to keep the cell references from changing. To make them **static** we put dollar signs around the column name, for example, writing $A$1 instead of A1. If we write =$A$1+A2 in B1, then on copying to other cells we find that the A2 reference changes but the $A$1 reference does not.

	A	B	C
1	25	=$A$1+A2	=$A$1+B2
2	50	=$A$1+A3	=$A$1+B3
3	75	=$A$1+A4	=$A$1+B4

$\longrightarrow$

	A	B	C
1	25	75	125
2	50	100	50
3	75	25	25

Quick Review

After reading this section, you should be able to:

- Carry out arithmetical operations in a spreadsheet.
- Use spreadsheet functions such as Sum and Product.
- Use a spreadsheet to create a frequency table from raw data.
- Use both dynamic and static cell references while copying formulas.

EXERCISE 4.2

1. Use a spreadsheet program to calculate.

(a) $\sqrt{10}^3$
(b) $\dfrac{2/3}{3/5}$
(c) $\log_{10}(500)$
(d) $\sin(60°)$
(e) $1 + 2 + \cdots + 10$
(f) $100!$

2. Write a spreadsheet command for applying the Sum function to the numbers in the following cells: A1, A2, A3, B1, B2, B3.

3. Consider the following Excel spreadsheet:

	A	B	C
1	25	75	125
2	50	100	50
3	75	25	25

Calculate the output of the following commands:

(a) Sum(A1:C3) (b) Sum(A1:A3, C1:C3) (c) Sum(A1:A3,-150)

4. Consider the Excel spreadsheet in the previous exercise. Use the Frequency function to group the data into the classes 0–19, 20–39, 40–59, . . . , 100–119, 120 and above.

5. The list below gives the number of hours a student has spent studying each day in April:

$$4, 3, 4, 6, 8, 5, 5, 0, 1, 8, 8, 2, 3, 4, 5, 6, 7, 8, 0, 0, 9, 1, 1, 1, 9, 2, 2, 9, 3, 4$$

   Create a single frequency table which can be used to calculate both the number of days on which the student has studied less than 4 hours and the number of days on which he has studied at least 7 hours.

4.3 Bar Charts, Histograms and Pie Charts

Bar Charts

A **bar chart** provides a visual representation of a frequency table. Each class is represented by a column (a vertical bar) and the height of the column gives the frequency of the class. All the columns have equal width (Figure 4.3.1).

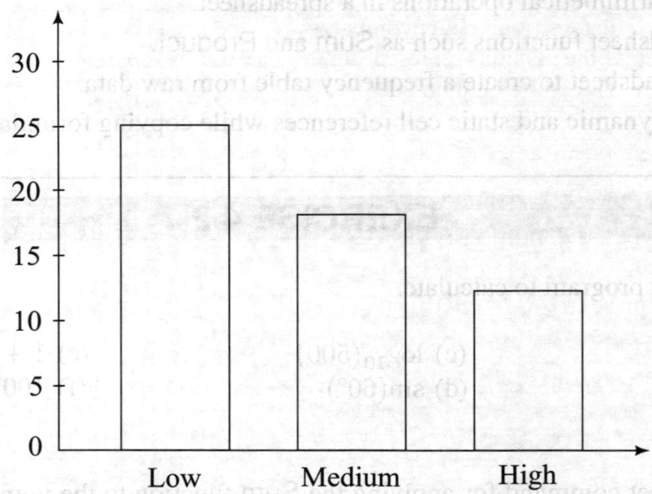

Figure 4.3.1 Bar chart for the grouped data in Table 4.1.3. In a bar chart, the number of items in a particular class is represented by a vertical column whose height equals that number.

Bar charts may also be used to compare different sets of grouped data provided, of course, that they use the same classes. For example, we may have a list of patients who were admitted to a hospital for a certain epidemic disease and they could be labelled as 'critical', 'stable' and 'recovering'. Suppose our data for the hospitalised patients covers two dates, a week apart. Then, by comparing the frequencies, we obtain an overview of the progress of the epidemic.

	Critical	Stable	Recovering
June 1	15	25	10
June 8	5	15	20

Here is a bar chart that helps us visualise the situation:

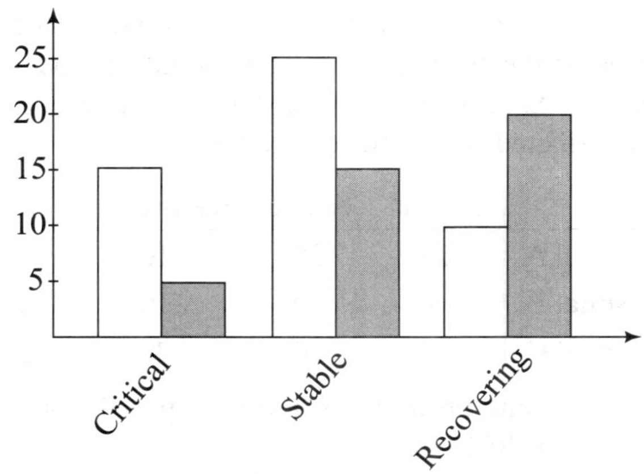

Figure 4.3.2 This **clustered bar chart** compares the state of patients one week apart, by placing together the columns representing the numbers in the same category but on different dates. The white columns depict the situation on June 1 and the grey columns on June 8.

The depiction in Figure 4.3.2 makes it easy to see that over a week, the state of the patients has generally improved: the numbers have shifted away from 'Critical' and 'Stable' towards 'Recovering'. There is, however, one further positive development—the decrease in the total number of patients from 50 to 40—and this is not so immediately seen from the clustered bar chart.

Figure 4.3.3 uses a single column for each date to enable a clear view of the change in the total number of patients, as well as the changes in each category. This would be an especially useful approach if we had a longer series of data over several weeks.

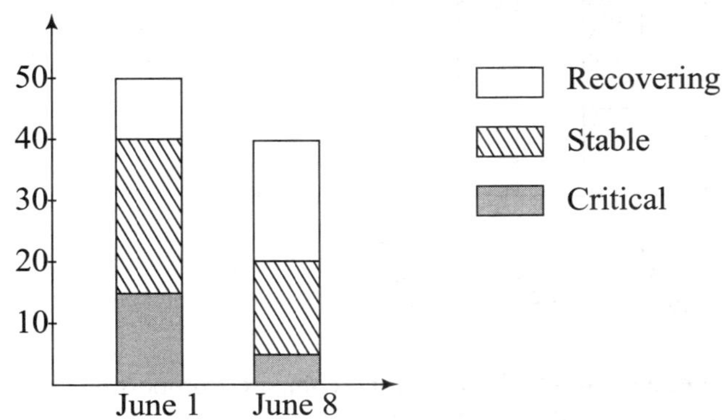

Figure 4.3.3 This **stacked bar chart** uses single columns to depict the situations on June 1 and June 8. Each column is divided into three parts which show the distribution of the patients among the three categories.

We will now take up our last variation on the basic bar chart. Imagine that the data for patients have come from various hospitals scattered over a city, and we are interested in the geographical variation in the epidemic pattern. In this situation, we would not be too concerned with the total number of cases from a particular hospital, since this could depend simply on the size of the hospital or the population size that it serves. It would be more relevant to look at the proportions in which the patients are distributed among the categories.

	Critical	Stable	Recovering	Total
Hospital A	15	25	10	50
Hospital B	5	10	10	25
Hospital C	40	50	70	160

First, we convert these numbers into proportions. For example, the proportion of critical patients in Hospital A is $15/50 = 0.30 = 30\%$.

	Critical (%)	Stable (%)	Recovering (%)
Hospital A	30	50	20
Hospital B	20	40	40
Hospital C	25	31	44

We now use a single column for each hospital, with each column having the same height. Each column is then divided into parts according to the proportion of patients in each category.

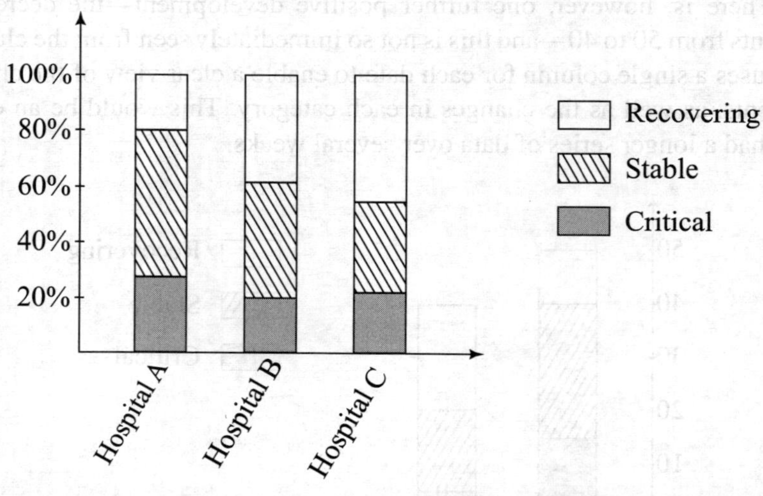

Figure 4.3.4 This **percent stacked bar chart** compares proportions or percentages in each category across different datasets.

Figure 4.3.4 shows that the situation in Hospital A is significantly worse than the others, with a much smaller proportion of recovering patients. Fortunately, the number of critical cases is not much higher than in the others—the difference is in the stable category, and so one may hope that soon this hospital too will have a greater number of recovering patients.

Making Bar Charts with Spreadsheets

Consider the following grouped data:

	Critical	Stable	Recovering
Hospital A	15	25	10
Hospital B	5	10	10
Hospital C	40	50	70

To make a clustered bar chart for this data in Excel, we take the following steps:

1. Open a worksheet and enter the table in Cells A1:D4. Select these cells.
2. Open the Insert tab.
3. In the Charts group, click on Column. A menu opens up giving various types of charts that can be created. The first row of options has the ones we want: Clustered, Stacked and Percent Stacked.
4. Click on Clustered Column to create the clustered bar chart.

Figure 4.3.5 shows these steps and the result.

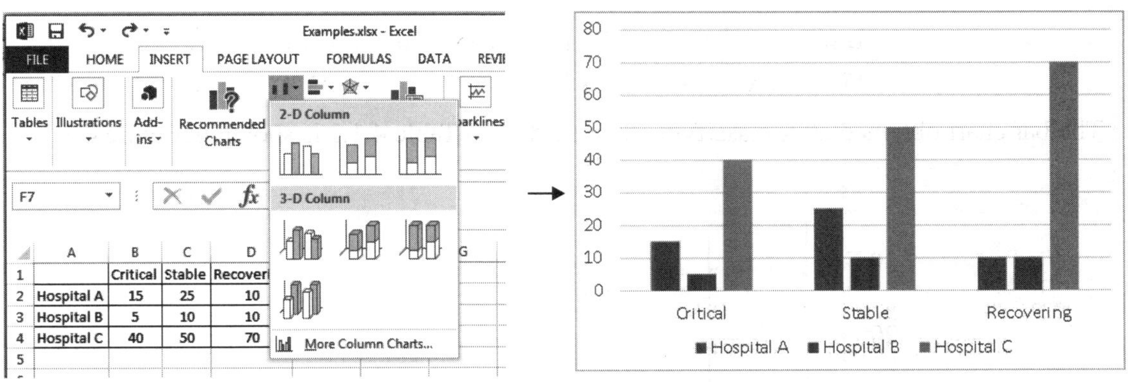

Figure 4.3.5 Creating a clustered bar chart in Excel.

The clustered bar chart we just created has grouped all the frequencies of each state of health (such as Critical). We might instead wish to group together all the frequencies for each hospital. We can switch to that through the following steps (Figure 4.3.6):

1. Right click on an empty part of the chart.
2. In the menu that opens, click on Select Data.
3. Another box opens up. Click on Switch Row/Column.

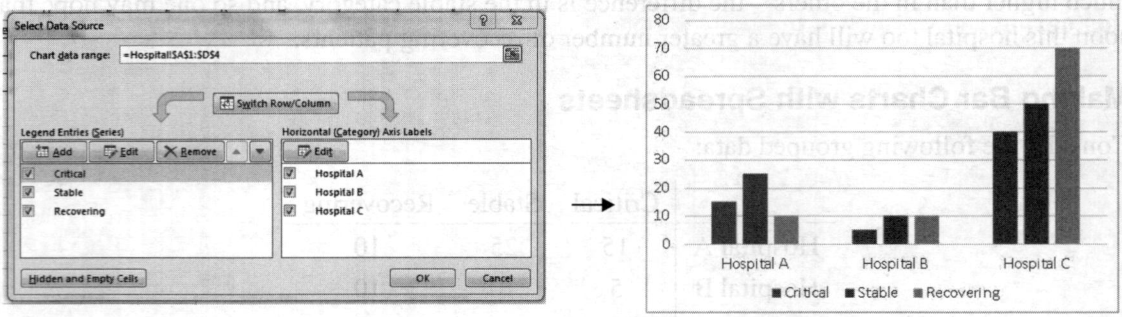

Figure 4.3.6 Switching the clustering in an Excel bar chart.

Histograms

The categories on which a bar chart is based can come about in various ways. An especially important case is when the categories are based on the value of a particular number. As an example, let us take up the cricket scores in Table 4.1.1. The scores vary from 0 to 179 and so we can conveniently distribute them among four ranges of equal width: 0–50, 50–100, 100–150 and 150–200. We use the convention that the category '50–100' includes 50 but not 100. Thus, there is no overlap between these categories, which are also called **class intervals** or **bins.** The frequency table for this choice of bins is:

Bin	0–50	50–100	100–150	150–200
Frequency	37	13	7	3

The bar chart obtained after classifying the raw data in Table 4.1.1 is shown in Figure 4.3.7.

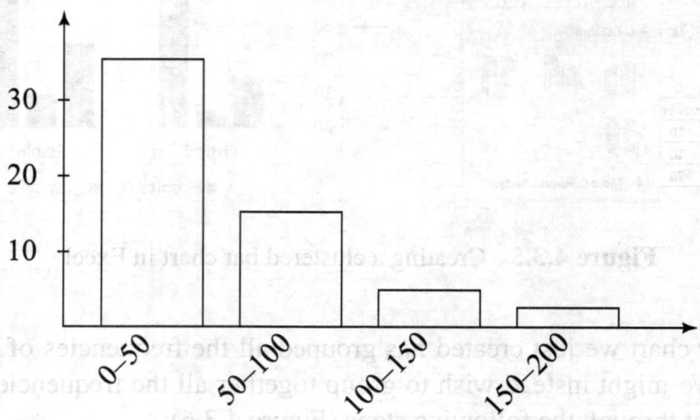

Figure 4.3.7 Bar chart obtained after classifying the raw data in Table 4.1.1.

The fact that the bins are intervals (like '0-50') can be used to simplify the presentation of the bar chart. We treat the x-axis as a number line and mark the end-points 0, 50, 100, 150 and 200 on it. Each bar is then drawn between the corresponding end-points. The resulting figure is called a **histogram** and we show it in Figure 4.3.8.

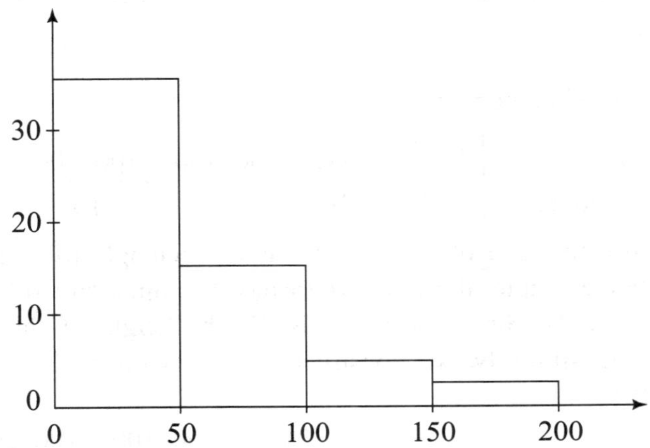

Figure 4.3.8 Histogram obtained after classifying the raw data in Table 4.1.1.

Now, we shall take up the aspects of histograms which distinguish them from bar charts. We note that it is desirable to allow the class intervals to have different lengths. We may then have smaller intervals where more precision is required (or more data are present) and larger intervals elsewhere. For example, we may settle on the following bins for Table 4.1.1:

Bin	0–25	25–50	50–100	100–200
Frequency	27	10	13	10

If we plot these frequencies as the heights of the corresponding columns, we get the following chart (Figure 4.3.9). This gives a rather misleading picture of the situation.

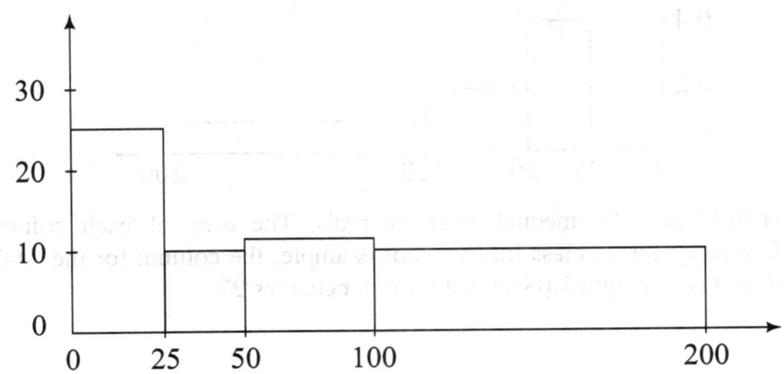

Figure 4.3.9 Chart of frequencies of class intervals. Plotting frequencies as heights gives a distorted picture when the class intervals have unequal lengths. In this example, it makes it look as though the batsman is more likely to score over 100 than under 25!

The reason that plotting frequencies as heights fails to work well for unequal class intervals is that the human eye judges the size of a rectangle by its area rather than by a single dimension such as height or width. For this reason, we will plot the frequencies as areas rather than heights, when making a histogram for class intervals of unequal length. As usual, we clarify this statement by means of an example.

Let us revisit the data used in the previous figure:

Bin	0–25	25–50	50–100	100–200
Frequency	27	10	13	10

This time around, we use the area of the corresponding rectangle to represent the number of points in each range. For example, the range 0-25 has 27 points and so we need to construct over it a rectangle of area 27. Since the width is 25, the height of this rectangle should be $27/25 = 1.08$. Proceeding similarly, we construct the following table and the corresponding histogram (Figure 4.3.10):

Bin	0–25	25–50	50–100	100–200
Area = Frequency	27	10	13	10
Width	25	25	50	100
Height = Area/Width	1.08	0.4	0.26	0.1

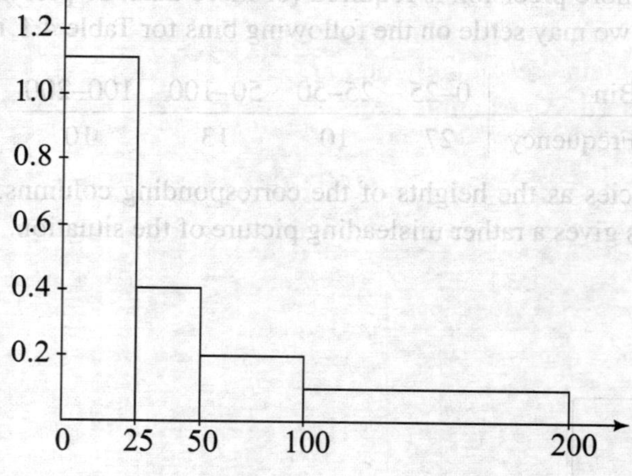

Figure 4.3.10 Histogram with unequal class intervals. The *area* of each column equals the frequency for its class interval. For example, the column for the '0–25' interval is drawn with height 1.08 so that its area becomes 27.

The Normal Shape

Take a look at the histogram drawn in the figure below. It has a symmetric shape distinguished by a single central peak and values that die off to zero on either side of the peak. It is quite common for histograms to have such a shape and so it has come to be called **normal.** A more descriptive name is **bell-shaped** due to the resemblance to the profile of a church bell (Figure 4.3.11).

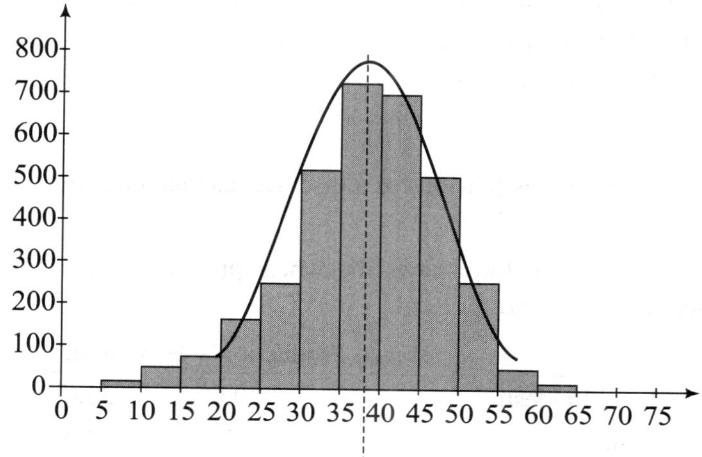

Figure 4.3.11 This histogram shows the distribution of marks for about 3,500 students in an exam with a maximum possible score of 75. Note the approximately *bell-shaped* or *normal* form of the histogram. The vertical dashed line represents the average marks.

The normal shape, though common, is not universal. One variation is that the peak may be off-centre so that the histogram dies quickly on one side but has a long tail on the other. Such a histogram is said to be **skewed.** An example is drawn below (Figure 4.3.12).

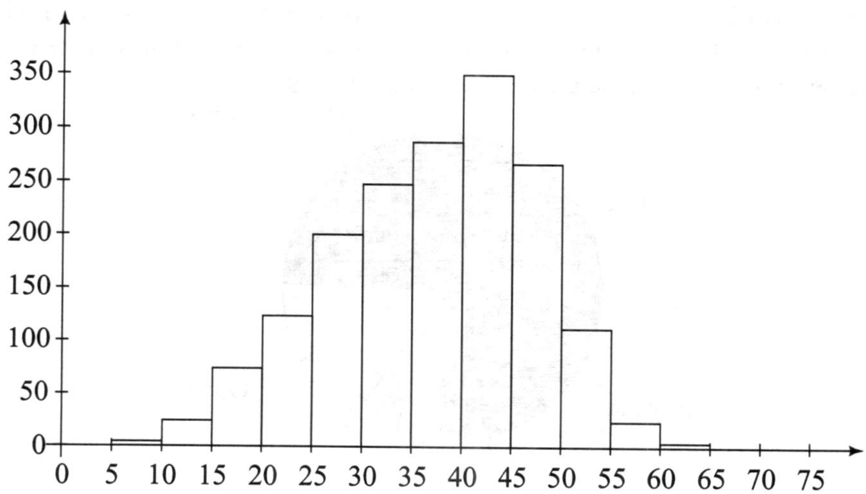

Figure 4.3.12 This histogram of marks for about 1,500 students in an exam with a maximum possible score of 75 is **skewed to the left**—more of the data is amassed on the right and there is a tail on the left.

Making Histograms with Spreadsheets

Unfortunately, there is no provision in Excel and Calc for drawing histograms with unequal class intervals. All we can do is create class intervals of equal lengths and calculate the corresponding frequencies through the Frequency function as described earlier. The output can then be converted into a histogram using the bar chart menu.

Pie Charts

A **pie chart** represents quantities as portions of a disc. Let us take up a small example by way of illustration.

Example 4.3.1 The table given below shows the wheat production (in millions of tonnes) of various states of India in 2006–2007:

State	Wheat Production	% of Total
Uttar Pradesh	25.03	33.0
Punjab	14.60	19.3
Haryana	10.06	13.3
Madhya Pradesh	7.33	9.7
Rajasthan	7.06	9.3
Others	11.73	15.4
Total	75.81	

Each data point is depicted by a sector of the circle whose area is proportional to the data value. For example, since Uttar Pradesh accounts for 25.03 units out of a total of 75.81, it represents $25.03/75.81 = 0.33$ or 33% of the total production. So it is represented by a sector whose area is 33% (almost exactly one-third) of the area of the circle. To draw this sector, we note that its central angle will also be 33% of the total angle of 360°, which is 119°.

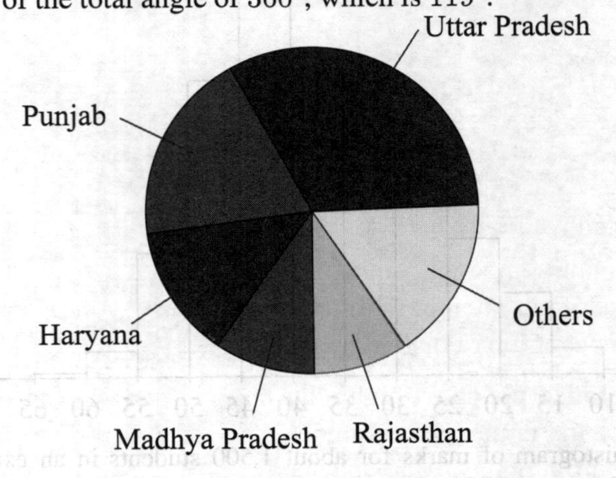

□

A pie chart represents a quantity by a sector whose area is proportional to the fraction of the total that the quantity constitutes. It is visually appealing; yet it is not suited for fine comparisons, as it can be hard to estimate and compare areas of sectors. It also quickly becomes cluttered if many quantities are involved.

Example 4.3.2 Let us incorporate more data in our last example:

State	Wheat Production	% of Total
Bihar	3.91	5.2
Gujarat	3.00	4.0
Maharashtra	1.63	2.1
Uttaranchal	0.80	1.1
West Bengal	0.80	1.1
Others	1.59	1.9

The pie chart for the combined data is shown below.

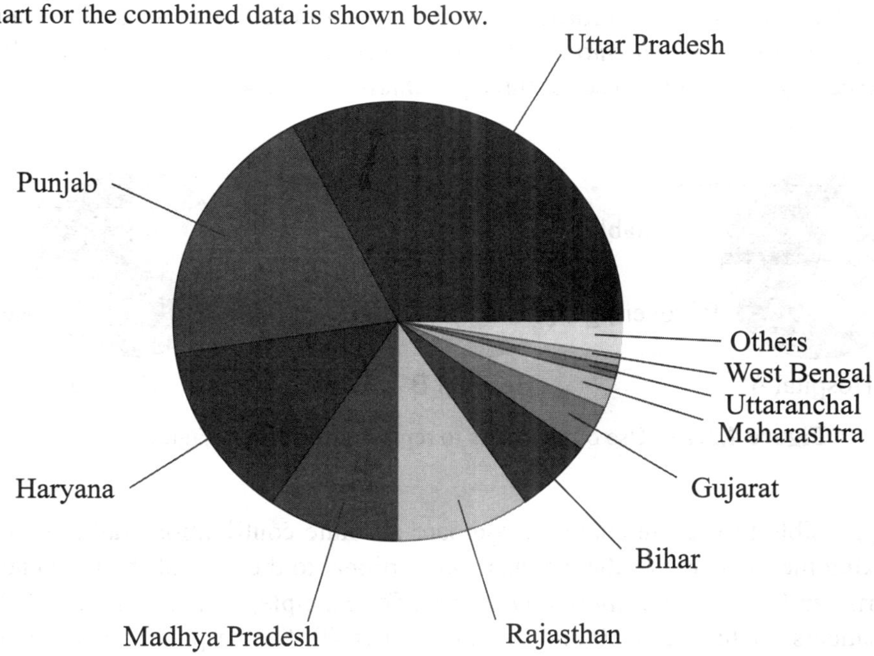

The sectors corresponding to small values have areas that are too small for safe comparisons with each other, in spite of our attempt to improve matters by working with a larger circle. A bar chart would do a much better job of distinguishing them. □

Example 4.3.3 An **exploded pie chart** highlights one or more values by slightly separating their sectors from the rest of the circle. For example, the following chart guides the viewer's eye to compare the relative contributions of Haryana and Rajasthan:

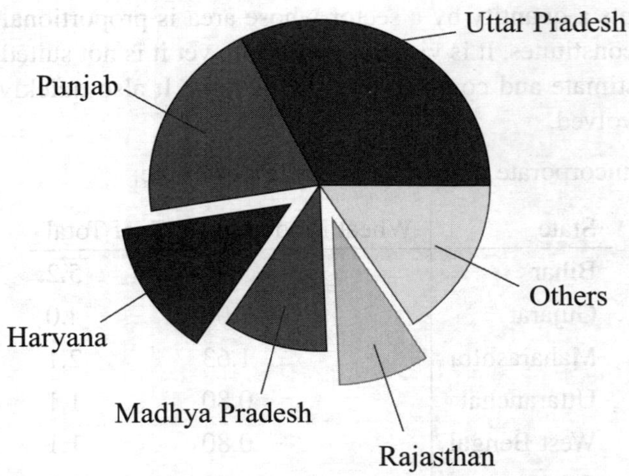

On pages 122–126, we took up various types of bar charts and examples where they are of use. The pie chart, because it shows the fractional contribution of each category to the total, is closely related to the *percent stacked bar chart*. For example, the percent stacked bar chart in Figure 4.3.4 could be replaced by the following set of three pie charts (Figure 4.3.13).

□

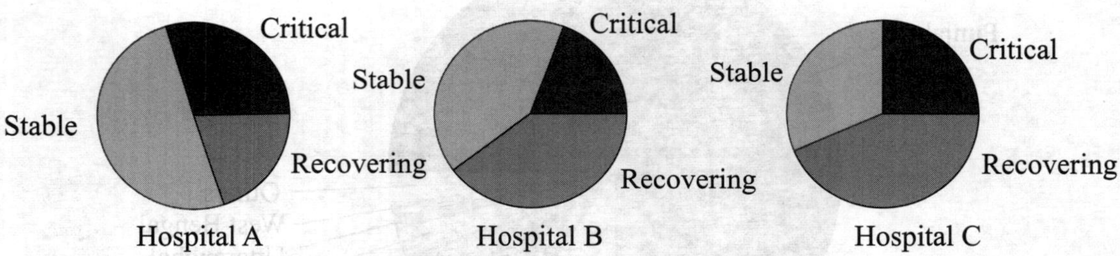

Figure 4.3.13 Use of pie charts to represent the data in Figure 4.3.4.

It is even possible to use pie charts to compare absolute contributions rather than fractional ones, by making the total area of the pie chart proportional to the total of the frequencies. In this way, pie charts can function like *stacked bar charts*. For example, consider Figure 4.3.3. The total number of patients on June 1 is 50 and on June 8 it is 40. So the pie chart for June 1 should have an area which is $50/40 = 1.25$ times that of the pie chart for June 8. Since the area of a circle is proportional to the square of its radius, the radius of the June 1 pie chart should be about $\sqrt{1.25} = 1.12$ times that for June 8.

Task 4.3.1 Let one circle have radius R and area A, while another circle has radius R' and area A'. If the ratio $A' : A$ is k, show that the ratio $R' : R$ is $\sqrt{k}$.

We have seen that pie charts can be used in many situations. However, they do not have the same ease of use as bar charts. They get cluttered if there are many categories, and it is also harder to make accurate visual comparisons of the sectors (Figure 4.3.14).

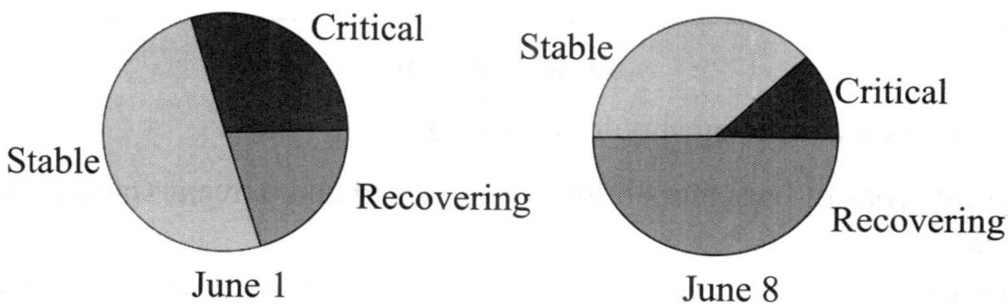

Figure 4.3.14 Use of pie charts to represent the data in Figure 4.3.3.

Pie Charts in Spreadsheets

The procedure for making a pie chart in Excel is similar to that of making a bar chart. Under the Insert tab, select the Pie option (Figure 4.3.15).

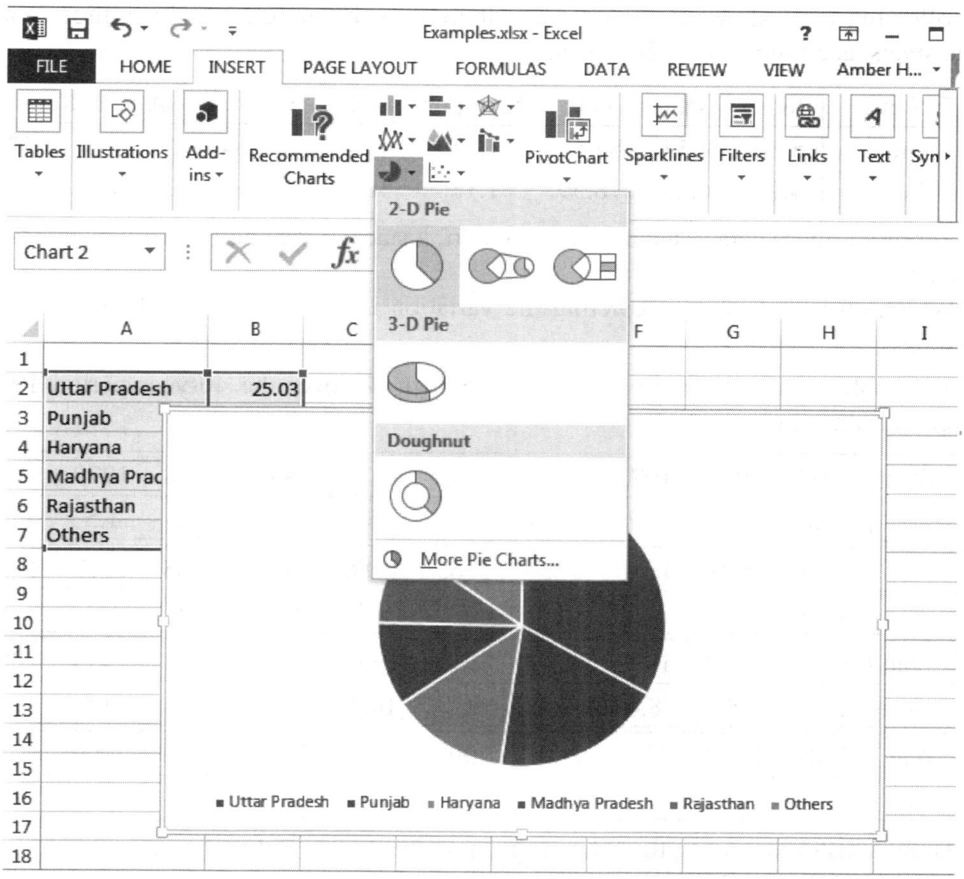

Figure 4.3.15 Options for creating a pie chart in Excel.

Quick Review

After reading this section, you should be able to:

- Recognise types of bar charts (clustered, stacked and percent stacked) and when to use them.
- Distinguish histograms from bar charts, and describe common histogram shapes (normal, left skewed, right skewed).
- Use pie charts to represent proportions.
- Create all these charts using spreadsheets.

EXERCISE 4.3

1. The following table shows the gender distribution in the population (in millions) of certain Indian states, according to the 2001 Census[6]:

State	J&K	Delhi	U.P.	Manipur	W.B.	Gujarat	A.P.	Kerala
Male	5.30	7.57	87.47	1.21	41.49	26.34	38.29	15.47
Female	4.77	6.21	78.59	1.18	38.73	24.25	37.44	16.37

(a) Represent these data in a clustered bar chart, a stacked bar chart and a percent stacked bar chart.

(b) Which chart is best for depicting the variation in the gender distribution among these states?

2. The next table shows the annual inflation rates (in %) over the 40-year period 1971–2010 (Source: inflation.eu):

Year	1971	1972	1973	1974	1975	1976	1977	1978	1979	1980
%	5.0	4.9	7.7	23.8	25.4	−6.2	0	7.9	1.5	11.7

Year	1981	1982	1983	1984	1985	1986	1987	1988	1989	1990
%	12.7	8.1	12.5	5.2	7.1	9.2	9.3	8.8	5.4	13.7

Year	1991	1992	1993	1994	1995	1996	1997	1998	1999	2000
%	13.1	8	8.6	9.5	9.7	10.4	6.3	15.3	0.5	3.5

Year	2001	2002	2003	2004	2005	2006	2007	2008	2009	2010
%	5.1	3.2	3.7	3.8	5.6	6.5	5.5	9.7	15.0	9.5

(a) Draw a histogram using the following class intervals: −10 to −5, −5 to 0, 0 to 5, 5 to 10, 10 to 15, 15 to 20, 20 to 25, 25 to 30.

(b) Describe the shape of this histogram.

[6] Source: Cyber Journalist website [W3].

3. Can we talk of a bar chart being normally shaped?
4. Use a pie chart to represent the following data on average annual fatalities involving trains of the Indian Railways during 1992–2002 (Source: indianrailways.gov.in):

Type	Collision	Derailment	Level Crossing (Road Users)	Fire
Fatalities	148	52	176	8

5. Use suitable charts to depict the given data on an average value of total assets (AVA) in rupees, owned by households of different social groups in 1991 (source: National Sample Survey Organization) :

Group	AVA (Rural)	AVA (Urban)
Scheduled Tribes	52,660	68,763
Scheduled Castes	49,189	57,908
Others	1,34,500	1,59,745
All	1,07,007	1,44,330

4.4 Tracking Trends: Line Plots

Consider the following data for the population of India (Table 4.4.1)[7]:

Table 4.4.1 Population of India

Year	1871	1881	1891	1901	1911	1921	1931	1941
Population (Millions)	255	257	282	285	303	305	338	389

We can plot these data to get a visual depiction which will make its key features more obvious. We mark the years at constant intervals along the horizontal axis (which, by convention is called the X-axis). We similarly mark population (in millions) on the vertical axis (the Y-axis). Thus, we first create the following frame in which we will plot the actual data:

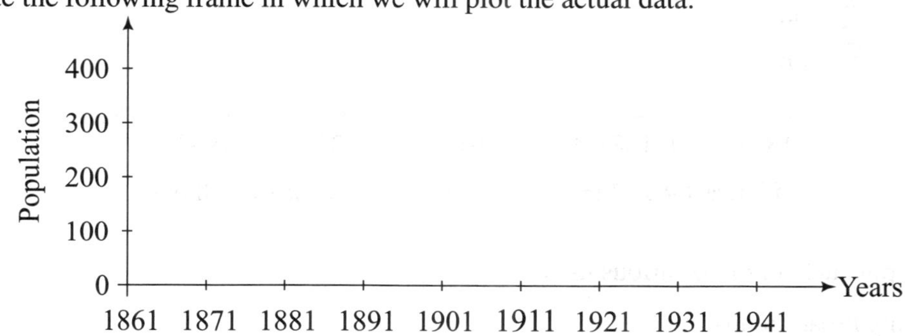

[7] Estimates by Kingsley Davis. Source: Irfan Habib, *Man and Environment* [2].

Now, we place the data inside this frame. In 1871, the population in millions was 255, so we mark 255 on the Y-axis and, at that height, put a dot above the location of 1871 on the X-axis.

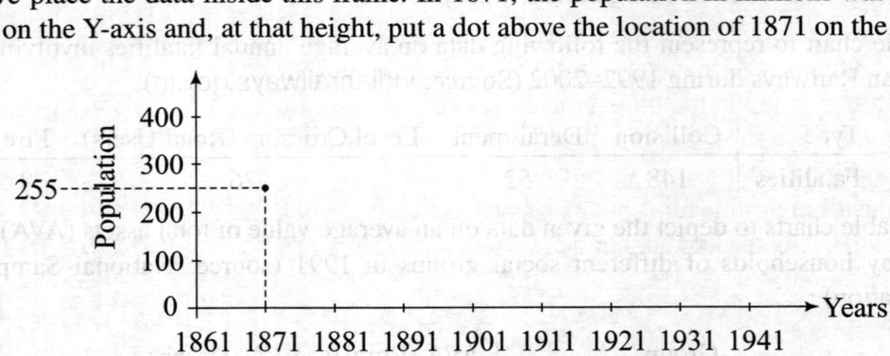

We repeat this process for each data value and get the following plot:

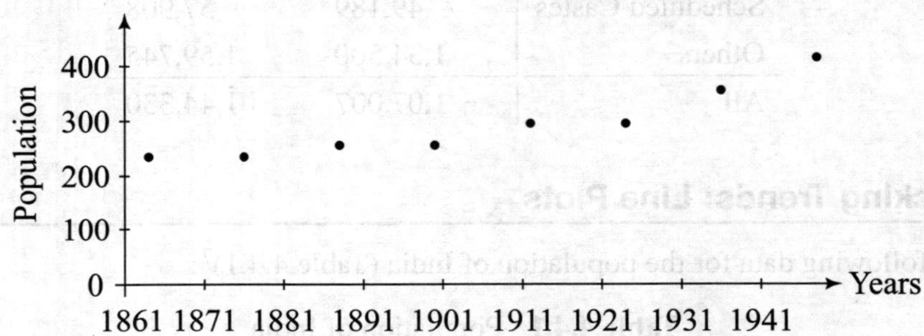

We can give the plot a greater impact by drawing a curve that passes through its points—the resulting figure is called a **graph.** A simple way to do this is to join the points by straight line segments, in which case we call it a **line plot.**

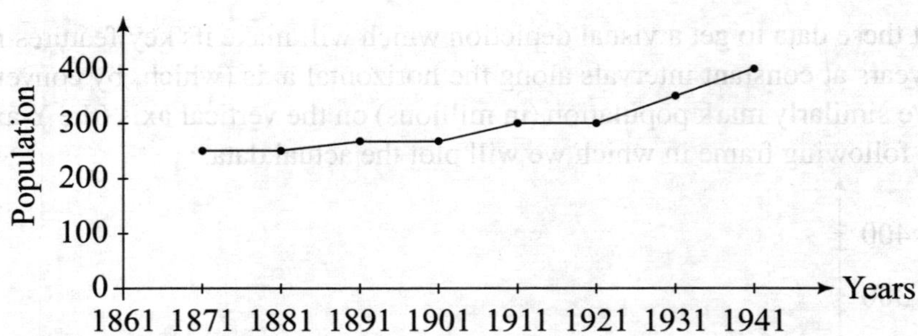

Figure 4.4.1 Line plot of population of India (in millions).

One can put such a plot to various uses:

1. Identifying broad trends.
2. Estimating the value at other than the observed occasions.
3. Comparing two different quantities to illuminate the relationship between them.

Line Plots and Graphs in Spreadsheets

The process of creating a line plot in Excel is similar to that of a bar chart. We first put the paired data in two columns. We then select the values that are to be plotted along the vertical axis and, under the Charts group in the Insert menu, we click on Line.

In Figure 4.4.2, we show a spreadsheet with the population data we were discussing earlier. The highlighted option enables us to draw multiple plots on the same graph. The other options allow us to stack the different line plots on top of each other (as in a stacked bar chart), or to further their totals (as in a percent stacked bar chart).

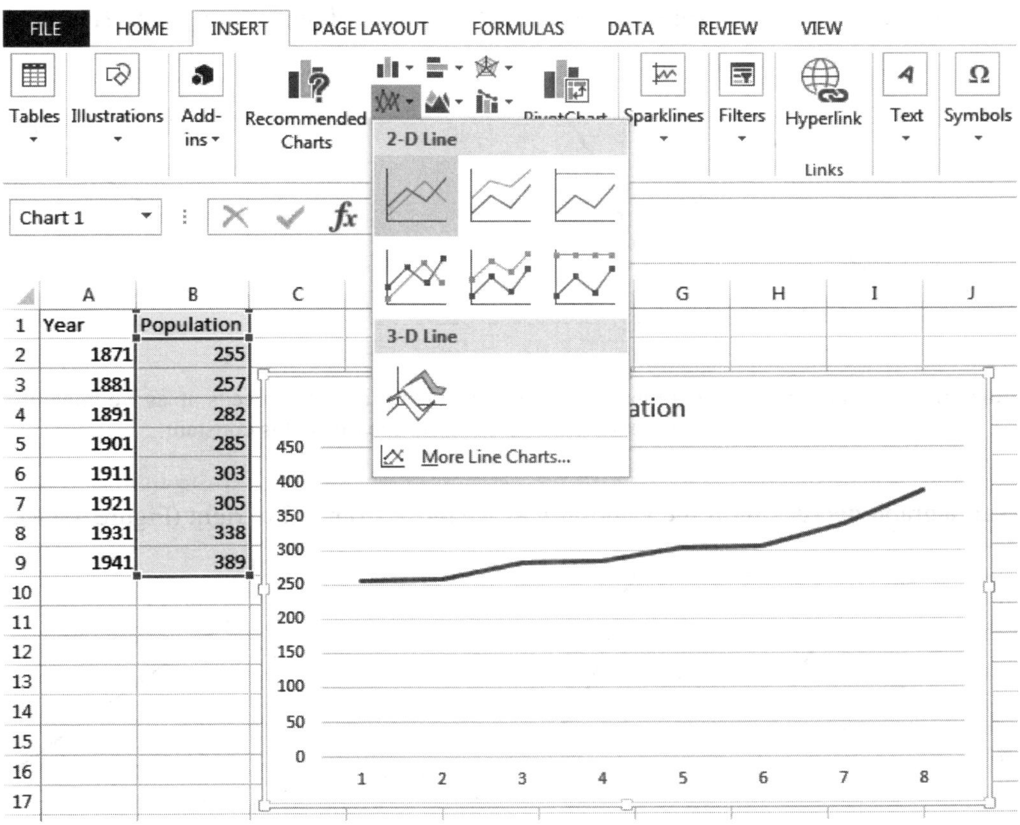

Figure 4.4.2 Creating a line plot in Excel.

One defect in this plot is that the years are not marked on the X-axis, the points being simply labelled as 1, 2, . . . , 8. This can be fixed as follows:

1. Right click on the horizontal axis and select Select Data.
2. Click on Edit in the column named Horizontal (Category) Axis Labels.
3. Enter the range containing the desired axis labels.

An Excel line plot always places the successive X-axis values at equal intervals, regardless of their actual gaps. If the gaps are uneven, then we need to use the scatter plot option, which we will discuss shortly.

Identifying Trends from Graphs

From the graph of Figure 4.4.1, it is evident that the Indian population began a period of sustained high rate of growth from about 1921. We can often use graphs in this manner to identify trends as well as changes.

We call a graph **increasing** if the values rise as we move from left to right. Figure 4.4.1 is an example of an increasing graph. Here are some other examples of increasing graphs (Figure 4.4.3):

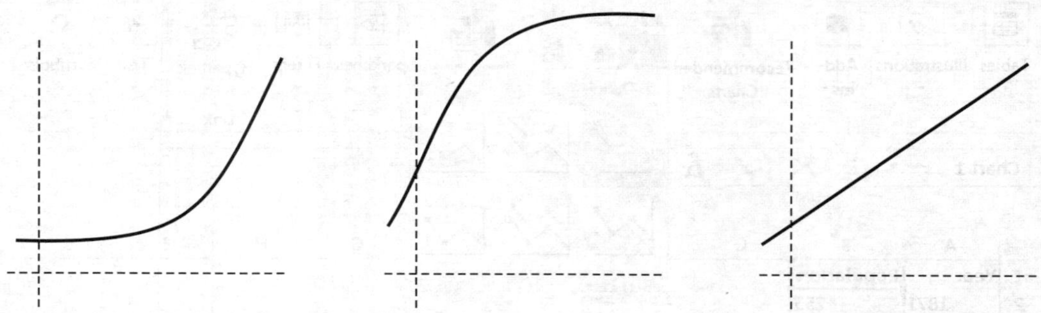

Figure 4.4.3 Three *increasing* graphs. In the first, the rate of increase speeds up as we move from left to right. In the second, it slows down. In the third, it is constant.

We call a graph **decreasing** if the values *fall* as we move from left to right (Figure 4.4.4).

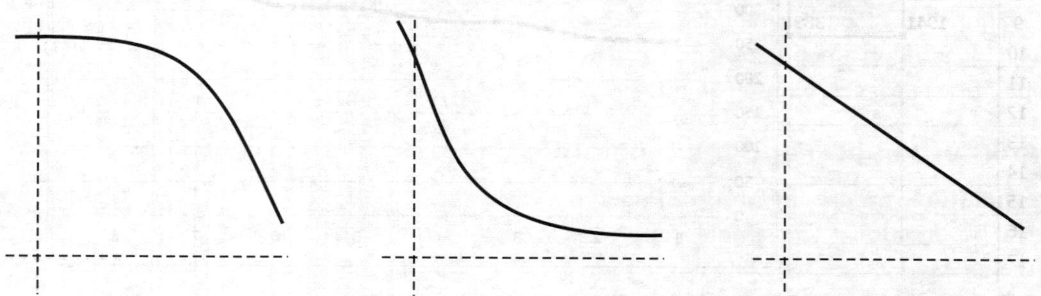

Figure 4.4.4 Three *decreasing* graphs. In the first, the rate of decrease speeds up as we move from left to right. In the second, it slows down. In the third, it is constant.

Some graphs are **cyclic** they alternately increase and decrease. The table below shows the number of snowshoe hares caught per year by the Hudson Bay Company in Canada between 1845 and 1873. They can be assumed to reflect the total population of these hares.[8]

[8] Source: BioTopics website [W1].

Year	Hares (1000s)	Year	Hares (1000s)	Year	Hares (1000s)
0	20	10	5	20	25
2	55	12	15	22	50
4	65	14	50	24	70
6	95	16	75	26	30
8	55	18	20	28	15

Here is a line plot of this data (Figure 4.4.5):

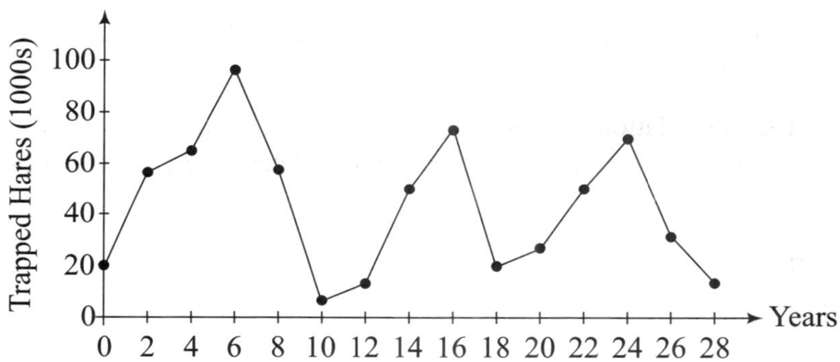

Figure 4.4.5 *Cyclic* line plot of variation in number of hares.

This plot shows a periodic rise and fall in the population, with a period of about 10 years. Hare population is affected mainly by two factors—food supply and hunting by predators. The cyclic pattern has been attributed to the hares breeding so quickly that they use up their food resources, leading to a population crash. After the crash, the habitat gradually recovers—and then the population starts to zoom up again!

Linear Interpolation and the Line of Best Fit

In **interpolation,** we use the observed data to estimate values at points for which we do not have empirical observations. The simplest approach is called **linear interpolation.** In this approach, we fill in the missing values by using a line plot.

Let us consider the population data in Figure 4.4.1. How can we estimate the population for the year 1926, when we do not have a census number for that year? We carry out the following steps:

1. Make a line plot for the population data.
2. Mark 1926 at the appropriate place on the X-axis.
3. Draw a vertical line passing through the mark for 1926, and put a dot where it cuts the line connecting the 1921 and 1931 values.
4. Draw a horizontal line through this dot and mark the point where it cuts the Y-axis. The population corresponding to this point is our estimate of the 1926 population. (We call it an **interpolated** value of the population.)

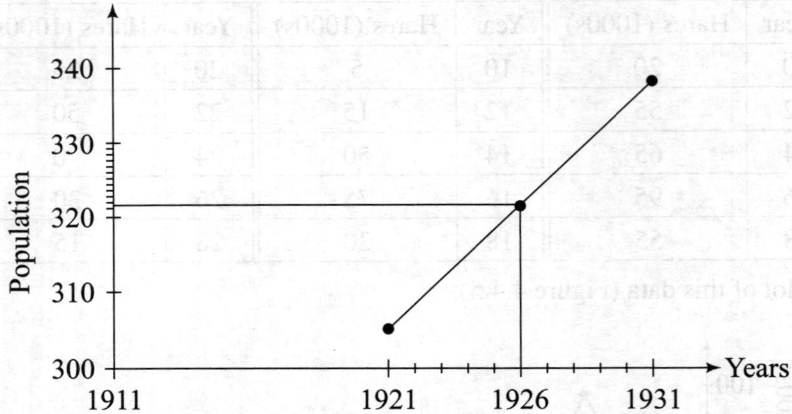

We see that the 1926 population falls between 321 and 322 million. If we had drawn the graph on a larger area, with the points further apart, then we would have seen that the population estimate is 321.5 million.

Line of Best Fit

Consider the following data:

X	1	2	3	4	5	6	7	8
Y	5.6	6.9	8.3	10.3	12.2	12.6	14.5	15.4

We plot the X values on the horizontal axis and the Y values on the vertical axis to get the following plot:

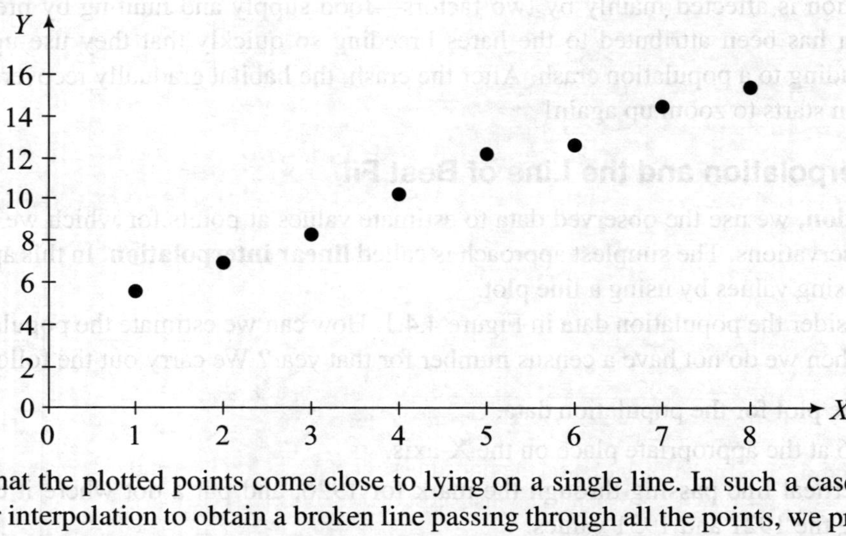

We see that the plotted points come close to lying on a single line. In such a case, rather than doing linear interpolation to obtain a broken line passing through all the points, we prefer drawing a single line which *almost* passes through them. Such a line gives a simple description which captures the essence of the relationship between the two quantities. For example, we draw the following line:

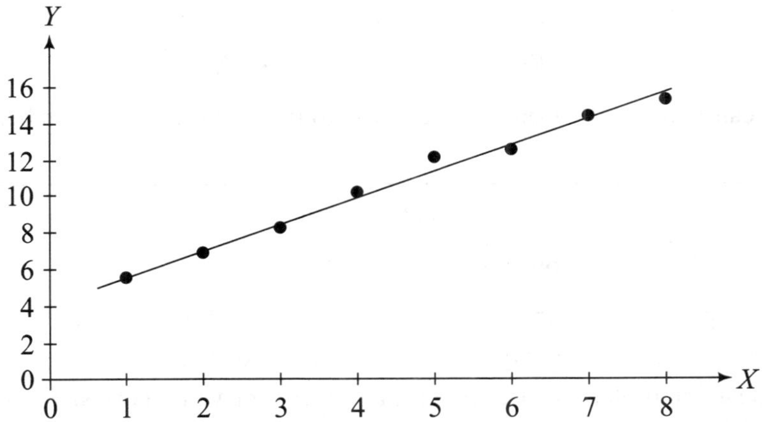

This particular line is the **line of best fit** for the data—it minimises the average vertical distance from the given points to the line. We will not explain the formula for this line. A convenient way to obtain it is to make a line plot of the data in Excel, right-click on one of the plotted points and select Add Trendline followed by Linear.

The line of best fit gives another way of estimating unobserved data values. For example, to estimate the Y value corresponding to $X = 4.5$, we proceed as follows (Figure 4.4.6):

1. Mark 4.5 at the appropriate place on the X-axis.
2. Draw a vertical line passing through the mark for 4.5, and put a dot where it cuts the line of best fit.
3. Draw a horizontal line through this dot and mark the point where it cuts the Y-axis. This is our estimate of Y.

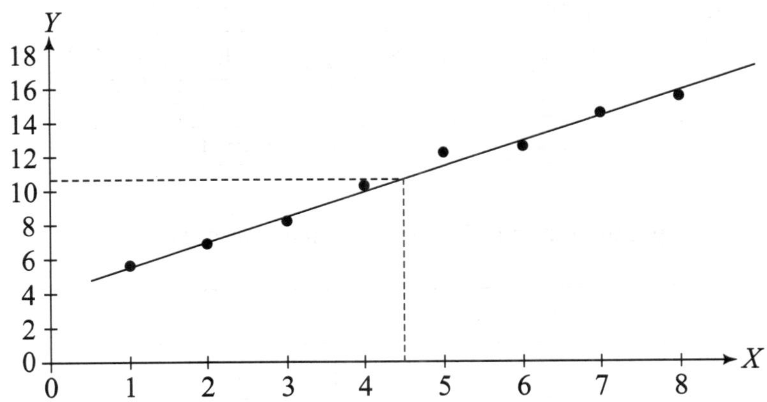

Figure 4.4.6 Interpolation based on the line of best fit. The interpolated value of Y, corresponding to $X = 4.5$, is seen to be just short of 11. (In fact, it is 10.7.)

The capability to interpolate values from the best-fit line is also present in Excel. Suppose the X values are in cells A2:A9 and the corresponding Y values are in B2:B9. To find the interpolated value of Y corresponding to $X = 4.5$, we execute the following command:

$$= \text{Forecast}(4.5, \text{B2:B9}, \text{A2:A9})$$

Task 4.4.1 How can you use the Forecast function to do linear interpolation between two given data points?

> **Quick Review**
>
> After reading this section, you should be able to:
>
> - Draw line plots for data in which one quantity depends on another.
> - Describe graphs qualitatively in terms of trends and changes.
> - Use linear interpolation to estimate the dependent value corresponding to a new value of the base quantity.
> - Draw a line of best fit to paired data and use it for interpolation.

EXERCISE 4.4

1. Classify the following graphs as increasing, decreasing or cyclic:

(a) Electrocardiogram recording of normal heart beat:

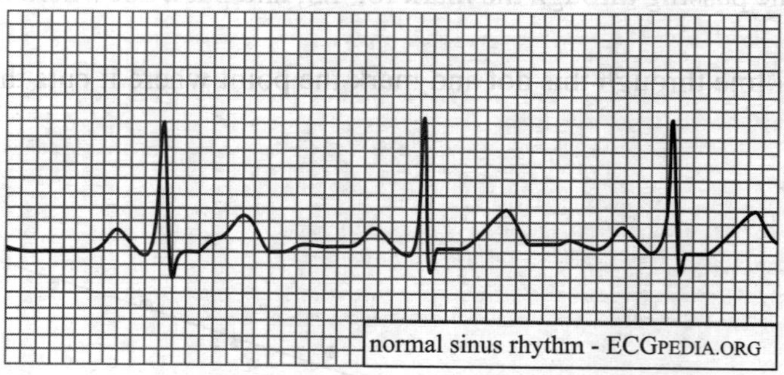

(b) Estimated world population in millions over three centuries:

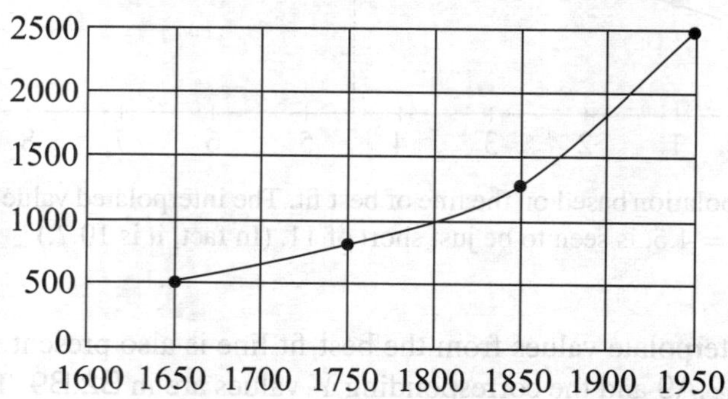

(c) Speed of an object free falling in air:

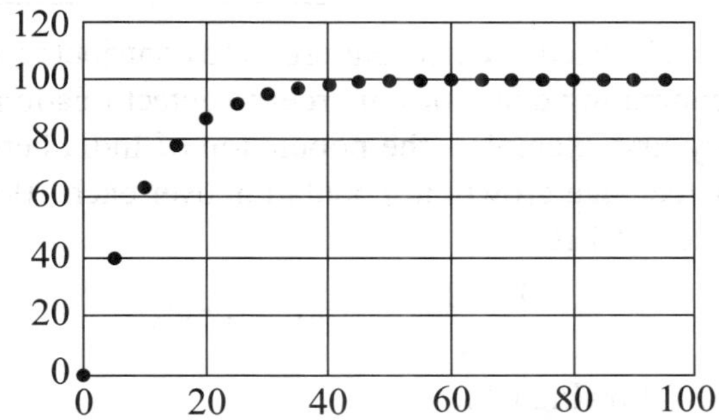

(d) India and China GDP per capita, compared to the world average:

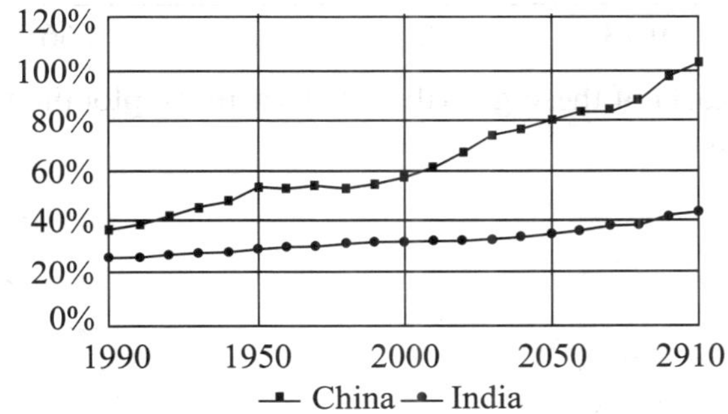

2. Make a line plot of the following data:

X	57	62	67	72	77
Y	20	19	17	19	12

Estimate the value of Y corresponding to $X = 70$ using

(a) Linear interpolation (b) Line of best fit

3. We return to the example of the snow shoe hare population in Arctic Canada. The table below shows the number of trapped lynx, a predator which mainly feeds on hares, during the same years.

Year	Hares (1000s)	Lynx (100s)	Year	Hares (1000s)	Lynx (100s)	Year	Hares (1000s)	Lynx (100s)
0	20	10	10	5	15	20	25	5
2	55	15	12	15	10	22	50	25
4	65	55	14	50	60	24	70	40
6	95	60	16	75	60	26	30	25
8	55	20	18	20	10	28	15	5

Use Excel's line plot option to plot the hare and lynx numbers together, in the same diagram, against the years. What conclusions can you draw?

4.5 Finding Relationships: Scatter Plots

We have seen how graphs can be used to detect patterns and contribute to making predictions. Now we shall take up ways of comparing quantities in order to detect relationships and causes.

Let us return to our very first example—the population of India between 1871 and 1941. We begin by calculating the percentage growth in population over each 10-year period. For example, the growth rate from 1871 to 1881 is

$$\frac{257 - 255}{255} \times 100 = 0.78\%$$

The calculations lead to the following table:

Year	1871–81	1881–91	1891–1901	1901–11	1911–21	1921–31	1931–41
% Growth	0.78	9.73	1.06	6.32	0.66	10.82	15.09

Here is a graph (Figure 4.5.1) of these growth rates (where we plot the final year of each 10-year period on the horizontal axis):

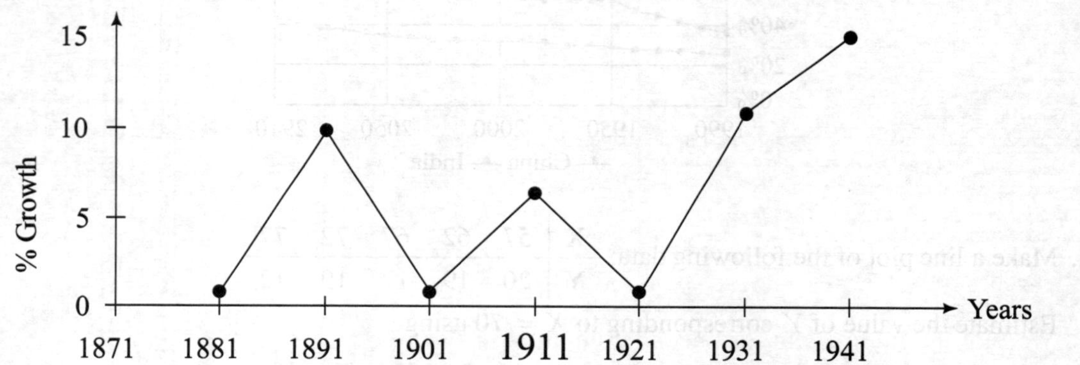

Figure 4.5.1 Line plot of population growth rates of India (in percentage) over 10-year periods.

We see that while the rates have fluctuated tremendously, there appears to be an overall downward trend during 1871–1921, after which the rates shoot up. What can explain these trends? The immediate factors affecting population growth are birth rates and life expectancy. These in turn are affected by various causes—weather, epidemics, prosperity levels, etc. Table 4.5.1 shows the mean life expectancy (in years) during the corresponding time periods.[9]

Table 4.5.1 Life Expectation Estimates

Year	1871–81	1881–91	1891–1901	1901–11	1911–21	1921–31	1931–41
L.E.	24.6	25.1	23.8	22.9	20.2	26.7	31.7

[9] Source: Irfan Habib, *Man and Environment* [2].

The graph (Figure 4.5.2) of this data shows the same trend as the population growth rates!

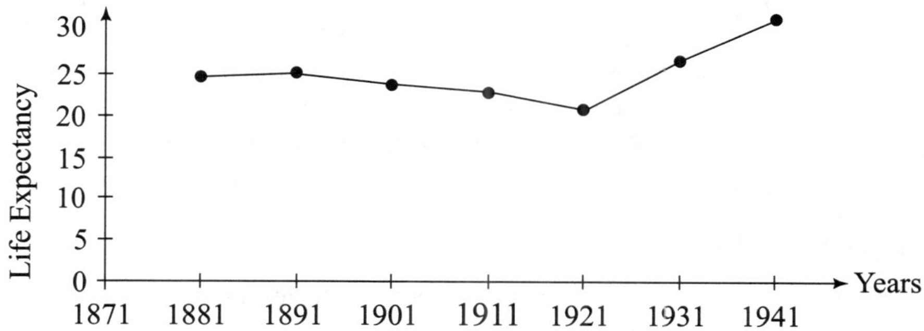

Figure 4.5.2 Line plot of mean life expectancy in India (in years) over 10-year periods.

It appears that fluctuations in life expectancy are the essential explanation for the changes in growth rate. We can directly look at the relationship between these two quantities by plotting one against the other. We start by combining all the data in a single table:

Period	Life Expectancy	Growth Rate
1871–81	24.6	0.78
1881–91	25.1	9.73
1891–1901	23.8	1.06
1901–11	22.9	6.32
1911–21	20.2	0.66
1921–31	26.7	10.82
1931–41	31.7	15.09

We plot the numbers in the second column along the horizontal axis, and the corresponding numbers in the third column along the vertical axis. This gives a plot (Figure 4.5.3) which shows how the growth rate varies with the life expectancy.

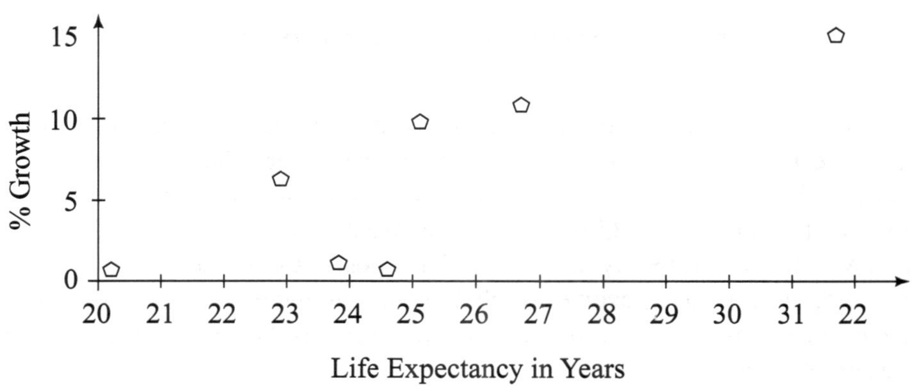

Figure 4.5.3 Scatter plot of population growth rate versus mean life expectancy.

This is an example of a **scatter plot.** It is used when we explore possible connections between two quantities which may not directly depend on each other. A line plot, on the other hand, is suitable when one quantity depends on the other. In a scatter plot, it may well happen that the same X value corresponds to two different Y values, nor are the pairs in sequence from left to right. So in a scatter plot we may not join the points by line segments.

On the other hand, it *would* make sense to ask if the relationship between X and Y could be approximated by a line, and so the concept of a line of best fit can be applied to a scatter plot.

Reading a Scatter Plot

Scatter plots are used to explore relationships between quantities and to answer questions like: If one quantity rises, then will the other also rise? If yes, then by how much? What is the strength of the connection?

Consider a scatter plot arising out of observations of a pair of quantities X and Y. Perhaps we measure the height (X) and the weight (Y) of each student in a classroom. Or else, we observe the prices of two shares on the stock market at 5-minute intervals. We want to know whether information of one quantity helps to predict the other. Can we give a range for the possible weight of a boy who is 4-feet tall? Is the range narrow (in which case, the height is helpful in predicting the weight) or wide (the height is not helpful)? Are we able to describe how the range changes with the height?

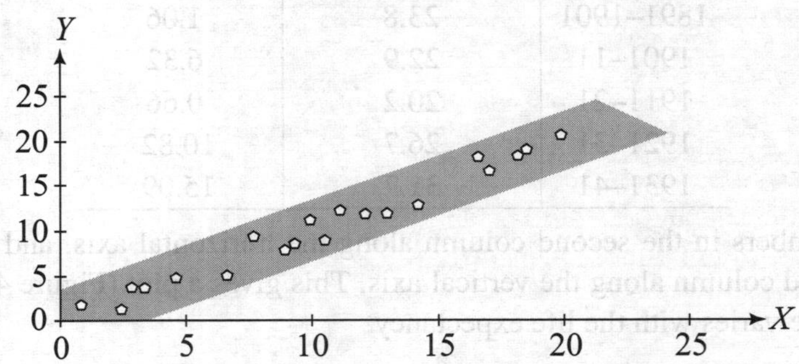

Figure 4.5.4 Scatter plot with high positive correlation.

The data points in the scatter plot in Figure 4.5.4 lie in a narrow band. This reveals a strong relationship between the two quantities. If we take any particular value of X, say 10, then we see that the part of the band above it is quite thin compared to the overall variation in Y: the technical term for this is that X and Y have **high correlation.** We also see that a rise in X tends to cause a rise in Y, and we express this by saying that they have **positive correlation.**

In Figure 4.5.5, we still have positive correlation since an increase in X tends to be associated with an increase in Y. But the connection is relatively weak as the band is wide. This is a case of **low positive correlation.**

Figure 4.5.6 shows an instance of **no correlation.** The variation in Y is the same for any value of X.

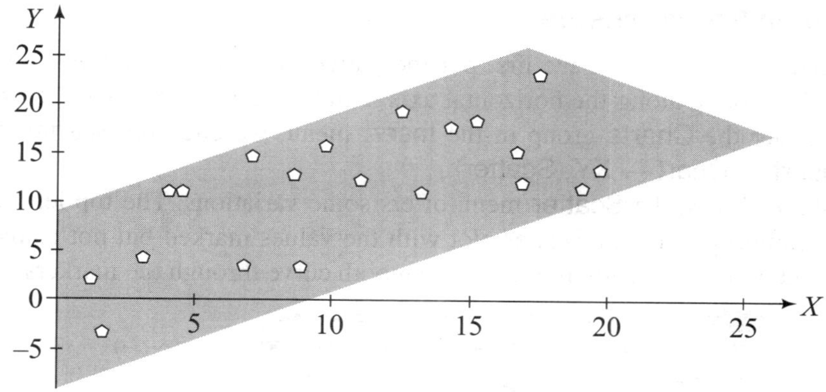

Figure 4.5.5 Scatter plot with low positive correlation.

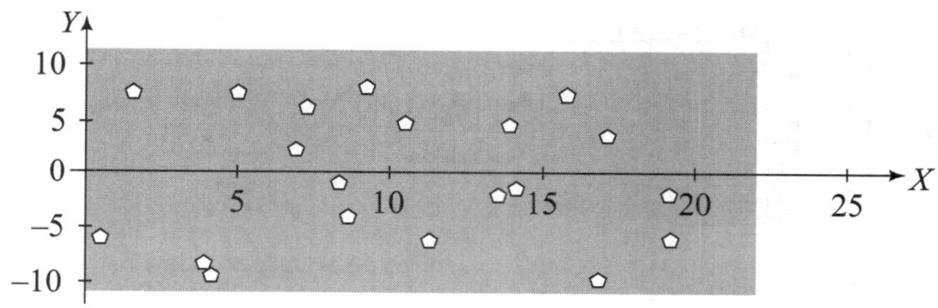

Figure 4.5.6 Scatter plot with no correlation.

The relation between X and Y could be an inverse one, with a rise in X tending to cause a *fall* in Y. In this case, we say that there is **negative correlation.** Figure 4.5.7 gives examples of high and low negative correlation.

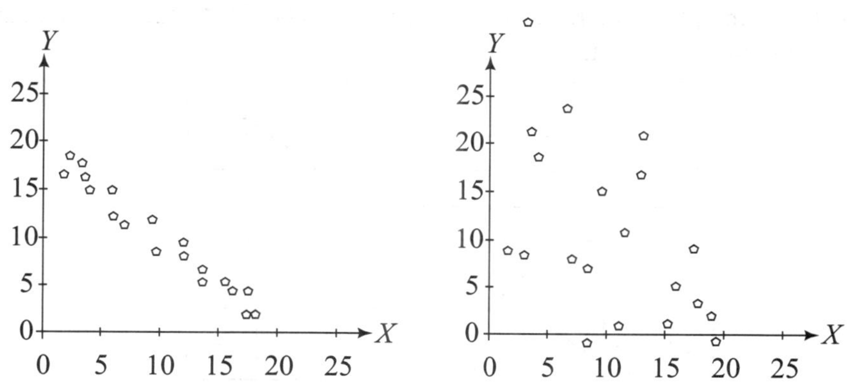

Figure 4.5.7 Scatter plots with high and low negative correlation.

Scatter Plots in Spreadsheets

To create a scatter plot in Excel, we first put the paired data in two columns, taking care that the quantity to be plotted along the horizontal axis is in the first column. We then select the two columns and, under the Charts group in the Insert menu, we click on Scatter. (In Calc, the sequence is Insert → Chart → XY (Scatter).)

As Figure 4.5.8 shows, the Scatter menu offers some variations. The top left option (shown selected in the figure) produces a scatter plot with the values marked but not connected to each other. The second option in the top row draws a smooth curve through the markers.

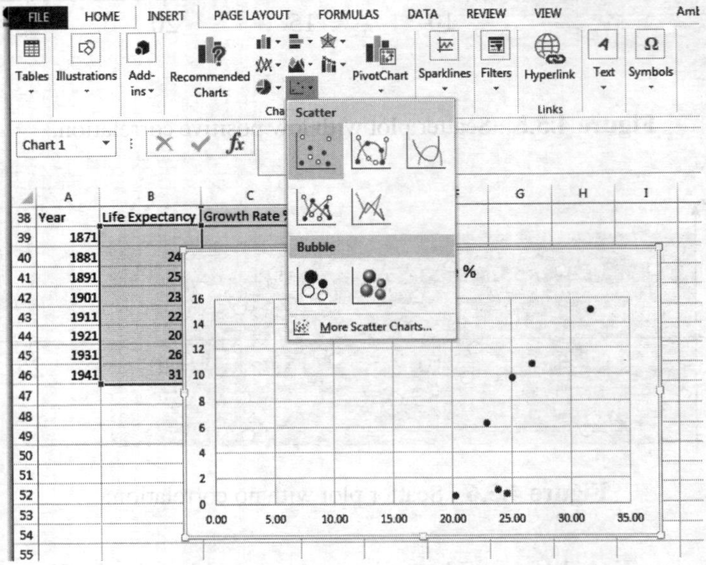

Figure 4.5.8 Creating a scatter plot in Excel.

As remarked earlier, we can add a line of best fit to the scatter plot by using the Add Trendline option, just as for a line plot. The result is shown below. The Forecast function can be used to interpolate using the line of best fit.

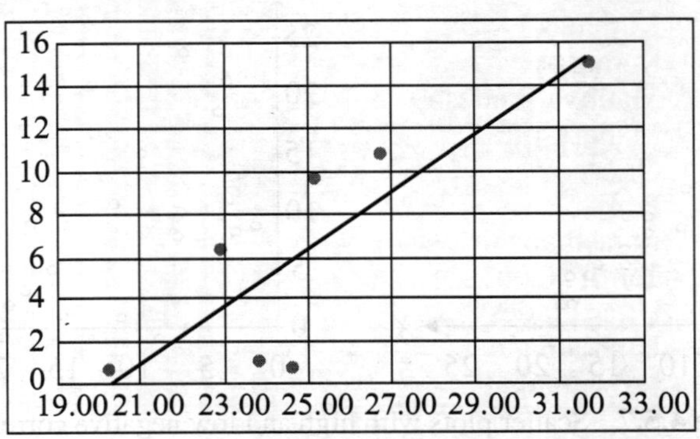

Quick Review

After reading this section, you should be able to:

- Draw scatter plots for paired data of quantities which may be related to each other, but perhaps do not directly depend on each other.
- Use scatter plots to analyse the strength and direction of the relationship between two quantities.
- Draw a line of best fit to a scatter plot and use it for interpolation.

EXERCISE 4.5

1. Use the Scatter option in Excel to make a line plot of the following data.

X	57	62	67	74	82
Y	20	19	17	19	12

Estimate the value of Y corresponding to $X = 70$ using linear interpolation.

2. How would you describe the correlation between the two quantities in Figure 4.5.3?

3. The table below gives the GDP per capita (in ₹) during 2000–01, and the sex ratio (number of women per 100 men) according to the 2001 census, for the 20 most populous states in India.

State	GDP per capita	Sex ratio
Andhra Pradesh	19,159	97.8
Assam	13,820	93.2
Bihar	6,911	92.1
Chhattisgarh	12,707	99.0
Delhi	44,419	82.1
Gujarat	21,966	92.1
Haryana	27,553	86.1
Jammu & Kashmir	16,584	90.0
Jharkhand	11,926	94.1
Karnataka	19,524	96.4
Kerala	22,659	105.8
Madhya Pradesh	13,116	92.0
Maharashtra	25,906	92.2
Orissa	11,849	97.2
Punjab	30,758	87.4
Rajasthan	14,597	92.2
Tamil Nadu	23,645	98.6
Uttar Pradesh	10,932	89.8
Uttarakhand	17,339	96.4
West Bengal	17,892	93.4

Use Excel to draw a scatter plot and the line of best fit. What can you say about the relationship?

4.6 Locating the Centre: Mode, Median and Mean

Mode

We shall now take up various ways of finding and expressing the information contained in the data. We can ask various questions of raw numerical data, of which the simplest is perhaps, 'Which is the most frequent value?' The answer to this question is called the **mode** of the data.

Example 4.6.1 Consider the following data: 1, 2, 1, 3, 1, 5, 4, 3, 4, 1, 4, 5, 1. The data value 1 has the highest frequency, hence the mode of this data is 1. □

It could happen that the highest frequency is achieved by more than one data value. In such a case, we cannot talk of *the* mode of the data, and the term is not very useful.

Task 4.6.1 A group of cricket commentators voted as follows for their favourite test bowler:

Bowler	Hadlee	Imran Khan	Lillee	Muralitharan	Warne	Kapil Dev
Votes	9	6	7	12	10	6

If we treat the name in each vote as one data entry, then what is the mode?

Mode Function in Spreadsheets

Excel and Calc provide a Mode function which can calculate the mode of a list of numbers. Suppose the numbers 1, 2, 3, 1, 2, 1 are entered in cells A1:A6. Then =Mode(A1:A6) will give the result as 1. However, there isn't a built-in function for directly handling non-numerical data.

Median

Suppose we have a collection of numerical data. We seek to find the centre of this data in the sense of a number which splits the data into two equal parts: half the data is above this number and half is below. This number will be called the **median** of the dataset.

Since we are interested in the size of the numbers, the first step in any example is to reorder the data so that the entries occur in an increasing order. If the original data is $1, 3, 1, 5, 2, -2$, then we will first rearrange it as $-2, 1, 1, 2, 3, 5$.

Example 4.6.2 Suppose the data is $-1, 0, 2, 3, 4, 5, 5$. Then the number 3 can serve as median, since there are 3 data values below it (–1,0,2) and three above it (4,5,5). □

Example 4.6.3 Suppose the data is $-1, 0, 2, 4, 4, 5, 5$. Which of these can serve as median? Since there are seven data points, the one in the fourth position is the median, as there are three numbers to its right as well as three to its left. Therefore the median is 4. □

Task 4.6.2 Consider a dataset with an odd number of data points

$$x_1,\ x_2,\ \ldots,\ x_{2N+1}$$

arranged in increasing order: $x_1 \leq x_2 \leq \cdots \leq x_{2N+1}$. Show that the median is x_{N+1}.

Example 4.6.4 Suppose the data is 1, 2, 2, 3, 4, 5. Here, there are six data values so there isn't a single central value. Instead, we have a middle pair: 2 and 3. In such a case, we take the median to be the mid-point of the two central values—in this case it is 2.5. Note that now the median is not itself a member of the original dataset. □

Task 4.6.3 Consider a dataset with an even number of data points

$$x_1,\ x_2,\ \ldots,\ x_{2N}$$

arranged in increasing order: $x_1 \leq x_2 \leq \cdots \leq x_{2N}$. Show that the median is $(x_N + x_{N+1})/2$.

Median of Raw Data with Spreadsheets

The Median function in Excel and Calc calculates the median of raw data by the rules explained above. For example, if the data is entered in cells A1:A8, then the median can be calculated by evaluating =Median(A1:A8).

Percentiles

The median attempts to split the data into two parts of equal size. We can also attempt to split the data into unequal proportions.

Given a collection of ordered data, and a percentage p, the ***p*-percentile** of the data is a number that splits the data into the bottom p percent and the top $(100 - p)$ percent. For example, the 90-percentile splits the data into the bottom 90 percent and the top 10 percent. Similarly, the 50-percentile splits the data into two equal halves (50 percent each) and therefore equals the median.

Example 4.6.5 Consider the data 0, 2, 4, 6, ..., 20. These 11 data points create 10 intervals between them, and each interval represents 10 percent of the data. So 0 is the 0-percentile, 2 is the 10-percentile, 4 is the 20-percentile, and so on until we end with 20 which is the 100-percentile.

Data:	0	2	4	6	8	10	12	14	16	18	20
Percentile:	0%	10%	20%	30%	40%	50%	60%	70%	80%	90%	100%

Now, suppose we want the 85-percentile. We simply take the mid-point of the 80-percentile (16) and 90-percentile (18): thus, its value is 17. Similarly, if we need the 83-percentile, we divide the gap between 16 and 18 into 10 equal parts and measure off three of these parts—this gives the value 16.6. □

The procedure shown in this example is applied to any kind of data, even when the data values are irregularly spaced and have repetitions.

Example 4.6.6 Consider the data 1, 1, 2, 3, 4, 4, 4, 6, 7, 10, 10. We have again taken 11 data points to keep the calculations simple. We plot them at equal spaces along a line as before, treating each data point as a separate number.

Data:	1	1	2	3	4	4	4	6	7	10	10
Percentile:	0%	10%	20%	30%	40%	50%	60%	70%	80%	90%	100%

Let us carry out some percentile calculations based on this diagram:

Percentile	Procedure	Value
40%	Read directly from diagram	4
45%	Mid-point of 4 and ④	4
83%	Three-tenths of the way from 7 to ⑩	7.9

□

In general, if we have $N+1$ data points, we mark them on a line as done above, creating N intervals. Each interval is viewed as containing $(100/N)$percent of the data.

The words *percentage* and *percentile* can be confused with each other. We have to keep in mind that percentage measures relative amount, while percentile measures relative position or rank.

Percentiles in Spreadsheets

If the raw data is in cells A1:A10 and we want the 90-percentile, we evaluate =Percentile(A1:A10,90%) in Excel and =Percentile(A1:A10;90%) in Calc.

Quartiles

The 25-, 50-, and 75-percentiles cut the data into four equal parts and, hence, are known as **quartiles.** The 25-percentile is called the **first quartile,** the 50-percentile is the **second quartile** and the 75-percentile is the **third quartile.**

Excel provides a Quartile function whose syntax is =Quartile(data, type) where type is 1, 2 or 3 corresponding to the first, second and third quartiles respectively (Figure 4.6.1).

	A	B	C	D	E
1	120		90-Percentile	=PERCENTILE(A1:A10,90%)	170.5
2	125				
3	130		First Quartile	=QUARTILE(A1:A10,1)	130
4	145				
5	145		Second Quartile	=QUARTILE(A1:A10,2)	147.5
6	150				
7	150		Third Quartile	=QUARTILE(A1:A10,3)	157.5
8	160				
9	170				
10	175				
11					

Figure 4.6.1 Excel commands for percentiles and quartiles, and their outputs.

Example 4.6.7 Percentiles can be used to methodically explore the distribution of certain features across a population. In this example, we will look at income changes among the poorer and richer parts of society. According to data released by the World Bank,[10] we can estimate that the gross national income per capita (the gross national income divided by the population) of the USA grew by about 7 percent between 2001 and 2010, after adjusting for inflation. How was this general increase in wealth distributed among the population?

The table below is based on data from the US Bureau of Labor Statistics.[11] It shows the change in weekly wages at different percentile levels of full-time wage earners over the period 2001–2010 (also adjusted for inflation).

Percentile	10	25	50	75	90
Change in Wages	−2%	1%	1%	4%	6%

The benefits are almost entirely confined to the richer half of the population, and the gap between the rich and poor has increased. □

Percent Rank

Suppose a number P is the p-percentile of some data. Then we say that p percent is the **percent rank** of P.

The PercentRank function calculates percent rank in our spreadsheet programs. The syntax in Excel is =PercentRank(data, number).

Mean

The **mean** or **average** of a list of numbers $\{x_1, x_2, \ldots, x_N\}$ is simply their total value divided by their total frequency:

$$\text{Mean} = \frac{x_1 + x_2 + \cdots + x_N}{N}$$

The mean of data $\{x_1, \ldots, x_N\}$ is usually denoted by $\overline{x}$. You may recall the notation $\sum_{i=1}^{N} x_i$ for the sum $x_1 + \cdots + x_N$. In this notation, our definition can be written as follows:

$$\overline{x} = \frac{1}{N}\sum_{i=1}^{N} x_i$$

Suppose we counted the money in five friends' wallets and found ₹40, 110, 55, 600 and 205. Then the mean is

$$\frac{40 + 110 + 55 + 600 + 205}{5} = 202$$

[10] See World Bank website [W12].
[11] See Bureau of Labor Statistics website [W11].

The mean is the number each gets if we pool the money and then redistribute it equally among the five. As this example illustrates, the mean is useful in getting a sense of the overall amount of a quantity, but it does not reveal anything about how it is actually distributed. In contrast, the median gives little sense of the total amount of the quantity, but it does give some idea of how it is distributed—half the data values are above the median and half are below.

Task 4.6.4 Suppose the data $\{x_1, \ldots, x_N\}$ is evenly spaced with gaps of size a: $x_2 = x_1 + a$, $x_3 = x_2 + a, \ldots, x_N = x_{N-1} + a$. Show that the mean is the mid-point of the extreme values x_1 and x_N: $\overline{x} = (x_1 + x_N)/2$.

Calculating Mean with Spreadsheets

Excel and Calc provide the Average and AverageA functions for calculating the mean. The syntax is simple: =Average(data) and =AverageA(data). We will now explain the details of how they work and how they differ from each other:

1. Both Average and AverageA calculate the mean of the data | 10 | 0 | 5 | 9 | as $(10+0+5+9)/4 = 6$.
2. Both Average and AverageA treat a blank cell as missing data and ignore it. So, they calculate the mean of | 10 | | 5 | 9 | as $(10+5+9)/3 = 8$.
3. Average also treats text as missing data and therefore calculates the mean of the data | 10 | A | 5 | 9 | as $(10+5+9)/3 = 8$.
4. AverageA treats text as zero, and therefore calculates the mean of | 10 | A | 5 | 9 | as $(10+0+5+9)/4 = 6$ (Figure 4.6.2).

	A	B	C	D
1		10	10	10
2		0		A
3		5	5	5
4		9	9	9
5	AVERAGE	6	8	8
6	AVERAGEA	6	8	6

Figure 4.6.2 The effect of using Average and AverageA on data with blank cells or text.

Task 4.6.5 In the following situations, will you use Average or AverageA?

1. You are the instructor of a course and wish to calculate the students' average marks in the final exam to assess the difficulty level of the exam. Some students missed the exam and their marks are entered as X.

2. You are the instructor of a course and wish to find a particular student's average marks in the assignments as part of the grade calculation. He had failed to submit some assignments and those marks are entered as X.
3. You sent out requests for donations on behalf of a charitable organisation. Now you have the list of people to whom you had written, and against each name you have entered the amount they contributed, or a blank space if they have not responded yet. You wish to compare the cost of the mails you sent out with the average donation received.

Remark The implementation of AverageA in Excel has one additional quirk: It evaluates the word TRUE as 1. In Calc, both Average and AverageA evaluate TRUE as 1.

Mean versus Median

Both mean and median attempt to provide a 'central value' for the set of data. So, which one does a better job? As always, the answer depends on the use you have in mind for this central value.

Example 4.6.8 In Major League Baseball in the USA, the performance of a team at any stage of the season is evaluated by the fraction of games it has won: If it has won 14 out of 20, then we say it is at 0.7 $(= 14/20)$. This is equivalent to giving it 1 point for a win, 0 points for a loss and then taking the mean of these scores.

The median of these points would not be very useful as it could only be 0 (when there are more losses than wins), or 0.5 (equal number of losses and wins) or 1 (more wins than losses). □

The median only notices the location of the centrally ranked data and is therefore not affected by the actual values of the extreme data points. Thus, the following three datasets have the same median:

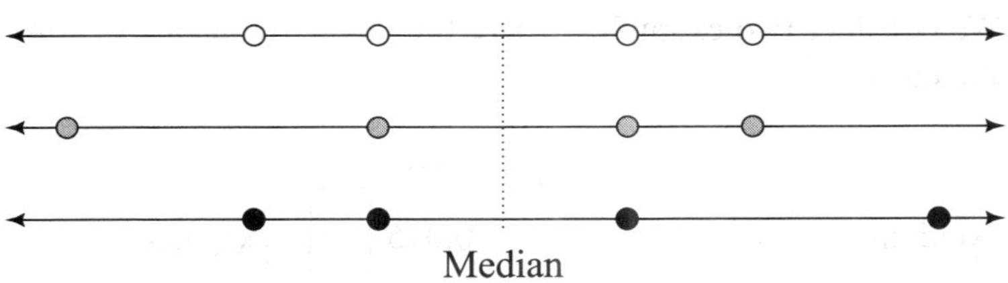

What is noteworthy here is that the middle values of the datasets are the same and, so, all three have the same median. The fact that their largest and smallest values vary does not matter.

The mean *is* affected by extreme values since every data point contributes to it. In fact, since each data point contributes in proportion to its value, extreme values affect the mean the most. Thus, the mean implements the first half of the ideal, 'From each according to his ability, to each according to his needs'!

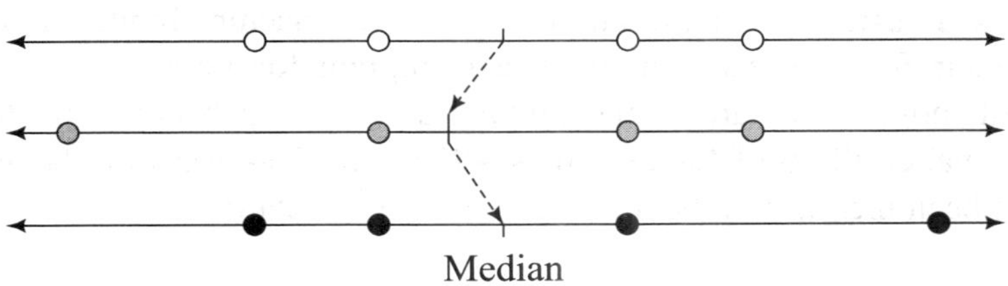

Example 4.6.9 In assigning letter grades to the results of an exam or course, one can use either *absolute* or *relative* grading. In absolute grading, there are fixed levels for achieving various grades (e.g., 85 percent is a commonly used cut-off for an A grade). In principle, under absolute grading, it is possible for every student to get an A, or for everyone to fail. Under a pure relative grading system, on the other hand, what matters is your rank rather than your marks. If you got 40 percent but topped the class, you would still get an A.

A relative grading system is often based on the mean of the student scores—typically the mean marks form the cut-off between a B grade and a C. However, such a grading system is affected by outliers. If there are a few extremely good students, then they will raise the mean and lower the grades of most of the students. If we wish to obtain a pure relative grading system, in which only rank matters, then we use the median (and other percentiles) instead of the mean in determining the cut-offs. □

We also need to understand the influences that can lead to median and mean being significantly different. If the data is symmetric about some value, then that value is both mean and median. It is when data is asymmetric, clustering more towards one end, that the mean and median differ in value.

Task 4.6.6 Suppose the data is completely symmetric about some number a: After excluding any occurrences of a itself, the remaining data can be split into two equal parts $\{x_1, \ldots, x_n\}$ and $\{y_1, \ldots, y_n\}$ such that $a - x_i = y_i - a$ for each i value. Then a is both the mean and the median of the data.

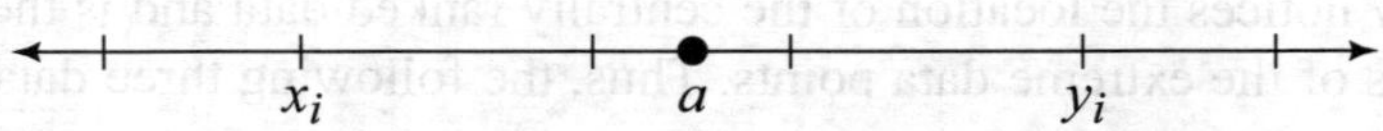

Example 4.6.10 We consider two examples where the larger frequency of low scores makes the median less than the mean.

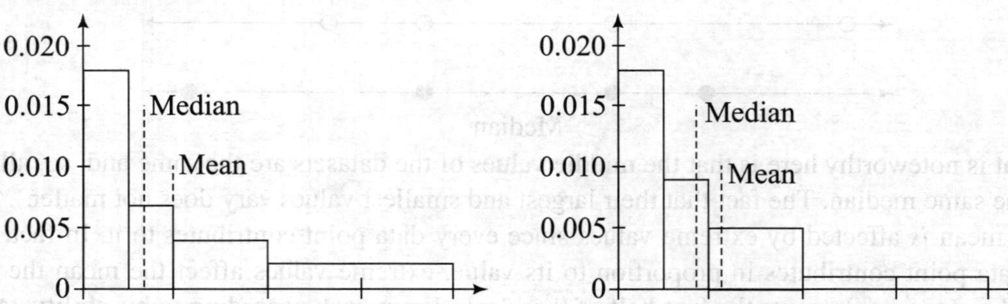

These normalised histograms represent the scores of Sachin Tendulkar and Brian Lara, respectively, in their first 60 test innings. The histogram for Lara's scores is slightly more symmetric in that more scores are in the middle, and the gap between median and mean is correspondingly smaller. (Two of Lara's scores—277 and 375—were too large to be shown in the chart but have been included in the mean and median calculations.) □

Trimmed Mean

Sometimes we wish to ignore the most extreme values, usually because we consider them as instances of poor measurement or exceptional circumstances. If we calculate the mean of the remaining data, we say we have obtained a **trimmed** or **truncated mean.**

Example 4.6.11 Suppose we drop the four most extreme values (two lowest and two highest) from the data in Example 4.6. For Tendulkar, this leads to a removal of scores of 0, 0, 177, 179 and for Lara the removal of 0, 0, 277, 375. The following table shows the trimmed mean along with the median and the original mean:

	Median	Mean	Trimmed Mean
Tendulkar	32.5	48.5	45.6
Lara	41.5	54.5	46.8

□

Trimmed means can be calculated in spreadsheets with the TrimMean function. The syntax in Excel is =TrimMean(Data, $P\%$), where P percent controls the number of dropped data points. For example, if there are 50 data points, =TrimMean(Data, 12%) will drop six (= 12% of 50) points—the lowest three and the highest three.

Example 4.6.12 The trimmed means in Example 4.6.11 were calculated by evaluating =TrimMean(Data, 7%). Since there were 60 data points, the number of dropped points at each end was

$$\frac{3.5}{100} \times 60 = 2.1$$

which was rounded down to 2. □

Quick Review

After reading this section, you should be able to:

- Calculate the mode, median, percentiles and mean of a dataset.
- Correctly select either Average or AverageA to calculate the mean of mixed data.
- Judge the appropriate measure—median or mean—of the 'centre' of a dataset in a particular context.

EXERCISE 4.6

1. Consider data $x_1, \ldots, x_N$ with mean $\overline{x}$ and data $y_1, \ldots, y_N$ with mean $\overline{y}$. Define numbers z_i by $z_i = x_i + y_i$. Show that the mean of $z_1, \ldots, z_N$ is given by $\overline{z} = \overline{x} + \overline{y}$.

2. Consider data $x_1, \ldots, x_N$ with mean $\overline{x}$ and data $y_1, \ldots, y_M$ with mean $\overline{y}$. Show that the mean of the pooled data $x_1, \ldots, x_N, y_1, \ldots, y_M$ is given by

$$\frac{N}{N+M}\overline{x} + \frac{M}{N+M}\overline{y}.$$

3. Consider the percentile calculations in Example 4.6.6. Verify that the Percentile spreadsheet function produces the same results.
4. The following marks were obtained by 20 students in a mathematics test:

$$0, 3, 4, 10, 11, 13, 19, 19, 20, 22, 24, 30, 31, 31, 33, 38, 39, 45, 46, 49$$

Use Excel to allocate grades to each student, according to the following rules:
   (a) 'A' if the percent rank is 85 or above.
   (b) 'B' if the percent rank is 65 or above but below 85.
   (c) 'C' if the percent rank is 45 or above but below 65.
   (d) 'D' if the percent rank is 25 or above but below 45.
   (e) 'F' if the percent rank is below 25.
   (f) If the score is below 15, then the student gets an 'F' regardless of the percent rank.

4.7 Measuring Diversity: Range and Deviation

The median and mean of data are called **measures of central tendency** as they are roughly in the middle of the data. We use them as indicators of where the data is located. Now, they are more representative of the total data if the data is clustered tightly about them. If the data is widely dispersed, then the import of these measures is weakened (Figure 4.7.1).

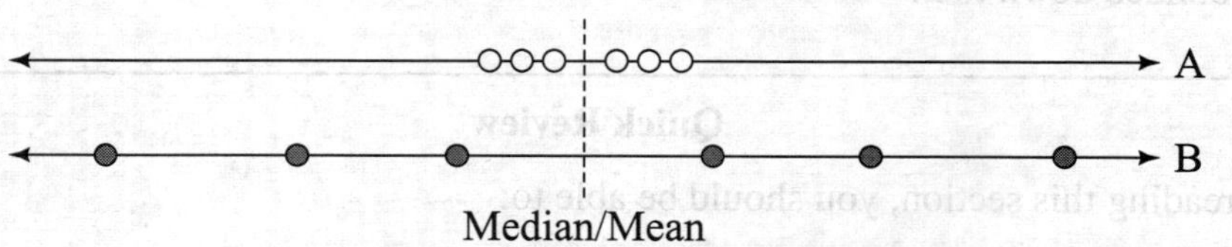

Figure 4.7.1 The mean and median are more representative of the dataset in case A, since all the data points are close to them. In B, the data points are highly scattered and the mean and median convey little information about their values.

It becomes important to also measure the extent of dispersion of the data. We not only want to judge if one dataset is more dispersed than another, we want to quantify the amount. We want to be able to say, 'This data has sufficiently low dispersion for the mean to be a useful quantity,' or 'The first dataset is twice as dispersed as the second.'

Range

The simplest measure of dispersion is the **range**—this is the gap between the largest and the smallest values in the dataset.

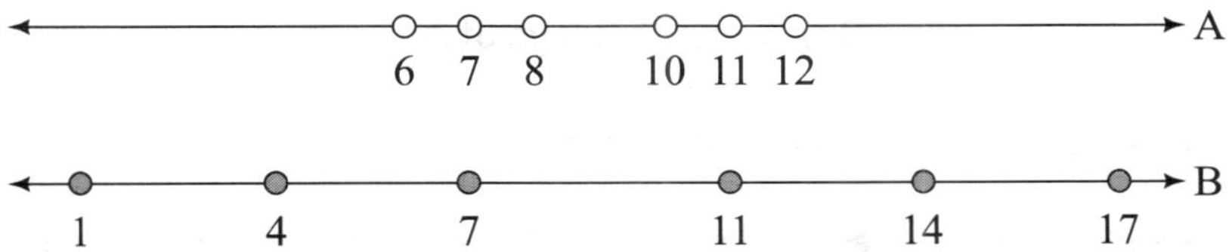

Dataset A has smallest value 6 and largest value 12, so its range is $12 - 6 = 6$. On the other hand, dataset B has range $17 - 1 = 16$. The larger range of B reflects its more dispersed nature.

Unfortunately, the range is rather unreliable. Suppose an IQ test is applied to a class. Most scores will be about 100, but there may be the odd low or very high score. The high score is particularly unstable. Experience shows that if a high-scoring person is subjected to repeated testing, then the scores vary significantly. Thus, the range for the class could be low on one day and high on another! Moreover, by concentrating on the most extreme values, the range ignores the distribution of the moderate values (Figure 4.7.2).

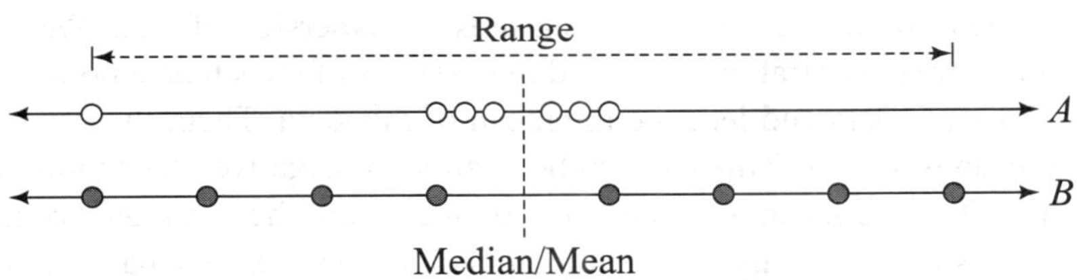

Figure 4.7.2 The points in dataset A are in general nearer to the central location than those in dataset B, but both have the same range.

In spreadsheets, the Max function gives the largest value in the dataset, while the Min function gives the least. So, we can find the range by evaluating = Max(data) – Min(data).

Interquartile Range

One way to avoid the trouble created by extreme values is to drop them. For example, the **interquartile range** is the range of the middle half of the (ordered) data points. It is formally defined as the difference between the third and first quartiles.

Example 4.7.1 The datasets depicted in Figure 4.7.2 are as follows:

$$A = \{4.5,\ 10.5,\ 11,\ 11.5,\ 12.5,\ 13,\ 13.5,\ 19.5\}$$

$$B = \{4.5,\ 6.5,\ 8.5,\ 10.5,\ 13.5,\ 15.5,\ 17.5,\ 19.5\}$$

Suppose we enter the data A in cells A1:A8 and the data B in B1:B8. We calculate the interquartile ranges as follows:

Interquartile range of A = Quartile(A1:A8, 3)−Quartile(A1:A8, 1) $= 2.25$

Interquartile range of B = Quartile(B1:B8, 3)−Quartile(B1:B8, 1) $= 8$

So, the interquartile range does detect that B is more dispersed than A.

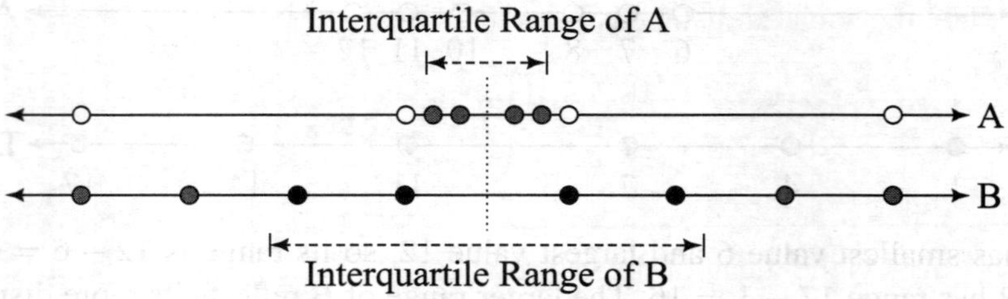

□

While the interquartile range is a marked improvement on the range, it has certain defects. The choice of working with the central half of the data is arbitrary—why not take the middle 75 percent or 90 percent instead? Another issue is that it completely ignores the distribution of the outer half of the data.

Variance and Standard Deviation

We now come to the most well-established measures of dispersion of data. We start by adopting the mean as our measure of central location, and then see how far each data point is from the mean. Let $\{x_1, \ldots, x_N\}$ be our data and let $\overline{x}$ be the mean of this data. Then, the gap between the data point x_i and the mean is $x_i - \overline{x}$. This gap can be positive or negative, depending on whether x_i is to the right or the left of the mean. One justification for taking the mean as a central value is that the sum of these gaps is necessarily zero: the points on the left exactly balance the points on the right.

Task 4.7.1 Show that $\sum_{i=1}^{N}(x_i - \overline{x}) = 0$.

To measure dispersion, we need to look at the *magnitude* of the gap between a data point and the mean. One way to do this is to just square the gap. This leads to the definition of the **variance** of the data, which we denote by σ^2:

$$\sigma^2 = \frac{1}{N}\sum_{i=1}^{N}(x_i - \overline{x})^2$$

Thus, the variance is the average squared gap between the data points and the mean.

If the data is in a certain unit, then the variance will be measured by the square of that unit. We therefore apply the square root to return to the original unit, and the new quantity is called the **standard deviation** of the data. It is denoted by σ:

$$\sigma = \sqrt{\frac{1}{N}\sum_{i=1}^{N}(x_i - \overline{x})^2}$$

Example 4.7.2 Consider the data from Figure 4.7.2 and Example 4.7.1. We calculate the variance and standard deviation of the dataset A below. First, we find the mean:

$$\overline{x} = \frac{1}{8}(4.5 + 10.5 + 11 + 11.5 + 12.5 + 13 + 13.5 + 19.5) = 12$$

Then, we calculate the squared distances from the mean:

x_i	4.5	10.5	11	11.5	12.5	13	13.5	19.5
$x_i - \overline{x}$	−7.5	−1.5	−1	−0.5	0.5	1	1.5	7.5
$(x_i - \overline{x})^2$	56.25	2.25	1	0.25	0.25	1	2.25	56.25

The variance σ_A^2 of A is the mean of the last row:

$$\sigma_A^2 = \frac{1}{8}(56.25 + 2.25 + 1 + 0.25 + 1 + 2.25 + 56.25) = 14.94$$

The standard deviation of A is the square root of the variance:

$$\sigma_A = \sqrt{14.94} = 3.86$$

We leave it to you to verify that the variance and standard deviation of B are given by:

$$\sigma_B^2 = 25.25, \qquad \sigma_B = 5.02$$

□

The spreadsheet functions for evaluating variance and standard deviation of data are Var.P and StDev.P respectively.

Quick Review

After reading this section, you should be able to:

- Distinguish between measures of central location and measures of dispersion.
- Calculate different measures of dispersion—range, interquartile range, variance and standard deviation.

EXERCISE 4.7

1. Suppose each value in the data $x_1, \ldots, x_n$ is shifted by a constant c to create new data $x_1 + c, \ldots, x_n + c$. Show that the variance and standard deviation do not change.
2. Suppose each value in the data $x_1, \ldots, x_n$ is scaled by a constant c to create new data $cx_1, \ldots, cx_n$. Show that the variance is multiplied by c^2 and the standard deviation by $|c|$.
3. Consider the following two sets of data:

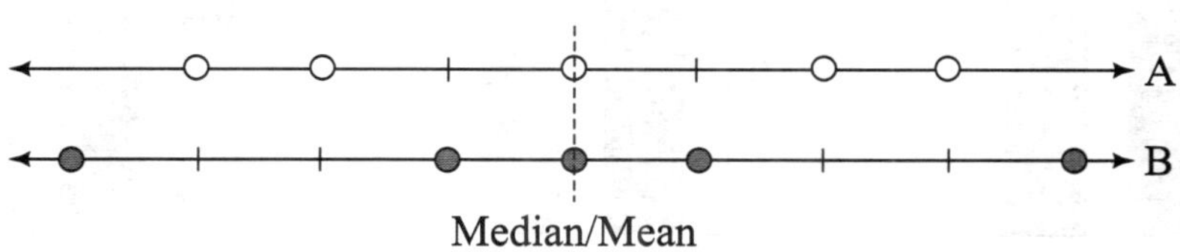

(a) Which has larger interquartile range?
(b) Which has larger variance?

4. The following table shows the monthly percentage changes in the BSE Sensex stock index from February 2006 to May 2009. The first 20 months of this period were a 'boom' in which the index rose from 9,920 to 17,291. The next 20 months featured a crash with the index falling all the way to 9,708, before a recovery to 14,625.

Month	1	2	3	4	5	6	7	8	9	10
% Change	4.5	6.8	8.8	−13.7	2.0	1.3	8.9	6.5	4.1	5.7
Month	11	12	13	14	15	16	17	18	19	20
% Change	0.7	2.2	−8.2	1.0	6.1	4.8	0.7	6.1	−1.5	12.9
Month	21	22	23	24	25	26	27	28	29	30
% Change	14.7	−2.4	4.8	13.0	−0.4	−11.0	10.5	−5.0	−18.0	6.6
Month	31	32	33	34	35	36	37	38	39	40
% Change	1.5	−11.7	−23.9	−7.1	6.1	−2.3	−5.7	9.2	17.5	28.3

Calculate the standard deviation for each 10-month period above. Is there a pattern?

4.8 How to Lie with Statistics

Statistical data, presented in various forms, pervades every argument and claim—from the advertisements of competing news channels to the relative effectiveness of two brands of medicine or the magnitude of injustices suffered by rival religious or ethnic groups at each other's hands. It is natural that much of the time the data is not presented as correctly as it could be, and this leads to a popular distrust of statistics itself. We will therefore supplement our earlier discussions of tools that help us to summarise and depict data, with examples of misuse of these tools. The pioneering reference on these matters is the book *How to Lie with Statistics* by Darrell Huff [3] and we have borrowed its title for this section.

A common way of misrepresenting statistical data to make it favour a particular cause is to manipulate its graphical representation so that differences appear much more important than they are.

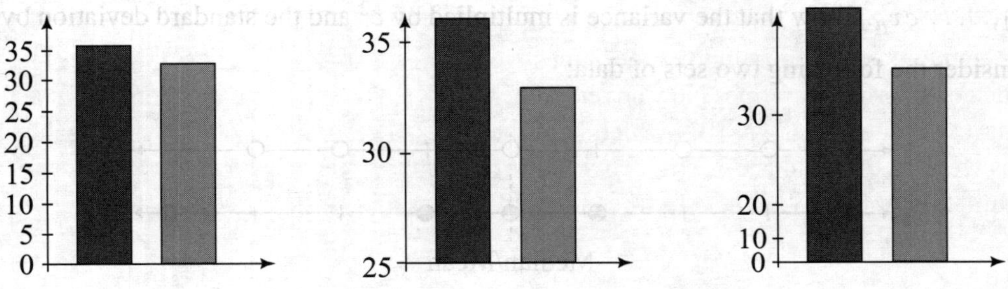

All three of these charts represent the same data. In the second chart, a slight difference in two quantities is made to look quite large by simply changing the baseline to 25 from 0. The third chart achieves a similar result by using a non-linear vertical scale. The example below (Figure 4.8.1) shows a particularly creative use of bar charts—it is hard indeed to conceive what rules were used to draw it! Note the absence of a vertical scale that would give the game away.

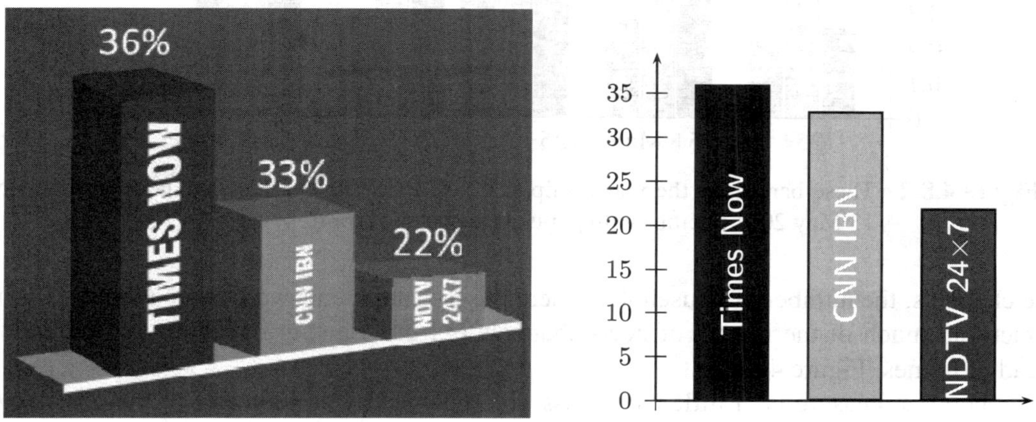

Figure 4.8.1 The bar chart on the left appeared in an advertisement in the 20 May 2011 edition of *Hindustan Times*. It purports to show the viewer-ship numbers for some TV news channels during election coverage on 13 May. The chart on the right shows the columns as they *should* have been drawn. [C1]

The other popular technique is to start with a mass of data and present only that which is favourable. For example, the advertisement described in Figure 4.8.1 included the following fine print: All India 1mn+, CS 25+, Males AB. This means that the viewership percentages refer to the population segment of males of age over 25, with a college education, with careers as businessmen or professionals, living in households with cable or satellite connections, situated in cities with population of over 1 million. While this is doubtless a very important segment for marketing purposes, it is far less satisfactory as a proxy for the total viewership.

When we see a particular segment of data being picked out like this, it is natural to wonder about what has *not* been picked. The chart below shows the distribution of the channel shares across various population segments. Perhaps the first thing to appreciate is their small overall share of the viewership—just 1–2 percent over the different segments. The other is that, of these similar segments, only one supports the Times Now advertisement.[12]

While considering the Times Now advertisement and its context, we have seen that in all but one segment it was actually CNN IBN which held the lead. Another observation is that, whoever led, the gaps were on the small side. This brings up the issue of whether the gaps were even significant. An argument that they were *not* significant could be based on factors such as the low reach of all

[12] TAM data sourced from article by Rajat Arora [7].

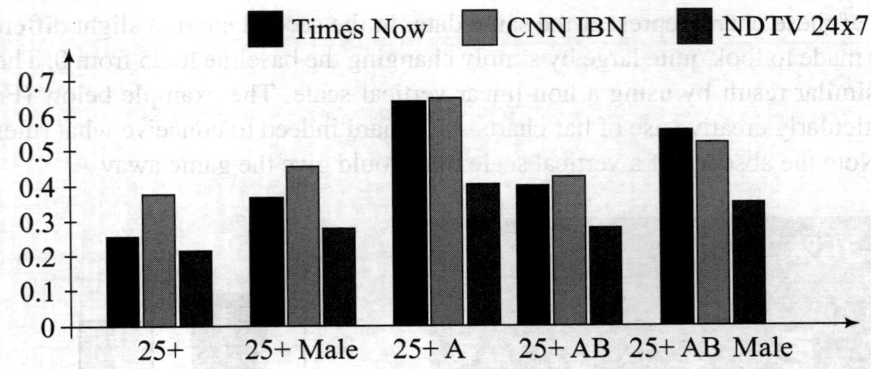

Figure 4.8.2 These bars show the viewership share of three English-language news channels on 13 May 2011 in cities with population over 1 million.

these channels, the number of households where the measurements were taken (about 9,000), and the fact that much of the impact of news channels is from viewing in offices and public places instead of homes (Figure 4.8.2).

In general, a measure is of little use unless accompanied by an indicator of the possible error. Absence of such indicators is one sign of possible misrepresentation. In the next chapter, on probability, we will look into some aspects of estimating errors in surveys.

Quick Review

After reading this section, you should be able to:

- Detect the use of distorted charts to misrepresent data.
- Detect the selective use of data to promote a claim.

4.9 Advanced Examples

Radioactivity

We will now consider an example of interpolation which will introduce us to a very useful technique of detecting patterns by suitably transforming data.

The following data, from a 1905 paper by Meyer and von Schweidler,[13] shows radioactive activity of a sample material over a month.

[13] Source: Connected Curriculum Project at Duke [W4].

Time (days)	Relative Activity
0.2	35.0
2.2	25.0
4.0	22.1
5.0	17.9
6.0	16.8
8.0	13.7

Time (days)	Relative Activity
11.0	12.4
12.0	10.3
15.0	7.5
18.0	4.9
26.0	4.0

Here is a plot of this data:

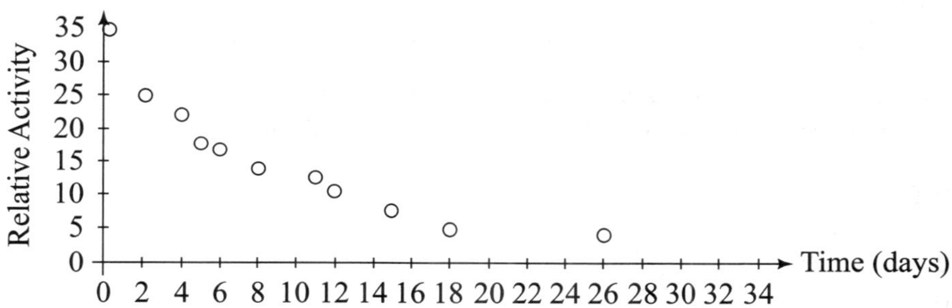

We see that this plot is like the middle graph in Figure 4.4.4 and represents a process whose rate of decrease is slowing down. For such examples, with a wide range of values, it is often helpful to take the logarithm of the data values. If we do so with our data, we get the following data:

Time (days)	Log (base 10) of Relative Activity
0.2	1.54
2.2	1.40
4.0	1.34
5.0	1.25
6.0	1.23
8.0	1.14
11.0	1.09
12.0	1.01
15.0	0.88
18.0	0.69
26.0	0.6

When we plot this transformed data, something fantastic happens—it looks almost linear. Below, we plot this data and its line of best-fit:

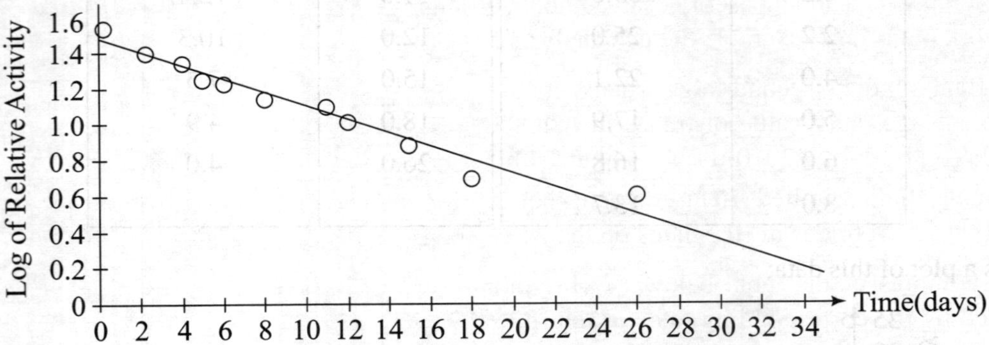

We can use the line of best-fit to get estimates for the relative activity on other days. The diagram below shows how to obtain an estimate for the activity on the 34th day:

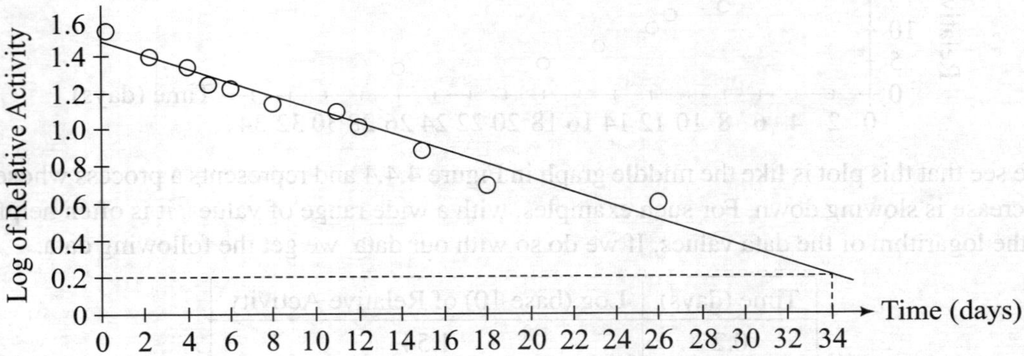

The predicted value of the log of the relative activity is 0.2. Therefore, the prediction for the relative activity itself is $10^{0.2} = 1.58$.

Market Risk

A **share** represents part of the capital of a company. The owner of a share is thus a part-owner of the company and takes part in the rise and fall of its fortunes. Shares are traded on stock exchanges, and this trading leads to a constant fluctuation in the share price. These fluctuations depend on various factors—such as events in the general market, changes in the company's strategy or ownership, and its latest financial statement or forecast, etc. Their magnitude is described by the **rate of return.** If the price of a share changes from U to V over some time, then we define the corresponding rate of return r to be

$$r = \frac{V - U}{U}$$

For example, suppose the price changes from ₹100 to ₹120. Then the rate of return is $r = (120 - 100)/100 = 0.20$ or 20 percent. If the price falls, we get a negative rate of return. For example, if the price changes from ₹100 to ₹80, we have $U = 100$, $V = 80$, and $r = (80 - 100)/100 = -0.20$ or -20 percent.

We can use data about past prices to compare different shares in terms of profit and risk. We will now work out a small example of such analysis, based on data for the following two companies:

ICICI Bank: India's second-largest bank. Its assets on 31 March 2010 totalled over ₹3 *trillion*. The average daily value of trades in ICICI Bank shares on the National Stock Exchange of India (NSE) during 2010–11 was about ₹4 billion.

Insecticides India Limited (IIL): A young company manufacturing agrochemicals. The average daily value of trades in IIL shares on the NSE during 2010–11 was about ₹44 million.

In the columns of the table below, we list monthly values (from the financial year 2010–11) followed by the corresponding rates of return.[14] For example, the first entry in the rate of return column for ICICI shares is calculated as follows:

$$r_{ICICI} = \frac{\text{Value on 3 May } - \text{ Value on Apr 1}}{\text{Value on 1 Apr}} = \frac{937 - 953}{953} = -0.017 = -1.7\%$$

Date	ICICI	r_{ICICI} (%)	IIL	r_{IIL} (%)
1 Apr	953		123	
3 May	937	−1.7	139	13.1
1 Jun	838	−10.6	147	5.3
1 Jul	841	0.4	239	63.1
2 Aug	940	11.7	220	−7.9
1 Sep	995	5.9	219	−0.7
1 Oct	1,135	14.1	213	−2.4
1 Nov	1,232	8.5	225	5.3
1 Dec	1,167	−5.2	217	−3.7
3 Jan	1,145	−1.9	224	3.6
1 Feb	995	−13.1	237	5.5
1 Mar	1,026	3.2	303	28.1

Investors are concerned with two aspects of any asset—possible profit and risk. The rate of return provides a measure of the first. To see how well a share has generally done in the past, we take the average of its rates of return over many consecutive periods. For ICICI Bank, the average of the given monthly rates of return was

$$\overline{r}_{ICICI} = \frac{1}{11}(-1.7 + \cdots + 3.2) = 0.010 = 1\%$$

while for IIL it was

$$\overline{r}_{IIL} = \frac{1}{11}(13.1 + \cdots + 28.1) = 0.099 = 9.9\%$$

[14] Data obtained from the NSE website [W9].

So, over this period, IIL was a much better performer—-on average—than ICICI. Is it a good bet for the future? Now we need to evaluate risk, and to measure it we use the standard deviation of the rates of return. Higher standard deviation corresponds to higher fluctuation and hence to higher risk. We find the two standard deviations to be as follows:

$$\sigma_{ICICI} = 0.082 = 8.2\%$$

$$\sigma_{IIL} = 0.192 = 19.2\%$$

Thus, the higher promise of IIL is at the cost of a much greater risk!

We can go further and ask about the nature of the risk. How much of it is from general trends, and how much from uncertainties peculiar to the company? Due to the nature and size of the two companies, we would expect ICICI share value to be closely tied to general trends in the market, while the risks for IIL should come more from particular events such as the monsoon and new products from rivals.

A **stock index** enables investors to track general trends by creating an imaginary portfolio of a number of shares of different companies, and tracking their total value. A typical index may have anywhere from 30 to 500 different shares. One obvious concern of an investor in a company is the exposure of that company to general trends in the market, and we can gain insights into this by comparing the rates of return of the company shares to the rates of return from the index portfolio over the same time periods. We will use the following index:

S&P CNX 500 index: A stock index maintained by the NSE. It has shares of 500 companies and covers above 90 percent of the daily trading on NSE, thus providing a broad-based summary of market trends.

Here is monthly data for the value of this index over the same time intervals as the two shares:

Date	S&P CNX 500	r_{SP} (%)
1 Apr	4,345	
3 May	4,341	−0.1
1 Jun	4,149	−4.4
1 Jul	4,386	5.7
2 Aug	4,524	3.1
1 Sep	4,603	1.8

Date	S&P CNX 500	r_{SP} (%)
1 Oct	5,019	9.0
1 Nov	5,053	0.7
1 Dec	4,876	−3.5
3 Jan	4,967	1.9
1 Feb	4,352	−12.4
1 Mar	4,392	0.9

To check out our intuition about the risk profiles of the two companies, we scatter plot their rates of return against the corresponding ones for the index (Figure 4.9.1).

The first plot shows a strong connection between r_{ICICI} and r_{SP}—they tend to move up or down together. On the other hand, there is no discernible relationship between r_{IIL} and r_{SP}. (The second plot also has two missing points, whose values were too high to fit in the chosen range.)

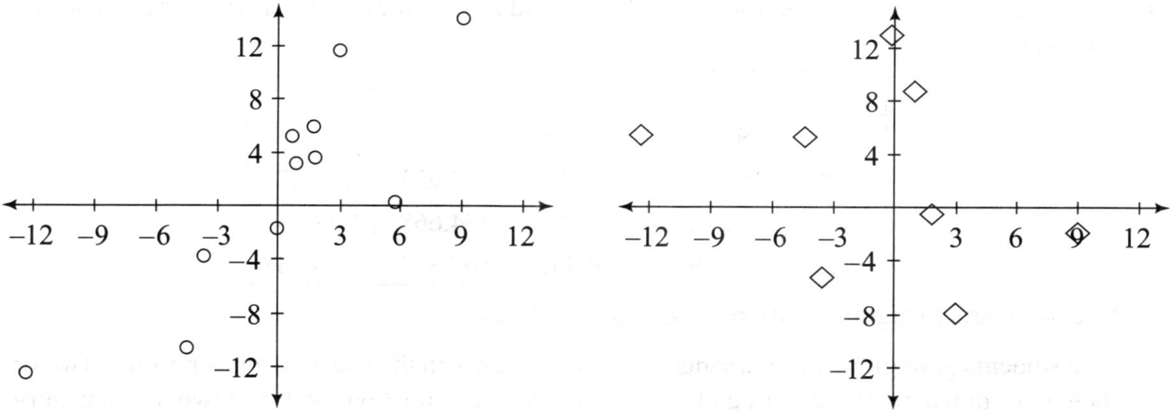

Figure 4.9.1 These scatter plots show the relations between the rates of return of the ICICI Bank and IIL share prices and the values of the S&P CNX 500 index.

Review Exercises

1. Suppose petrol prices increase by 10 percent every month for three months. What is the overall percentage increase?
2. Does a 15 percent decrease cancel a 15 percent increase?
3. What is the mean of the numbers $1, 2, \ldots, 100$?
4. Are the following statements true or false?
   (a) If the Sum function is applied to a single blank cell, then the result will be 0.
   (b) If data is symmetric, then the mean and median are equal.
   (c) If a value equal to its mean is added to a dataset, then the mean does not change.
   (d) If a dataset is skewed to the right, then its mean is greater than its median.
   (e) The standard deviation of a dataset is always less than its variance.
5. Are the following statements true or false?
   (a) If the Average function is applied to a single blank cell, then the result will be 0.
   (b) If the mean and median are equal, then the data is symmetric.
   (c) If a value equal to its median is added to a dataset, then the median does not change.
   (d) If a dataset is skewed to the left, then its mean is less than its median.

6. The table below shows the performance of boys and girls in the CBSE Class XII exams during 2009–11.[15]

Year	Boys		Girls	
	Appeared	Passed	Appeared	Passed
2009	354,151	274,108	257,952	221,674
2010	388,249	294,697	284,668	242,769
2011	427,795	328,442	307,814	266,488

Use bar charts to highlight different aspects of this data.

7. Five students gave glowing comments about their teacher in the course feedback forms. Two of them gave him a numerical rating of 4 out of 4, one gave him 3 out of 4, and two left that entry blank. Yet his average rating was reported as a mere 2.2 out of 4. What could have happened?

8. (Adapted from Huff [3]) A real estate agent tells you that the typical income in the colony where you are thinking of buying a house is ₹20 lakhs per year. You move in, and a little later the agent approaches you to sign a petition to keep certain municipality charges down because, after all, 'the typical income is only ₹2 lakhs per year'. Has the agent lied, or is he just clever with statistics? What interpretation of his words could lead to both statements being true?

The next exercise provides additional understanding of why mean can be considered as a central location.

9. The variance σ^2 of data $x_1, \ldots, x_N$ is the average squared deviation from the mean $\overline{x}$. Let $V(a)$ be the average squared deviation from the number a:

$$V(a) = \frac{1}{N}\sum_{i=1}^{N}(x_i - a)^2$$

Show that $V(a)$ has least value when $a = \overline{x}$ by verifying the following identity:

$$V(a) = \sigma^2 + (\overline{x} - a)^2$$

10. In 2016, the University Grants Commission recommended a relative grading policy in which a student's grade depends on the mean $\overline{x}$ and standard deviation σ of the marks of the entire class. For example, to get the top 'O' grade, a student needs to earn $\overline{x} + 2.5\sigma$ marks, while to avoid a failing 'F' grade the student needs to earn $\overline{x} - \sigma$ marks. In each of the following cases, identify which students will fail, and who will get an 'O':
    (a) The marks out of 100 are 5, 20, 35, 40, 55, 62, 63, 66, 71, 95.
    (b) The marks out of 100 are 20, 35, 40, 63, 75, 84, 85, 89, 99, 100.
    (c) The marks out of 100 are 5, 8, 11, 12, 17, 22, 33, 35, 40, 80.

    Analyse what kind of marks distribution (symmetric, skewed to the left, skewed to the right) will lead to higher grades for the students.

[15] Source: CBSE Annual Reports [W2].

11. Identify the anomalies produced by the grading policy in the previous exercise. What would be the relative advantages and disadvantages of a grading policy based on percentiles instead of mean and standard deviation?

12. The table below is from a 1940 article by D. D. Kosambi [5], who was unique in being eminent as a mathematician as well as a historian. He took up the problem of a type of ancient coin whose reverse carried a variable number of marks. In almost all cases, this number varied from zero to six. To understand the significance of the marks, Kosambi studied their relation to two other quantities—the weights and numbers of the coins. One part of his data is reproduced below:

Number of marks	0	1	2	3	4	5
Number of coins	224	128	132	85	64	46
Average weight (in 'grains')	53.26	52.93	52.74	52.47	52.53	52.17

Plot the second and third rows against the first and generate the lines of best-fit. Is there a pattern? Is it improved by using the log transformation (as we did for the radioactivity example in Section 4.9)? What does the number of marks appear to reflect?

Bibliography

Books and Articles

[1] Brian Albright. 2009. *Mathematical Modeling with Excel*. Sudbury, MA: Jones and Bartlett.

[2] Irfan Habib. 2011. *Man and Environment: The Ecological History of India*. A People's History of India. Vol. 36. New Delhi: Tulika.

[3] Darrell Huff. 1993. *How to Lie with Statistics*. Reissue of 1954 edition. New York: W W Norton and Company.

[4] Gerald Knight. 2006. *Analyzing Business Data with Excel*. O'Reilly.

[5] D. D. Kosambi. 1940. 'A statistical study of the weights of old Indian punch-marked coins'. *Current Science*. 9(7).

[6] Mathew Macdonald. 2013. *Excel 2013: The Missing Manual*. O'Reilly.

[7] Rajat Arora. 2011. 'Assembly elections 2011: CNN-IBN gets viewers' mandate `http://www.bestmediainfo.com/2011/05/assembly-elections-2011-cnn-ibn-gets-viewers-mandate/` (accessed 12 January 2017).

Websites

[W1] BioTopics. *Predator-prey interdependence*. `http://www.biotopics.co.uk/newgcse/predatorprey.html` (accessed 12 January 2017).

[W2] Central Board of Secondary Education. *Annual Report*. `http://cbseacademic.in/annualreport.html` (accessed 12 January 2017).

[W3] Cyber Journalist. *Census of India 2001—State wise population totals*. http://cyberjournalist.org.in/census/cenindia.html (accessed 12 January 2017).

[W4] W. H. Barker and D. A. Smith. *Radioactive Decay*. Connected Curriculum Project at Duke University. https://services.math.duke.edu/education/ccp/materials/precalc/raddec/ (accessed 29 January 2017).

[W5] Michael Friendly. *Data Visualization Gallery*. http://www.datavis.ca (accessed 12 January 2017).

[W6] HowSTAT! http://www.howstat.com (accessed 12 January 2017).

[W7] LibreOffice. *LibreOffice Calc Help*. https://help.libreoffice.org/Calc/Welcome_to_the_Calc_Help (accessed 12 January 2017).

[W8] Microsoft. *Excel Help*. https://support.office.com/en-us/excel (accessed 27 January 2017).

[W9] National Stock Exchange of India. https://www.nse-india.com (accessed 27 January 2017).

[W10] Statistics Canada. *Statistics: Power from Data*! http://www.statcan.gc.ca/edu/power-pouvoir/toc-tdm/5214718-eng.htm (accessed 27 January 2017).

[W11] US Bureau of Labor Statistics. https://www.bls.gov (accessed 29 January 2017).

[W12] World Bank. http://data.worldbank.org (accessed 29 January 2017).

Image Credit

[C1] Times Now Advertisement, *Hindustan Times*, May 20, 2011.

5 CHAPTER

Probability

Almost every action that we take represents a gamble. Even the writing of this book reflects the taking of a chance, that the energy spent would be compensated by its being useful to someone someday. Certainly, any significant decision we take is based on estimates of what is more or less likely, as well as the consequences of each. Mathematicians have taken up the challenge of quantifying this intuitive notion of the likelihood of an event actually coming to pass. They seek to be able to say not only that one team is more likely than another to win, but that it is twice as likely to win.

What do we gain by attaching numbers to chance? Suppose a card game is being played by two players, and the first player to reach 100 points wins ₹100. The game has to be suddenly stopped when one player has 90 points and the other has 60. We could say that as there is no winner, each player should get half the prize, but the player with 90 points would surely object that she had almost won. While if we give her the entire prize, her opponent would complain that he had not yet lost. A fair distribution of the prize can be worked out if we are able to estimate how likely each player was to win from this position. For example, if the player on 90 points has a 70 percent chance of winning, we could give her 70 percent of the prize.

Also, since certainty is rarely available, we like to ask if something is 'almost certain'. This cannot have an absolute answer, but we may be reassured on hearing that the chances of a desired event occurring are, say, 95 percent. In today's world, we are constantly subjected to such estimates. Perhaps the Meteorological Department forecasts a 96 percent probability of the Monsoon rains being normal or better. Or an opinion poll mentions in its fine print that there is a 95 percent probability of its estimates being accurate to within 3 percent. What pieces of information are these numbers trying to convey, and how should they affect our actions? How do we come up with them? These are the questions this chapter addresses.

5.1 Measuring Chance

A friend offers you a bet on the number of heads that will show up when a coin is tossed twice. Would you bet on no heads or one or two? Chances are that you feel an attraction towards betting on one head (and one tail) as the most balanced and reasonable choice. Perhaps you even notice that if we track the individual tosses, there are four possible combinations:

1. First toss is head, and second is tail.
2. First toss is tail, and second is head.
3. Both tosses are tails.
4. Both tosses are heads.

Of the four possible combinations, two are favourable to the bet on one head. The bets on no head or two heads have only one favourable combination each. This seems to go in favour of betting on one head.

This little thought experiment indicates that evaluating the probability of a possible event has at least two components:

1. Breaking down the possible events in terms of combinations of simpler events.
2. Counting the favourable combinations for a specific event.

Our game of coin tossing consists of two simpler parts: the individual tosses of the coin. Each of these has two possible outcomes, head and tail, of which exactly one can occur. In each toss, we probably have no reason to favour head over tail or vice versa. We express this by calling these two possible outcomes 'equally likely' and giving each a probability of $1/2$. The value of $1/2$ indicates our belief that given a lack of bias between head and tail, if we were to toss the coin a large number of times, about half the time we would get head.

If head and tail are equally likely in a single toss, and if the result of the first toss does not influence the second toss, then each of the four possible combinations for two tosses should also be equally likely. We assign them a probability of $1/4$ each. This indicates our belief that if the game were played many times, each combination would occur in about onefourth of the total times the game is played.

We follow this informal discussion with an introduction of the standard terms that are used in studying Probability. The first one is **random experiment.** This is any process whose outcome cannot be predicted with certainty beforehand (a coin toss, an election, an exam), but which is well enough understood so that we do have an exhaustive list of all possible outcomes. The outcomes should be such that whenever the process is run, exactly one of them will occur. The list of all possible outcomes constitutes a set and we call it the **sample space** of the experiment. In this book, we will study only random experiments with a finite sample space. Finally, we have **events,** which are combinations of outcomes. Each event can be viewed as a subset of the sample space.

Example 5.1.1 Consider a random experiment of tossing a coin. Let us denote by H the outcome of heads, and by T the outcome of tails. Then, the sample space is $\{H,T\}$. The list of events is $\emptyset$, $\{H\}$, $\{T\}$ and $\{H,T\}$. □

Example 5.1.2 Consider a random experiment of tossing a coin twice. Let us retain the H and T notation of the previous example for each individual toss. If we get H in the first toss and T in the second, we denote the overall outcome by HT. Thus, the sample space for this experiment is $\{HH, HT, TH, TT\}$. The number of events, the possible combinations of these four outcomes, is $2^4 = 16$, so we will not list all of them. But here are a few:

- Getting only heads: $\{HH\}$.
- Getting no heads: $\{TT\}$.
- Getting one head: $\{HT, TH\}$.
- Both tosses have the same result: $\{HH, TT\}$. □

An outcome is called **favourable** for an event if it is a member of that event. For example, in the two toss experiment, the favourable outcomes for the event of getting one head are HT and TH. We consider that a random experiment has resulted in the **occurrence** of a certain event if the actual outcome of the random experiment is favourable for that event.

Two events are particularly extreme. The **empty event** $\emptyset$ has no outcomes at all and is therefore impossible. We assign it a probability of 0. The entire sample space constitutes the **sure event.** It will always occur and is assigned a probability of 1. All other events are assigned probabilities between 0 and 1.

How should we assign probabilities and what do they represent? If a random experiment can be indefinitely repeated, the probability of an outcome or event should reflect how often that outcome or event occurs over multiple runs of the experiment. When we say that the probability of H in a coin toss is $1/2$, repeatedly tossing a coin a large number of times should result in obtaining heads about half the time. We may not be too surprised if 30 tosses resulted in 20 heads, but if 300 tosses resulted in 200 heads we would surely suspect that something is amiss. In other words, as the number of tosses increases the fraction in which heads results should be nearer to half. Perhaps the first attempt to verify this was in the eighteenth century by Count Buffon, who obtained 2,048 heads from 4,040 tosses—a fraction of 0.507. To reflect this interpretation of probability, we often express probability as a percentage. For example, a probability of 0.7 may be given as 70 percent and on reading it, we visualise the event as being likely to occur 70 times out of 100.

Therefore, one way to estimate the probability of an event A is to run the experiment some number of times, say N, and count the number of times n_A that the outcome is favourable to the event A. Then, the **relative frequency** $\frac{n_A}{N}$ can be taken as an estimate of the probability of A.

Using relative frequency to estimate probabilities is often impractical and does a poor job in estimating the probabilities of rare events. It is more satisfying if we can directly obtain probabilities through a theoretical understanding of the experiment. For example, in the coin toss experiment, the symmetry of the coin suggests that we should assign equal probabilities to heads and tails (Figure 5.1.1).

In many other situations too, it is reasonable to assume that each outcome in the sample space has equal probability. If we do so, we are employing **classical probability.** In classical probability, if the sample space has N outcomes, then each outcome is assigned a probability of $1/N$. Further, if an event has k favourable outcomes, its probability is taken to be k/N.

Task 5.1.1 Use classical probability to obtain the probabilities of the four events in a single coin toss.

Task 5.1.2 Use classical probability to obtain the probabilities of the events listed in Example 5.1.2.

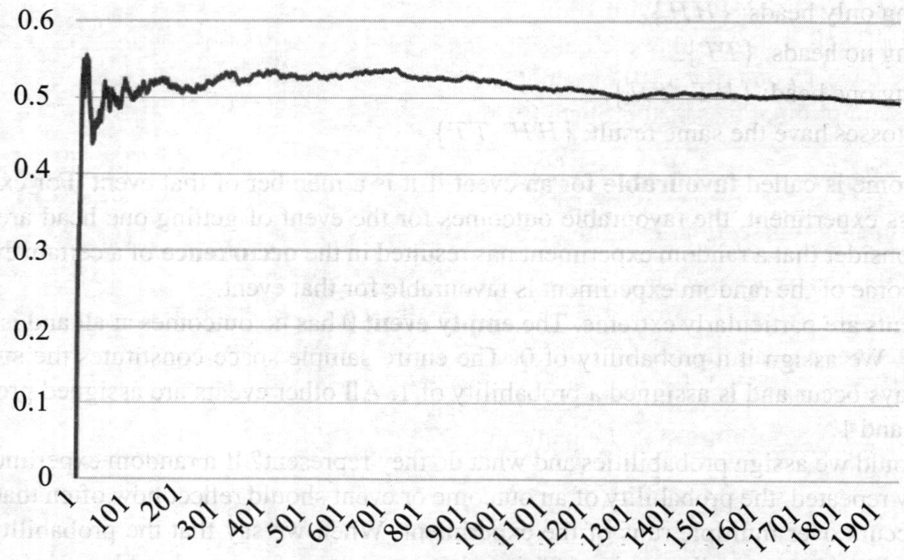

Figure 5.1.1 Two undergraduate students in California tossed a coin for an hour every day over a semester, achieving a total of 40,000 tosses. The chart above shows the changes in relative frequency of heads over the first 2,000 tosses where the coin initially had tails on top. [W1]

Example 5.1.3 Consider a die—a cube with the numbers 1, 2, 3, 4, 5 and 6 painted on its six faces. Our random experiment is to throw the die, and the outcome is the number that shows on the top face. Due to the symmetry of the shape, we expect each number to be equally likely to be the outcome. So we take this to be an instance of classical probability.

The sample space is $\{1, 2, 3, 4, 5, 6\}$ and has 6 members. Hence, each outcome is assigned the probability of 1/6. The number of possible events is $2^6 = 32$. We calculate the probabilities for some of them.

The event is that the outcome is an even number. The favourable outcomes are three: 2, 4 and 6. So the probability of this event is $3/6 = 1/2$.

The event is that the outcome is a factor of 6. The favourable outcomes are four: 1, 2, 3 and 6. The probability of this event is $4/6 = 2/3$.

Example 5.1.4 An urn is a tall vase or a container with a cover. Consider an urn containing a number of balls that are identical in size, weight and feel. If we pick out one of these balls from the urn, without knowing their arrangement inside the urn, all the balls can be taken as equally likely to be picked. Again, this becomes a case of classical probability.

Now suppose that these balls have other characteristics. Perhaps they have different colours. These colours may not be equally likely, but we could still apply classical probability to the balls themselves to solve for probabilities of different colours being picked.

For example, consider an urn with 10 black balls and 5 white balls. We want the probability of picking a black ball. Imagine that the balls are numbered 1–15, and the ones numbered 1–10 are black. So we have 15 outcomes, 10 of which are favourable for the event of picking a black ball. Under our assumption of classical probability, the probability of this event is $10/15 = 2/3$. □

Quick Review

After reading this section, you should be able to:

- Describe a process with uncertain outcomes formally as a random experiment with a sample space.
- Interpret probability in terms of repeated experiments and relative frequency.
- Identify when it is appropriate to use classical probability.
- Compute probabilities of events from those of single outcomes following classical probability.

EXERCISE 5.1

1. Describe the sample space for the following random experiments using set notation:
   (a) Tossing a coin thrice
   (b) Tossing a coin and a die
2. A high school student who has to choose between science and humanities is twice as likely to choose humanities (compared to science). What is the probability that she will choose humanities?
3. Two coins are tossed and the number of heads counted. They could be 0, 1 or 2. Would it be reasonable to consider the three outcomes to be equally likely and assign each a probability of 1/3?
4. Our experiment is to pick a ball from an urn which has a mix of two black and three white balls. If we wish to use classical probability, which of the following is a better way of describing the sample space?
   (a) Label each black ball B, each white ball W and take the sample space to be $\{B, W\}$.
   (b) Number the black balls 1 and 2, the white balls 3–5 and take the sample space to be $\{1, 2, 3, 4, 5\}$.
5. An urn has two black and three white balls. We remove four balls without looking at them. What is the probability that the remaining ball is black?
6. We toss a coin and a die. What is the probability of getting a head and an even number? If the experiment was repeated 1,000 times, about how many times would you expect this event to occur?

7. Let us explore further the comment that using relative frequencies as approximations for probabilities works poorly for rare events. Imagine an urn filled with a million identical beads, except that one is white and all the others black. The experiment is to pick a bead and note its colour. What is the probability of picking a white bead? If we repeat this experiment 1,000 times, would relative frequency counts be able to give us a reasonable idea of this probability?

5.2 Probability and Sets

If a random experiment has only finitely many possible outcomes, it is not difficult to set up a framework for studying the probabilities of events arising from this experiment. Let the sample space be written as $\{o_1, o_2, \ldots, o_N\}$. Let the probability assigned to outcome o_i be p_i, with $0 \le p_i \le 1$ and $p_1 + \cdots + p_N = 1$. (Classical probability is a special case with every $p_i = 1/N$.) The probability of any event is then defined to be the sum of the probabilities of all the outcomes that are favourable to the event. This definition is consistent with both classical probability and the relative frequency approach.

Let us introduce some convenient notation. First, we use $\mathcal{S}$ to denote the sample space of a random experiment. Then, we use $\mathcal{E}$ to represent the set of events, that is, the subsets of $\mathcal{S}$. We denote the probability of an event A by $P(A)$. We now express some obvious facts in this notation.

1. For every $A \in \mathcal{E}$, $0 \le P(A) \le 1$.
2. $P(\emptyset) = 0$ and $P(\mathcal{S}) = 1$.

When two events A and B have no common favourable outcome ($A \cap B = \emptyset$), they are called **mutually exclusive.** Mutually exclusive events can never happen together.

3. If $A, B \in \mathcal{E}$ and $A \cap B = \emptyset$, then $P(A \cup B) = P(A) + P(B)$.
   We can justify this as follows:
   Let $A = \{o_1, \ldots, o_M\}$ and $B = \{o'_1, \ldots, o'_N\}$.
   So $A \cup B = \{o_1, \ldots, o_M, o'_1, \ldots, o'_N\}$.
   Also, let $p_i = P(o_i)$ and $p'_j = P(o'_j)$. Then,

$$\begin{aligned} P(A \cup B) &= p_1 + \cdots + p_M + p'_1 + \cdots + p'_N \\ &= (p_1 + \cdots + p_M) + (p'_1 + \cdots + p'_N) \\ &= P(A) + P(B). \end{aligned}$$

This is a fundamental result from which many other consequences can be derived:

4. If $A \in \mathcal{E}$, then $P(A^c) = 1 - P(A)$.
   We have $A \cap A^c = \emptyset$ and $A \cup A^c = \mathcal{S}$. Hence, $1 = P(\mathcal{S}) = P(A) + P(A^c)$.
5. If $A, B \in \mathcal{E}$, then $P(A \cap B^c) = P(A) - P(A \cap B)$.
   $A \cap B$ and $A \cap B^c$ are disjoint, and their union is A. Hence, $P(A) = P(A \cap B) + P(A \cap B^c)$.
6. If $A, B \in \mathcal{E}$ and $B \subseteq A$, then $P(B) \le P(A)$.

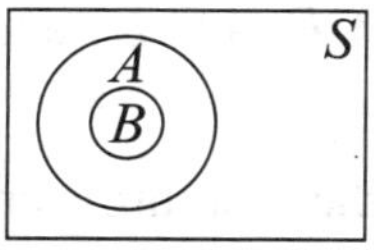

$$\begin{aligned} P(A) &= P(A \cap B) + P(A \cap B^c) \\ &= P(B) + P(A \cap B^c) \\ &\geq P(B). \end{aligned}$$

7. If $A, B \in \mathcal{E}$, then $P(A \cup B) = P(A) + P(B) - P(A \cap B)$.

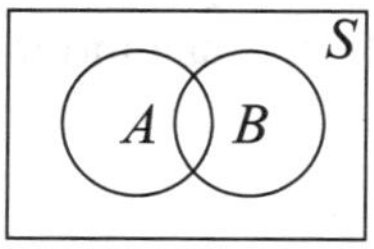

$A \cup B$ is the union of three pairwise disjoint sets:

$A \cup B = (A \cap B^c) \cup (B \cap A^c) \cup (A \cap B)$.

Hence,

$$\begin{aligned} P(A \cup B) &= P(A \cap B^c) + P(B \cap A^c) + P(A \cap B) \\ &= P(A) - P(A \cap B) + P(B) - P(A \cap B) + P(A \cap B) \\ &= P(A) + P(B) - P(A \cap B) \end{aligned}$$

Example 5.2.1 Consider an urn with black, white and red balls. Let B be the event of picking a black ball, W of picking a white one and R of picking red. It is given that $P(B) = 0.5$ and $P(W) = 0.3$. From this information, we can calculate all other probabilities. First, we note that B, W and R are pairwise mutually exclusive—no two can occur simultaneously. Hence, the probability that the selected ball is either black or white is:

$$P(B \cup W) = P(B) + P(W) = 0.8$$

The event R is complementary to $B \cup W$; hence,

$$P(R) = 1 - P(B \cup W) = 1 - 0.8 = 0.2$$

□

Quick Review

After reading this section, you should be able to:

- Describe events by using set operations.
- Compute probabilities using set operations and the corresponding properties of probability.

EXERCISE 5.2

1. Verify that under classical probability, the probability of an event is the sum of the probabilities of the favourable outcomes.
2. Verify that the relative frequency of an event is the sum of the relative frequencies of the favourable outcomes.

3. Let $P(A) = 0.4$ and $P(B) = 0.2$. What is the largest possible value of $P(A \cup B)$ and that of $P(A \cap B)$?

4. A die is tossed twice. Let A be the event of getting an even number on the first throw and B the event of getting an odd number on the second throw. Then, find $P(A)$, $P(B)$, $P(A \cap B)$ and $P(A \cup B)$.

5. The Venn diagram shows the coffee preferences of 100 people. E is the set of people who like espresso and C is the set of those who like cappuccino. A person is selected at random. Find $P(E)$, $P(C)$, $P(E \cup C)$ and $P(E^c)$.

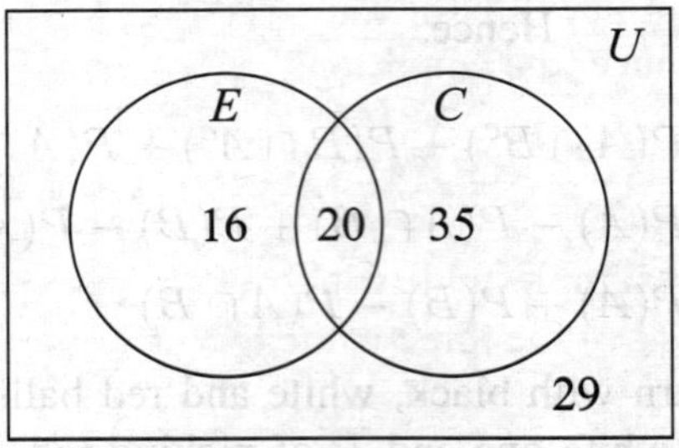

6. One-hundred fifteen children are allotted to overlapping sets D, M and L as shown. One child is selected at random. Find $P(M \cap L^c \cap D^c)$ and $P(L \cup M)$.

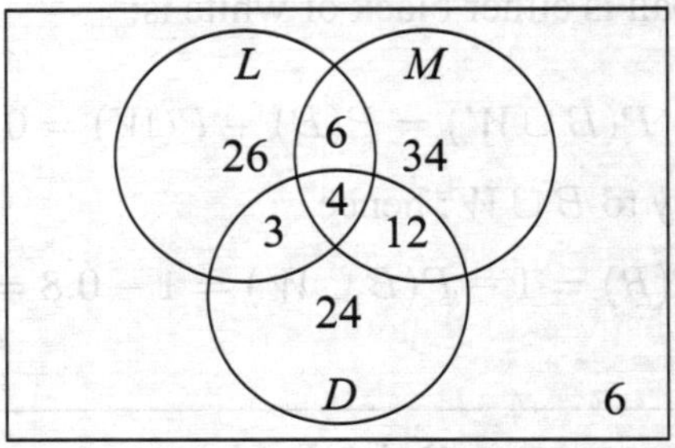

5.3 Conditional Probability

An urn contains three black balls labelled 1, 2 and 3, and three white balls labelled 4, 5 and 6. If we pick a ball at random from the urn, the probability of picking an even number is 1/2. Now, suppose we manage to sneak a look while picking the ball and see that the ball being picked is white. This extra information changes the probabilities. Since the ball is white, we know it is numbered 4 or 5 or 6. As two of these numbers are even, the probability that the ball carries an even number becomes 2/3.

In general, suppose we know that a random experiment has produced an event A and wish to know the probability that event B has also occurred. This is called the **conditional probability** of B, given that A has occurred and is denoted by $P(B|A)$. In the example given above, A is the event that the ball is white, B is the event of picking an even number and the conditional probability is $P(B|A) = 2/3$.

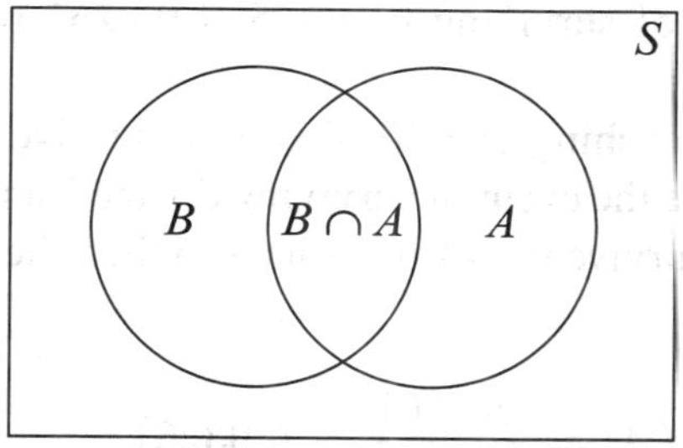

When we know that event A has occurred, in effect our sample space becomes A. From the outcomes that make up A, the ones that are favourable to B constitute $B \cap A$. Thus, for B to occur, $B \cap A$ should occur. Hence, the conditional probability of B, given A, is defined by

$$P(B|A) = \frac{P(B \cap A)}{P(A)} \tag{5.1}$$

Example 5.3.1 Consider the example with which we began this section. The event A which is known to have occurred consists of the three white balls ($A = \{4,\ 5,\ 6\}$) and thus $P(A) = 3/6 = 1/2$. The event B consists of the even numbered balls ($B = \{2,\ 4,\ 6\}$). Hence, $B \cap A = \{4,\ 6\}$ and $P(B \cap A) = 2/6 = 1/3$. Therefore,

$$P(B|A) = \frac{P(B \cap A)}{P(A)} = \frac{1/3}{1/2} = \frac{2}{3}$$

□

Example 5.3.2 A die is rolled twice. What is the probability that the total of the displayed numbers is greater than 6, given that one of the numbers is 2?

We are interested in the conditional probability $P(B|A)$, where B is the event that the total exceeds 6 and A is the event that one of the numbers is 3. Hence,

$$A = \{(2,1),\ (2,2),\ (2,3),\ (2,4),\ (2,5),\ (2,6),\ (1,2),\ (3,2),\ (4,2),\ (5,2),\ (6,2)\}$$
$$B \cap A = \{(2,5),\ (2,6),\ (5,2),\ (6,2)\}$$

Hence, $P(B|A) = \dfrac{P(B \cap A)}{P(A)} = \dfrac{4/36}{11/36} = \dfrac{4}{11}$.

□

Example 5.3.3 According to the life table prepared on the basis of the Sample Registration Survey carried out by the Government of India over 2006–10, out of a random collection of 100,000 births, the expected number of survivors to certain ages is as given as follow[1]:

Age	5	40
Survivors	93,261	87,233

[1] Source: Census of India website [W2].

Thus, at birth, the probability of surviving to age 5 is 93.261 percent and to age 40 is 87.233 percent.

What is the probability of reaching age 40 *if* you do survive to age 5? This is a conditional probability calculation. Let A be the event of surviving till age 5 and B be the event of surviving till age 40. Obviously, those who survive till 40 also survived till 5; hence, $B \subseteq A$ and so $B \cap A = B$. Therefore,

$$P(A) = \frac{93,261}{100,000} = 0.93261$$

$$P(B) = \frac{87,233}{100,000} = 0.87233$$

$$P(B \cap A) = P(B) = 0.87233$$

$$P(B|A) = \frac{P(B \cap A)}{P(A)} = \frac{0.87233}{0.93261} = 0.93536$$

How difficult are the first five years! The probability of surviving them is essentially the same as surviving the next 35 years. □

Multistage Experiments

Equation (5.1), which defines conditional probability, can be rearranged as

$$P(B \cap A) = P(B|A)P(A) \tag{5.2}$$

This rearrangement is very useful for analysing random experiments that have multiple stages.

Example 5.3.4 Two balls are picked from an urn that contains three black balls and three white balls. What is the probability that both are white?

We can view this as an experiment with two stages: First one ball is picked, and then another.

Let A be the event that the first ball is white and B be the event that the second one is white. If A is known to have occurred, then the second ball is picked from three black and two white balls. Hence $P(B|A) = 2/5$, and

$$P(\text{both balls are white}) = P(B \cap A) = P(B|A)P(A) = \frac{2}{5} \cdot \frac{3}{6} = \frac{1}{5}$$

□

Example 5.3.5 A company's share currently costs ₹100. A financial analyst models the possible changes in its value over the next two months as follows: each month the value will increase or decrease by 5 percent, with the probability of an increase being 0.6 and that of a decrease being 0.4. What is the probability that the price will not drop below ₹100?

We can depict the possibilities by the following 'tree diagram':

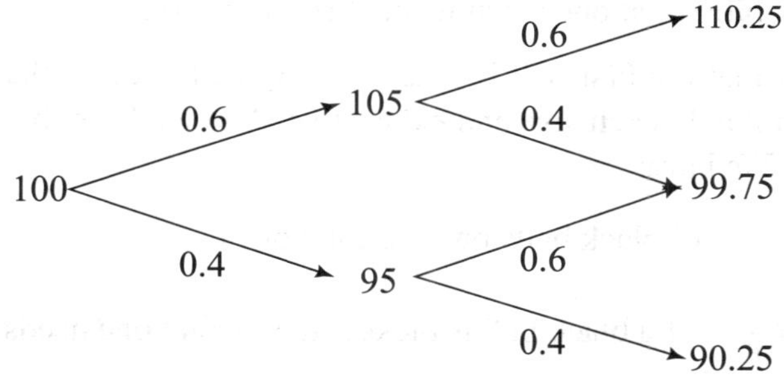

The event of interest to us is $B \cap A$, where

$$A = \{\text{Price after one month is ₹105}\}$$

$$B = \{\text{Price after two months is ₹110.25}\}$$

We are given that $P(A) = 0.6$ and $P(B|A) = 0.6$. Hence,

$$P(B \cap A) = P(B|A)P(A) = 0.6 \cdot 0.6 = 0.36$$ □

Equation (5.2) can be extended to cover combinations of three or more events. For example,

$$P(C \cap B \cap A) = P(C|B \cap A)P(B \cap A) = P(C|B \cap A)P(B|A)P(A)$$

Example 5.3.6 Three balls are picked from an urn that contains three black balls and three white balls. What is the probability that all three are white?

Let A be the event that the first ball is white, B be the event that the second one is white and C be the event that the third one is white. As in Example 5.3.4, $P(B|A) = 2/5$. Further, if A and B are known to have occurred, then the third ball is picked from three black balls and one white ball. Hence, $P(C|B \cap A) = 1/4$. Therefore,

$$P(\text{all balls are white}) = P(C \cap B \cap A) = P(C|B \cap A)P(B|A)P(A) = \frac{1}{4} \cdot \frac{2}{5} \cdot \frac{3}{6} = \frac{1}{20}$$

□

Exhaustive Events

Events $A_1, \ldots, A_n$ from a sample space $\mathcal{S}$ are called **exhaustive** if their union is $\mathcal{S}$. When the experiment is run, at least one of these events must occur. If the events are both mutually exclusive and exhaustive, they become very useful for calculating probability by enumerating cases. To see how, let $A_1, \ldots, A_n$ be mutually exclusive and exhaustive, and E be any event. Then, E is the union of $E \cap A_1, \ldots, E \cap A_n$, and these are pairwise disjoint. Hence,

$$P(E) = \sum_{i=1}^{n} P(E \cap A_i) = \sum_{i=1}^{n} P(E|A_i)P(A_i) \tag{5.3}$$

Example 5.3.7 Two balls are picked from an urn that contains three black balls and four white balls. What is the probability that one is white and the other is black?

Let A_1 be the event that the first ball is black and A_2 be the event that the first ball is white. Then A_1 and A_2 are mutually exclusive and exhaustive. And let E be the event that the two balls have different colours. We have:

$$P(A_1) = \frac{3}{7} \quad \text{(3 black balls out of a total of 7)}$$

$$P(E|A_1) = \frac{4}{6} \quad \text{(If a black ball is picked, the second one needs to be white)}$$

Similarly,

$$P(A_2) = \frac{4}{7} \quad \text{(4 white balls out of a total of 7)}$$

$$P(E|A_2) = \frac{3}{6} \quad \text{(If a white ball is picked, the second one needs to be black)}$$

Therefore,

$$P(E) = P(E|A_1)P(A_1) + P(E|A_2)P(A_2) = \frac{4}{6} \cdot \frac{3}{7} + \frac{3}{6} \cdot \frac{4}{7} = \frac{4}{7}$$

□

Independent Events

Two events A and B from the same experiment are considered **independent** if the knowledge of the occurrence of one of them does not affect the probability of the other: $P(B|A) = P(B)$ and $P(A|B) = P(A)$. Both requirements are captured by a single equation:

$$P(A \cap B) = P(A)P(B)$$

Task 5.3.1 Check that $P(A \cap B) = P(A)P(B)$ implies $P(B|A) = P(B)$ and $P(A|B) = P(A)$.

Example 5.3.8 A coin is tossed twice. A is the event that the first toss is H, B is the event that both tosses have the same result and C is the event that both tosses are H. Assuming that the four combinations HH, HT, TH and TT are equally likely, we have the following:

$$P(A) = P(HH, HT) = 1/2, \quad P(B) = P(HH, TT) = 1/2, \quad P(C) = P(HH) = 1/4$$

$$P(A \cap B) = P(HH) = 1/4, \quad P(A \cap C) = P(HH) = 1/4, \quad P(B \cap C) = P(HH) = 1/4$$

Hence, A and B are independent, A and C are not, and B and C are not. □

Example 5.3.9 Two balls are picked from an urn that contains three black balls and three white balls. Let A be the event that the first ball is white and B be the event that the second one is white. We would not expect these to be independent as the colour of the first ball would affect the proportions of the remaining ones. We can reach the same conclusion by the following calculation:

$$P(B|A) = \frac{2}{5}$$

$$P(B|A^c) = \frac{3}{5}$$

$$P(B) = P(B|A)P(A) + P(B|A^c)P(A^c) = \frac{2}{5} \cdot \frac{3}{6} + \frac{3}{5} \cdot \frac{3}{6} = \frac{1}{2} \neq P(B|A)$$

(Can you see a direct way to obtain $P(B) = 1/2$ without resorting to conditional probabilities?) □

If all probabilities are known, we can calculate conditional probabilities and establish if two events are independent. But there may be little point to just labelling events as independent or not. It is more useful to deduce independence from an understanding of the features of the experiment, and then use this to calculate probabilities.

A common application of the concept of independence is to experiments that are run repeatedly without any change in their parameters such that the outcome of a particular run does not affect the outcomes of future ones. Such repeated runs are called **independent trials** of the experiment. We usually assume that repeated tosses of a coin, or throws of a die, are of this type. If A is an event in one trial and B in another, then the assumption of independence implies $P(A \cap B) = P(A)P(B)$.

Consider N independent trials of an experiment with n equally likely outcomes. Suppose that the outcome of the ith trial is denoted o_i, and the sequence of outcomes is denoted $o_1 \cdots o_N$. Then,

$$P(o_1 \cdots o_N) = P(o_1) \cdots P(o_N) = \frac{1}{n} \cdots \frac{1}{n} = \frac{1}{n^N}$$

and so the experiment consisting of the N independent trials also has all outcomes as equally likely.

Example 5.3.10 A die is thrown thrice. What is the probability of getting exactly two sixes?

Assuming that the three trials are independent, each combination of numbers is equally likely and hence has probability $1/6^3$. There are 15 combinations which have exactly two sixes. Hence, the desired probability is $15/6^3$. □

Sampling without Replacement

When we want to learn about the distribution of some property among the members of a population (of people, or bacteria or balls in an urn and so on), the ideal experiment would involve observing or measuring that property vis-à-vis each member of the population. But this ideal is rarely attainable as the task of observing each member is too difficult or may just be undesirable for some reason. So we usually satisfy ourselves with observing only some members of the population, whom we select at random. This act of randomly selecting a subset of the population is called **sampling** and the selected members are called a **sample.** Sampling problems are of two kinds:

1. We know how some property is distributed in the population and we use that knowledge to make probability estimates regarding the sample.

2. We do not know how some property is distributed in the population, and we use our observations of the sample to make estimates about the population.

The first kind of problem is easier to solve, and we have already solved numerous instances. For example, all our problems of calculating probabilities for some balls picked from an urn have been of this kind. We shall take up the second kind in the section titled. 'Sampling'.

There are two basic types of sampling. In **sampling without replacement,** the observed member is removed from the population, and is not available for a repeat observation. Our examples of selecting multiple balls from an urn, such as Example 5.3.9, involve sampling without replacement.

Sampling with Replacement

In **sampling with replacement,** we observe a member of a population without altering or removing it. Thus, the sample space and the assigned probabilities are the same for each act of selection. We also assume that each selection is independent of the results of previous ones. Our examples involving repeated tosses of a coin or die can be viewed as instances of sampling with replacement, since the trials are taken to be independent and the sample space does not shrink.

Example 5.3.11 Consider the same urn and balls as in Example 5.3.9. Pick a ball, note its colour and return it to the urn. Now again pick a ball. In this case, the conditions are reset to the original situation after the first pick, and we take it to be a case of independent trials. If we again let A be the event that the first ball is white and B be the event that the second one is white, then $P(A \cap B) = P(A)P(B) = 1/4$. □

When we compare Examples 5.3.9 and 5.3.11, we see that the calculations are much simpler when we sample with replacement. When the size of the population is large, the probabilities of sampling with or without replacement hardly differ. Even in Examples 5.3.9 and 5.3.11, with a population of only 6, the answers of 1/5 and 1/4 are quite close. So, in practice, one may sometimes do the easier 'with replacement' calculation, even though the actual sampling is 'without replacement'.

Bayes' Formula

Consider the experiment of picking two balls from an urn that contains three black balls and four white balls. Let A be the event that the first ball is black, and E be the event that the two balls have different colours. Then $P(E|A)$ is easy to calculate, and we found it in Example 5.3.7. On the other hand, $P(A|E)$ is hard to visualise. Nevertheless, we can work it out as follows:

$$P(A|E) = \frac{P(A \cap E)}{P(E)} = \frac{P(E|A)P(A)}{P(E|A)P(A) + P(E|A^c)P(A^c)}$$

All the numbers in the last expression were already calculated in Example 5.3.7. Consequently,

$$P(A|E) = \frac{\frac{4}{6} \cdot \frac{3}{7}}{\frac{4}{6} \cdot \frac{3}{7} + \frac{3}{6} \cdot \frac{4}{7}} = \frac{2/7}{4/7} = \frac{1}{2}$$

In this example, we calculated the conditional probability $P(A|E)$ using the inverse conditional probabilities $P(E|A)$ and $P(E|A^c)$ which were much easier to obtain.

Such calculations can produce surprising results. We present an example, slightly modified from Wentzel [1].

Example 5.3.12 Two hunters, George and Fred, fired simultaneously at a bear and killed it. It was considered that George, the more experienced hunter, had a 75 percent chance of hitting the bear, while Fred had only a 50 percent chance. On inspection, it was found that only one of them had hit the bear. How should the reward of 100 rubles be divided among them?

It is tempting to split the award in proportion to the hunters' probabilities of hitting the bear, giving 60 rubles to George and 40 to Fred. But this does not take into account our knowledge that one of them missed. What we need are the conditional probabilities of George and Fred hitting the bear, given that exactly one of them has done so.

Let G be the event that George hit the bear, and E the event that exactly one hunter hit the bear. Then,

$$P(E|G) = 0.5$$

$$P(E|G^c) = 0.5$$

$$P(G|E) = \frac{P(E|G)P(G)}{P(E|G)P(G) + P(E|G^c)P(G^c)} = \frac{0.5 \times 0.75}{0.5 \times 0.75 + 0.5 \times 0.25} = 0.75$$

Similarly, if F is the event that Fred hit the bear, we find that

$$P(E|F) = 0.25$$

$$P(E|F^c) = 0.75$$

$$P(F|E) = \frac{P(E|F)P(F)}{P(E|F)P(F) + P(E|F^c)P(F^c)} = \frac{0.25 \times 0.5}{0.25 \times 0.5 + 0.75 \times 0.5} = 0.25$$

Therefore, the correct distribution is 75 rubles to George and only 25 to Fred! □

The next example illustrates one of the difficulties in dealing with rare diseases.

Example 5.3.13 A kind of cancer is estimated to occur in 0.01 percent of the population. A new method of detecting it early is advertised as highly reliable: if you have the cancer, there is a 99.9 percent that probability the test will give a positive result, and if you don't have it, there is a 99.9 percent probability that the test will turn out negative. You go for the test and the result is positive. What is the probability you actually have the cancer?

Let T be the event of a positive test and C be the event that you have the cancer. Then,

$$P(T|C) = 0.999, \qquad P(T|C^c) = 1 - 0.999 = 0.001$$

Hence,

$$P(C|T) = \frac{P(T|C)P(C)}{P(T|C)P(C) + P(T|C^c)P(C^c)} = \frac{0.999 \times 0.0001}{0.999 \times 0.0001 + 0.001 \times 0.9999} = 0.0908$$

There is only about a 9 percent chance that you actually have cancer. Time for caution, but not for despair! □

Our approach in this section can be given a more general setting. Let events $A_1, \ldots, A_n$ from a sample space $\mathcal{S}$ be exhaustive and mutually exclusive and let E be any event from $\mathcal{S}$. Then, we have **Bayes' formula:**

$$P(A_k|E) = \frac{P(E|A_k)P(A_k)}{\sum_{i=1}^{n} P(E|A_i)P(A_i)} \tag{5.4}$$

It is useful to think about Bayes' formula in terms of **hypotheses** and **evidence.** The exhaustive and mutually exclusive events A_i constitute competing hypotheses, exactly one of which can hold at a time. The event E constitutes the evidence. The probability of each hypotheses is known, as well as the conditional probability of the evidence given the occurrence of any hypothesis. Bayes' formula answers the question: If evidence E is observed, what is the conditional probability of each hypothesis?

Example 5.3.14 Consider the situation of Example 5.3.5, but now suppose the analyst wants to estimate the probability that the share's price was above ₹ 100 after two months, given that its price after three months is above ₹ 100.

Let P_i denote the price after i months. The tree diagram is

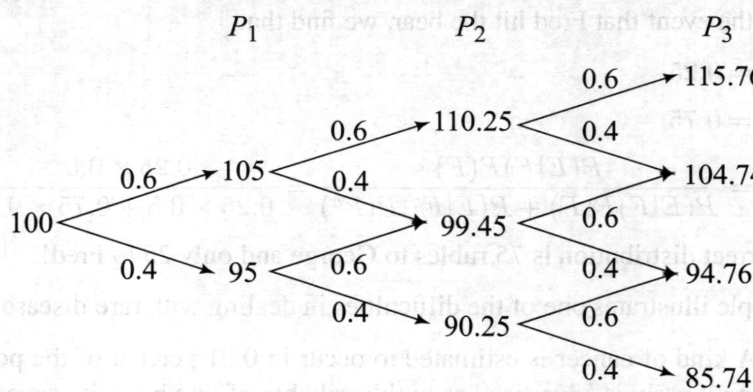

The 'evidence' is the event $E = \{P_3 > 100\} = \{P_3 = 115.76 \text{ or } 104.74\}$. The 'hypotheses' are the events $A_1 = \{P_2 = 110.25\}$, $A_2 = \{P_2 = 99.75\}$ and $A_3 = \{P_2 = 90.25\}$. We need to calculate $P(A_1|E)$.

A glance at the tree diagram gives the following conditional probabilities:

$$P(E|A_1) = 1, \quad P(E|A_2) = 0.6, \quad P(E|A_3) = 0$$

Also,

$$P(A_1) = 0.6 \cdot 0.6 = 0.36, \quad P(A_3) = 0.4 \cdot 0.4 = 0.16, \quad P(A_2) = 1 - P(A_1) - P(A_3) = 0.48$$

Hence,

$$P(A_1|E) = \frac{P(E|A_1)P(A_1)}{P(E|A_1)P(A_1) + P(E|A_2)P(A_2) + P(E|A_3)P(A_3)}$$
$$= \frac{0.36}{0.36 + 0.6 \cdot 0.48} = 0.56$$

□

Quick Review

After reading this section, you should be able to:

- Use the concept of conditional probability to incorporate partial information into a probability calculation.
- Identify independent events and use them to simplify calculations.
- Use Bayes' formula to find the probabilities of hypotheses, given the occurrence of certain evidence.

EXERCISE 5.3

1. A die is rolled twice. What is the probability that the total of the displayed numbers is greater than 6, given that
   (a) The first roll results in 2?
   (b) The total of the displayed numbers is less than 8?

2. Here are more data from the life table based on the 2006–10 Sample Registration Survey in India. These are the expected survival numbers for men and women from an initial cohort of 100,000 each:

Age	5	40	60	80
Men	93,595	86,879	71,429	24,862
Women	92,892	87,578	77,981	34,001

   Complete the following table showing the probability (rounded to three decimal places) of surviving from one age to the next:

Age	0 to 5	5 to 40	40 to 60	60 to 80
Men	0.936			
Women	0.929			

3. Find $P(C|E)$ and $P(E|C)$, where C and E are the events in the sample space S, and the numbers of elements of various parts of the sets are shown below, and assuming classical probability:

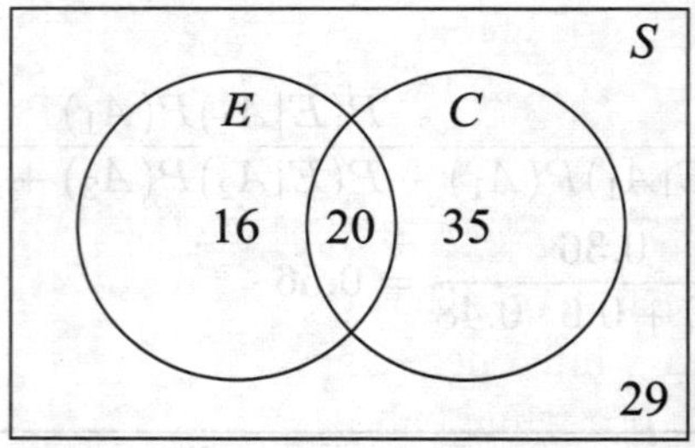

4. Show that $P(D \cap C \cap B \cap A) = P(D|C \cap B \cap A) \cdot P(C|B \cap A) \cdot P(B|A) \cdot P(A)$.
5. Are events A and B independent? The numbers denote the outcomes of that type. Assume classical probability.

(a)

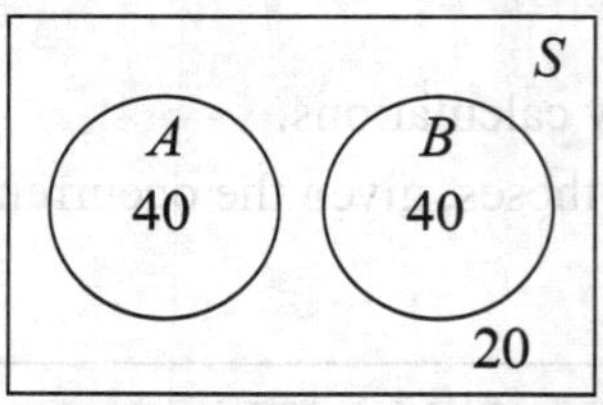

(b)

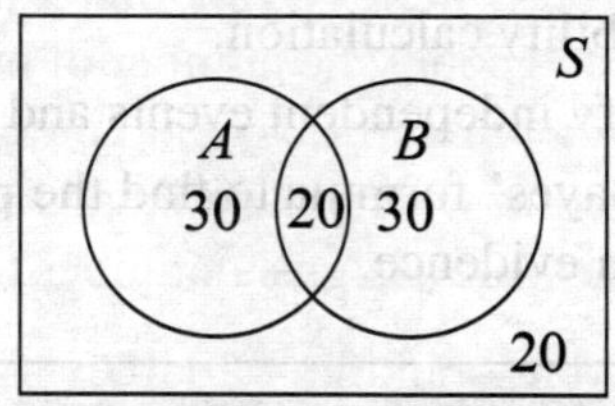

6. Are events A and B independent? The numbers denote the outcomes of that type. Assume classical probability.

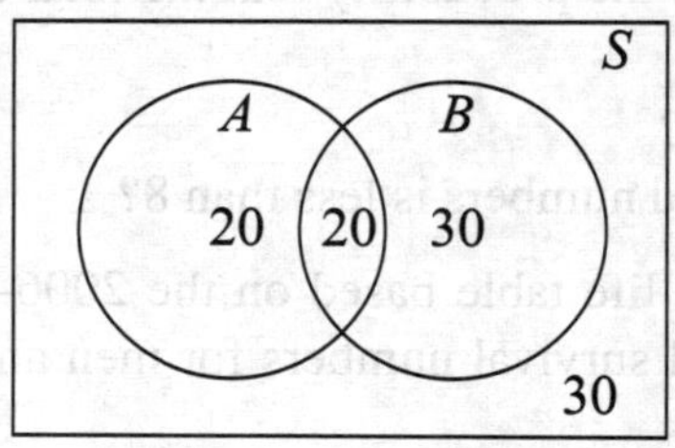

7. True or false?
   (a) $P(A^c|B) = 1 - P(A|B)$.
   (b) $P(A|B^c) = 1 - P(A|B)$.
   (c) If A and B are independent, so are A and B^c.
   (d) If A and B are independent, so are A^c and B^c.
8. A die is thrown four times. Would it be advantageous to bet on at least one occurrence of 6? (This was one of the questions that led Blaise Pascal and Pierre Fermat to methodically probe probability in the seventeenth century.)
9. A pair of dice is thrown 24 times. Would it be advantageous to bet on at least one occurrence of a pair of sixes? (This was another question that motivated Pascal and Fermat.)
10. In Example 5.3.12, suppose that the probabilities for George and Fred hitting the bear were 1 and 0.99, respectively, instead of 0.75 and 0.5. How should the reward be distributed?

11. Assume that births of boys and girls are equally likely. A family has two children. What is the probability of both children being boys, if we already know that:
    (a) The elder one is a boy.
    (b) One of them is a boy.

12. Two urns contain five balls each. Urn A has three black balls and two white ones, while urn B has one black ball and four white ones. One of the urns is selected at random, and then a random ball is picked from it. If a white ball is picked, what is the probability that urn A was the selected one?

5.4 Random Variables, Expectation and Variance

Quite often, the outcome of a random experiment is given by a number. For example, we toss a coin three times and count the number of heads. Or we pick a ball from an urn and read the number written on it. Even when the original outcomes are not numbers, we can label them by numbers. When we toss a coin, we can label heads as 1 and tails as 0.

A **random variable** assigns a number to each outcome in a sample space.

Example 5.4.1 Many random variables may be attached to the same sample space. Suppose that we toss a coin twice. The number of heads is a random variable. The number of tails is another. We could use these to create further examples. For example, the square of the number of heads is also a random variable (and we will see that it is a useful one). □

We will denote random variables by capital letters such as X, Y, Z. If a random variable assigns a number x to an outcome w, we will write $X(w) = x$.

Example 5.4.2 Let X be the number of heads when we toss a coin twice. Then,

$$X(HH) = 2, \quad X(HT) = 1, \quad X(TH) = 1, \quad X(HH) = 0$$

□

Our first interest is in the probabilities of the different values that a random variable can take. By the probability of a random variable X taking a particular value x, we mean the probability of the event E consisting of all the outcomes which are mapped to x by X. And the event E is denoted by $\{X = x\}$.

Example 5.4.3 Let X be the number of heads when we toss a coin twice. Then,

$$P(X = 2) = P(HH) = 1/4$$
$$P(X = 1) = P(HT, TH) = 1/2$$
$$P(X = 0) = P(TT) = 1/4$$

□

We can also be interested in the probabilities of collections of ranges of a random variable. We use some obvious notation for the events corresponding to these ranges:

$\{X \leq x\}$: outcomes mapped to numbers which are less than or equal to x

$\{X < x\}$: outcomes mapped to numbers which are less than x

$\{X \geq x\}$: outcomes mapped to numbers which are greater than or equal to x

$\{X > x\}$: outcomes mapped to numbers which are greater than x

$\{x_1 \leq X \leq x_2\}$: outcomes mapped to numbers which are greater than or equal to x_1 and also less than or equal to x_2

Example 5.4.4 Let X be the number of heads when we toss a coin twice. Then,

$$P(X \leq 2) = P(HH, HT, TH, TT) = 1$$
$$P(X \leq 1) = P(HT, TH, TT) = 3/4$$
$$P(1 \leq X \leq 2) = P(HT, TH, HH) = 3/4$$

□

We use the symbol f_X to stand for the rule which associates the probability $P(X = x)$ with the value x of X, expressing this by $f_X(x) = P(X = x)$. The rule f_X is called the **probability distribution function** of X, and this name is often abbreviated to **pdf.**

Example 5.4.5 Let X be the number of heads when we toss a coin thrice. Then, the possible values of X are 0, 1, 2 and 3:

$$f_X(3) = P(X = 3) = P(HHH) = 1/8$$
$$f_X(2) = P(X = 2) = P(HHT, HTH, THH) = 3/8$$
$$f_X(1) = P(X = 1) = P(HTT, THT, HTT) = 3/8$$
$$f_X(0) = P(X = 0) = P(TTT) = 1/8$$

□

Note that the values of the probability distribution function must sum to 1, since the sum gives the probability of the entire sample space.

Apart from probabilities of individual values, or ranges of values, of a random variable, we may be interested in getting an overall picture of how its values are distributed. This requires addressing questions such as: If the experiment has many independent trials, then what is likely to be the average of the observed values of the random variable? And will the various observed values be clustered close to this average or will they be widely dispersed?

Let us recall from the previous chapter the concepts of mean, variance and standard deviation of data. If some data consist of numbers $x_1, \ldots, x_n$, each with corresponding frequency $f_1, \ldots, f_n$, then with $N = \sum_{i=1}^{n} f_i$,

$$\text{Mean: } \overline{x} = \sum_{i=1}^{n} \frac{f_i}{N} x_i$$

$$\text{Variance: } \sigma^2 = \sum_{i=1}^{n} \frac{f_i}{N} (x_i - \overline{x})^2$$

$$\text{Standard deviation: } \sigma = \sqrt{\sigma^2}$$

Now, consider an experiment with N independent trials, and an associated random variable X with possible values $x_1, \ldots, x_n$. Then, the probability $f_X(x_i)$ can be viewed as a prediction of the relative frequency of the value x_i among the actually observed values of X. Hence, the following quantities may be viewed as predictions of the corresponding ones for the actually observed values:

$$\text{Mean or expectation: } E[X] = \sum_{i=1}^{n} f_X(x_i)x_i$$

$$\text{Variance: } \sigma_X^2 = \sum_{i=1}^{n} f_X(x_i)(x_i - E[X])^2$$

$$\text{Standard deviation: } \sigma_X = \sqrt{\sigma_X^2}$$

The expectation of X is also denoted by μ_X. The variance of X is the expectation of the mean squared distance between the values of X and their mean position, and thus measures the dispersal of the values of X.

Example 5.4.6 We toss a coin thrice. Let X be the fraction of throws which result in heads. The pdf of X is:

$$f_X(0) = 1/8, \quad f_X(1/3) = 3/8, \quad f_X(2/3) = 3/8, \quad f_X(1) = 1/8$$

Hence,

$$\begin{aligned} E[X] &= \frac{1}{8}\cdot 0 + \frac{3}{8}\cdot\frac{1}{3} + \frac{3}{8}\cdot\frac{2}{3} + \frac{1}{8}\cdot 1 = \frac{1}{2} \qquad \text{(as expected!)} \\ \sigma_X^2 &= \frac{1}{8}\left(0 - \frac{1}{2}\right)^2 + \frac{3}{8}\left(\frac{1}{3} - \frac{1}{2}\right)^2 + \frac{3}{8}\left(\frac{2}{3} - \frac{1}{2}\right)^2 + \frac{1}{8}\left(1 - \frac{1}{2}\right)^2 = \frac{1}{12} \\ \sigma_X &= \frac{1}{2\sqrt{3}} \end{aligned}$$

□

One can also calculate expectations of random variables that are derived from X. Three important instances are as follows:

1. For any real number k, $E[X + k] = E[X] + k$. (If every value of X is shifted by k, the expected value is also shifted by k.)
2. For any real number k, $E[kX] = k\,E[X]$. (If every value of X is multiplied by k, the expected value is also multiplied by k.)
3. $E[X^2] = \sum_{i=1}^{n} f_X(x_i)x_i^2$. (For the expectation of X^2, multiply each value of X^2 by the probability of the corresponding X value.)

Task 5.4.1 Show that $\sigma_X^2 = E[X^2] - (E[X])^2$.

Example 5.4.7 We toss a coin four times and let X be the fraction of throws resulting in heads. We leave it to you to verify that the pdf of X is given by

$$f_X(0) = 1/16, \quad f_X(1/4) = 4/16, \quad f_X(2/4) = 6/16, \quad f_X(3/4) = 4/16, \quad f_X(1) = 1/16$$

Hence,

$$E[X] = \frac{1}{16} \cdot 0 + \frac{4}{16} \cdot \frac{1}{4} + \frac{6}{16} \cdot \frac{2}{4} + \frac{4}{16} \cdot \frac{3}{4} + \frac{1}{16} \cdot 1 = \frac{1}{2} \quad \text{(again, as expected!)}$$

$$E[X^2] = \frac{1}{16} \cdot 0 + \frac{4}{16} \cdot \frac{1}{16} + \frac{6}{16} \cdot \frac{4}{16} + \frac{4}{16} \cdot \frac{9}{16} + \frac{1}{16} \cdot 1 = \frac{5}{16}$$

$$\sigma_X^2 = \frac{5}{16} - \frac{1}{4} = \frac{1}{16}$$

$$\sigma_X = \frac{1}{4}$$

Comparing this to the case of three throws, we see that the variance has decreased. This is in accordance with our intuition that with more throws, the relative frequency of heads should cluster more closely around 1/2. □

Quick Review

After reading this section, you should be able to:

- Use expectation to describe average behaviour.
- Use standard deviation and variance to describe dispersion from the mean.

EXERCISE 5.4

1. A pair of dice is thrown. The random variable X equals the number of prime numbers that show. What are the possible values of X? What is its pdf?
2. Two balls are picked from an urn that contains three black and three white balls. The random variable C counts the number of white balls that are picked. What is its pdf?
3. Find the mean and variance of the random variables in the first two exercises.
4. The probability of picking a white ball from an urn is known to be 0.6. The random variable W has the value 1 if a white ball is picked and the value 0 if a ball of another colour is picked. Find the mean and variance of W.
5. A share is currently priced at ₹100. Its value has a 0.6 probability of increasing by 100.4 probability of decreasing by 10 percent over the next month. What is its expected price after a month?

6. Two balls are tossed into three boxes, such that each ball is equally likely to fall into any box. The random variable X is the number of balls that lands in the first box. Find the pdf, mean and variance of X.

7. Check that the variance of $X + k$ is σ_X^2.

8. Check that the variance of kX is $k^2\sigma_X^2$.

9. The pdf of X is given by $f_X(-1) = f_X(1) = p$ and $f_X(0) = 1 - 2p$, with $0 <\leq p \leq 1/2$. Which value of p makes the variance as large as possible?

5.5 Jointly Distributed Random Variables

So far we have studied individual random variables. The next step is to study the relationship between a pair of random variables X and Y. For example, we may wish to know if the knowledge of values of X would improve our prediction of values of Y. Or we may want the probability that X and Y will obey a certain rule such as $X^2 + Y^2 \leq 1$. We will take up this joint study of random variables in this section.

Random variables X and Y are called **jointly distributed** if they are associated with the same random experiment. Their **joint probability distribution function,** or **joint pdf,** is defined by

$$f_{X,Y}(x, y) = P(X = x \text{ and } Y = y)$$

Example 5.5.1 A coin is tossed once. X is the number of heads and Y is the number of tails. Then, the joint pdf is given by

$$f_{X,Y}(0,0) = 0, \quad f_{X,Y}(1,0) = 1/2, \quad f_{X,Y}(0,1) = 1/2, \quad f_{X,Y}(1,1) = 0$$

□

Example 5.5.2 A die is tossed once, with outcome w. X and Y are defined as follows:

$$X(w) = \begin{cases} 1 & \text{if } w = 1,2,3 \\ 0 & \text{if } w = 4,5,6 \end{cases}, \qquad Y(w) = \begin{cases} 1 & \text{if } w = 2,4,6 \\ 0 & \text{if } w = 1,3,5 \end{cases}$$

Then the joint pdf is given by

$$f_{X,Y}(0,0) = 0, \quad f_{X,Y}(1,0) = 1/3, \quad f_{X,Y}(0,1) = 1/3, \quad f_{X,Y}(1,1) = 1/3$$

□

When X and Y are jointly distributed, their individual pdf's are called their **marginal distributions.** If the joint pdf is known, the marginal distributions can be recovered from it. Let X take values $x_1, \ldots, x_m$ and Y take values $y_1, \ldots, y_n$. Then the events $Y = y_j$ are exhaustive and mutually exclusive. Hence,

$$f_X(x_i) = P(X = x_i) = \sum_{j=1}^{n} P(X = x_i,\ Y = y_j) = \sum_{j=1}^{n} f_{X,Y}(x_i, y_j)$$

Similarly, $f_X(y_j) = \sum_{i=1}^{m} f_{X,Y}(x_i, y_j)$.

On the other hand, knowing the marginal distributions is not enough to know the joint pdf, as illustrated below.

Example 5.5.3 The tables below show two joint distributions for X and Y, leading to the same marginal distributions. We have left it for you to compute the marginal distributions and verify this fact.

	$Y=0$	$Y=1$
$X=0$	0	1/2
$X=1$	1/2	0

	$Y=0$	$Y=1$
$X=0$	1/4	1/4
$X=1$	1/4	1/4

□

The two joint distributions in this example capture very different relationships. In the first one, the value of X also determines the value of Y. For example, we have $f_{X,Y}(0,0) = 0$ and so if X takes the value 0, Y must take the value 1. In the second one, the value of X tells us nothing about Y: both values of Y stay equally likely, just as they are without the knowledge of X. At the most basic level, we would like to know if the values tend to move in the same direction, or in opposite directions, or if they are independent of each other. And if there is a connection, how strong is it?

To this end, we define the **covariance** of X and Y by

$$\mathrm{Cov}[X,Y] = E[(X-\mu_X)(Y-\mu_Y)] = \sum_{i,j} f_{X,Y}(x_i,y_j)(x_i-\mu_X)(y_j-\mu_Y)$$

A positive covariance indicates that the two variables tend to move in the same direction, while a negative covariance indicates that they tend to move in opposite directions.

Example 5.5.4 Consider jointly distributed X and Y, whose joint pdf is tabulated as follows:

	$Y=0$	$Y=1$	$Y=2$
$X=0$	1/3	1/6	0
$X=1$	0	1/6	1/3

By inspecting the table, we see that, when $X=0$, the lower Y values are more likely, and when $X=1$, the higher values of Y are more likely. To measure this connection, we calculate covariance. First, we calculate the marginal distributions of X and Y:

k	0	1
$P(X=k)$	$\frac{1}{3}+\frac{1}{6}+0=\frac{1}{2}$	$0+\frac{1}{6}+\frac{1}{3}=\frac{1}{2}$

k	0	1	2
$P(Y=k)$	$\frac{1}{3}+0=\frac{1}{3}$	$\frac{1}{6}+\frac{1}{6}=\frac{1}{3}$	$0+\frac{1}{3}=\frac{1}{3}$

Next, we obtain the expectations of X and Y:

$$E[X] = \frac{1}{2}\cdot 0 + \frac{1}{2}\cdot 1 = \frac{1}{2}, \qquad E[Y] = \frac{1}{3}\cdot 0 + \frac{1}{3}\cdot 1 + \frac{1}{3}\cdot 2 = 1$$

Now, we can calculate the covariance:

$$\begin{aligned}\text{Cov}[X,Y] &= \frac{1}{3}(0-\frac{1}{2})(0-1) + \frac{1}{6}(0-\frac{1}{2})(1-1) + \frac{1}{6}(1-\frac{1}{2})(1-1) + \frac{1}{3}(1-\frac{1}{2})(2-1)\\ &= \frac{1}{6} + \frac{1}{6} = \frac{1}{3}\end{aligned}$$

As expected, it is positive. □

Example 5.5.5 In the above example, suppose that the unit used to measure X were halved, so that the possible values of X became 0 and 2. Then, $E[X]$ becomes 1, and the covariance becomes $2/3$. □

The magnitude of the covariance depends on the choice of the units used to measure X and Y, with smaller units leading to larger values of covariance. Therefore, we modify covariance to make it independent of the unit. The new quantity is called **correlation** and is defined by

$$\rho_{X,Y} = \frac{\text{Cov}[X,Y]}{\sigma_X \sigma_Y}$$

Example 5.5.6 Consider the joint variables X and Y of Example 5.5.4. Their variances are as follows

$$\sigma_X^2 = E[X^2] - E[X]^2 = \frac{1}{2}\cdot 0^2 + \frac{1}{2}\cdot 1^2 - \frac{1}{2^2} = \frac{1}{4}$$

$$\sigma_Y^2 = E[Y^2] - E[Y]^2 = \frac{1}{3}\cdot 0^2 + \frac{1}{3}\cdot 1^2 + \frac{1}{3}\cdot 2^2 - 1^2 = \frac{2}{3}$$

Hence, the correlation is $\rho_{X,Y} = \dfrac{\text{Cov}[X,Y]}{\sigma_X\sigma_Y} = \dfrac{1/3}{(1/2)\sqrt{2/3}} = \sqrt{2/3} \approx 0.82.$ □

The correlation $\rho_{X,Y}$ takes values between -1 and 1. The extreme values of ± 1 occur when X and Y have a simple relationship of the form $Y = aX + b$ (in particular, one completely determines the other). When $a > 0$, we have $\rho_{X,Y} = 1$, and when $a < 0$, we have $\rho_{X,Y} = -1$.

We define jointly distributed random variables X and Y to be **independent** if their outcomes are independent, that is,

$$f_{X,Y}(x,y) = f_X(x) f_Y(y)$$

As was the case with events, independence of random variables is usually assumed when it seems plausible given our understanding of some situation. And then we use the assumption of independence to explore the joint behaviour of the two variables.

If X, Y are independent, their covariance (and hence correlation) is zero:

$$\text{Cov}[X,Y] = \sum_{i,j} f_{X,Y}(x_i, y_j)(x_i - \mu_X)(y_j - \mu_Y)$$

$$= \sum_{i,j} f_X(x_i) f_Y(y_j)(x_i - \mu_X)(y_j - \mu_Y)$$

$$= \sum_i f_X(x_i)(x_i - \mu_X) \sum_j f_Y(y_j)(y_j - \mu_Y)$$

$$= E[X - \mu_X]\, E[Y - \mu_Y] = 0$$

When the correlation between two random variables is zero, we call them **uncorrelated.** Independent random variables are uncorrelated. However, uncorrelated variables are not always independent, as illustrated in Exercise 1.5.2.

Sums of Random Variables

Suppose X and Y are jointly distributed random variables. Then, their sum $X+Y$ is another random variable. Its expectation is simply the sum of the individual ones:

$$E[X+Y] = E[X] + E[Y]$$

This is an expected result. If we expect to earn ₹2,000 from one investment, and ₹5,000 from another, then we expect to earn ₹7,000 from the combined investment. The nature of correlation of the investments does not matter. However, the correlation does matter when we consider the variance of $X+Y$. If X, Y are positively correlated, their fluctuations reinforce each other and $X+Y$ has higher variance. If they are negatively correlated, their fluctuations cancel each other and $X+Y$ has lower variance. We calculate the exact relationship in what follows. Since variance is unaffected by a constant shift in the variables, we can assume $E[X] = E[Y] = 0$, and this also gives $E[X+Y] = 0$.

$$\sigma_{X+Y}^2 = E[(X+Y)^2] = E[X^2] + E[Y^2] + 2E[XY]$$

$$= \sigma_X^2 + \sigma_Y^2 + 2\,\mathrm{Cov}[X,Y]$$

$$= \sigma_X^2 + \sigma_Y^2 + 2\sigma_X \sigma_Y \rho_{X,Y}$$

When X, Y are independent, this simplifies to $\sigma_{X+Y}^2 = \sigma_X^2 + \sigma_Y^2$ since covariance is zero.

The concepts of joint probability distribution and independence can be extended to any number of random variables. Suppose $X_1, \ldots, X_n$ are random variables associated with the same random experiment. Their joint pdf is defined by

$$f_{X_1,\ldots,X_n}(x_1, \ldots, x_n) = P(X_1 = x_1,\ \ldots,\ X_n = x_n)$$

The expectation of their sum is again the sum of their expectations:

$$E[X_1 + \cdots + X_n] = E[X_1] + \cdots + E[X_n] \tag{5.5}$$

The random variables are deemed **independent** if their joint pdf is the product of the marginal ones:

$$f_{X_1,\ldots,X_n}(x_1, \ldots, x_n) = f_{X_1}(x_1) \ldots f_{X_n}(x_n)$$

When $X_1, \ldots, X_n$ are independent, it can be shown that that each pair X_i, X_j is independent, the covariance of each pair is zero and the variance of the sum is the sum of the variances:

$$\sigma^2_{X_1+\cdots+X_n} = \sigma^2_{X_1} + \cdots + \sigma^2_{X_n} \tag{5.6}$$

Quick Review

After reading this section, you should be able to:

- Calculate marginal probability for one random variable when the value of a jointly distributed random variable is known.
- Use covariance and correlation to describe the joint behaviour of two random variables.
- Identify independent random variables.

EXERCISE 5.5

1. Let the joint pdf of X and Y be given by the following table:

	$Y=-1$	$Y=0$	$Y=1$
$X=0$	$1/8$	$1/4$	$1/8$
$X=1$	$1/8$	$1/4$	$1/8$

(a) Find the probability that $XY = 0$.
(b) Find the marginal pdfs of X and Y.
(c) Find the covariance and correlation of X and Y.
(d) Are X and Y independent?

2. Let the joint pdf of X and Y be given by the table below:

	$Y=-1$	$Y=0$	$Y=1$
$X=0$	0	$1/2$	0
$X=1$	$1/4$	0	$1/4$

(a) Find the marginal pdf's of X and Y.
(b) Find the covariance and correlation of X and Y.
(c) Are X and Y independent?

3. Consider the joint distribution in the previous exercise.

(a) Show that the pdf's of $X+Y$ are given by

k	-1	0	1	2
$P(X+Y=k)$	0	$3/4$	0	$1/4$

(b) Verify that $E[X+Y] = E[X] + E[Y]$
(c) Verify that $\sigma^2_{X+Y} = \sigma^2_X + \sigma^2_Y + 2\text{Cov}[X, Y]$

4. Two balls are drawn from an urn that contains three balls, numbered 1–3. X is the number on the first ball and Y is the larger of the two numbers drawn.
   (a) Find the joint pdf of X and Y.
   (b) Find the covariance of X and Y.

5. Show that $\text{Cov}[X+Y,Z] = \text{Cov}[X,Z] + \text{Cov}[Y,Z]$.

6. Let X, Y, Z be uncorrelated random variables with common variance σ^2. Find the correlation between $X+Y$ and $X+Z$.

5.6 Bernoulli and Binomial Random Variables

A random variable X is called a **Bernoulli variable with parameter** p if it takes only two values, 0 and 1, with p being the probability of $X = 1$. The outcome 1 is called 'success' and the outcome 0 is called 'failure'. A familiar example is the toss of a coin with $X = 1$ associated with heads, $X = 0$ associated with tails, and $p = 1/2$. Another example is the throw of a die, with success defined to be a result of 6. In this case, $p = 1/6$.

The expectation and variance of a Bernoulli variable X with parameter p are easy to calculate:

$$E[X] = 0 \cdot P(X=0) + 1 \cdot P(X=1) = 0 \cdot (1-p) + 1 \cdot p = p$$

$$\sigma_X^2 = E[X^2] - E[X]^2 = 0^2 \cdot P(X=0) + 1^2 \cdot P(X=1) - p^2 = p - p^2 = p(1-p)$$

Now, consider a random experiment with an associated Bernoulli variable X. If we run n independent trials of the experiment, we create n independent Bernoulli variables $X_1, \ldots, X_n$, with each having the same pdf as X. Let $B = X_1 + \cdots + X_n$. Then, B counts the total number of successes over the repeated experiments, and is called a **Binomial variable with parameters** n **and** p. We have

$$E[B] = E[X_1] + \cdots + E[X_n] = np$$

$$\sigma_B^2 = \sigma_{X_1}^2 + \cdots + \sigma_{X_n}^2 = np(1-p)$$

The random variable B/n counts the relative frequency of successes. We have

$$E[B/n] = E[B]/n = p$$

$$\sigma_{B/n}^2 = \sigma_B^2/n^2 = \frac{p(1-p)}{n}$$

Note that the variance of B/n gets smaller as n increases. This provides theoretical backing to the empirical observation that the relative frequency of heads clusters more tightly around $1/2$ as we increase the number of tosses. However, we would like to know more. We would like to have probability estimates about how close to p the relative frequency B/n is likely to be.

For these probability calculations, we need the pdf of the Binomial variables. We will not develop the formula for this pdf. Instead, in the spirit of Chapter 4, we will see how to use Microsoft Excel for such work.

The relevant Excel function is Binom.Dist(k, n, p, 0). It gives $P(B = k)$, where B is a Binomial random variable with parameters n and p. The number k can take values 0, 1, ... , n. A variation is Binom.Dist(k, n, p, 1), which gives $F_B(k) = P(B \leq k)$, the **cumulative distribution function** of B.[2]

Example 5.6.1 A coin is tossed 10 times. We wish to know the probability that the number of heads will be be between 4 and 6. Let B be the number of times the heads occur. Then, B is a Binomial random variable with parameters $n = 10$ and $p = 0.5$. We use Excel to calculate the probabilities of B:

Binom.Dist(4, 10, 0.5, 0)	Binom.Dist(5, 10, 0.5, 0)	Binom.Dist(6, 10, 0.5, 0)
0.205	0.246	0.205

Hence, $P(4 \leq B \leq 6) = P(B = 4) + P(B = 5) + P(B = 6) = 0.205 + 0.246 + 0.205 = 0.656$.

This method of finding the probability of a range of values by summing up the individual probabilities becomes inconvenient when the range is large. The function Binom.Dist(k, n, p, 1) provides an alternate approach (Figure 5.6.1):

$$\begin{aligned} P(4 \leq B \leq 6) &= P(B \leq 6) - P(B \leq 3) \\ &= \text{Binom.Dist(6, 10, 0.5, 1)} - \text{Binom.Dist(3, 10, 0.5, 1)} \\ &= 0.828 - 0.172 = 0.656 \end{aligned}$$

□

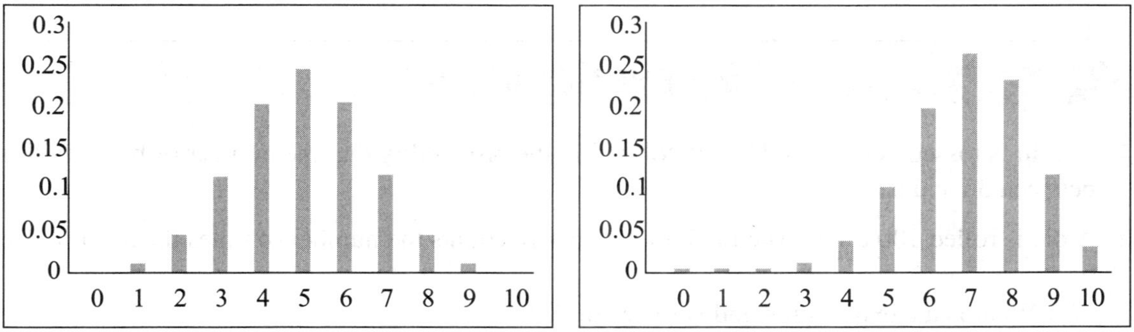

Figure 5.6.1 Probabilities for Binomial random variables with $n = 10$ calculated by the Binom.Dist function. The random variables take values 0, 1, ... , 10. The chart on the left has $p = 1/2$ and has a symmetric shape because failure and success are equally likely. The chart on the right has $p = 0.7$. In this chart, since success is more likely than failure, the bars are higher towards the right.

[2] The Binom.Dist function is also present in Calc, with the same syntax.

Excel also provides a specific function for calculating the probability of a range: Binom.Dist.Range(n, p, a, b) gives $P(a \leq B \leq b)$. In the setting of the above example, $P(4 \leq B \leq 6) =$ Binom.Dist.Range(10, 0.5, 4, 6).[3]

Example 5.6.2 A coin is tossed 2,000 times. What is the probability that the relative frequency of heads will be within 0.02–0.5?

Let B represent the number of heads that occur. Then, B is a binomial random variable with parameters $n = 2{,}000$ and $p = 0.5$. The relative frequency of heads is $S = B/n$. Now,

$$P(0.48 \leq S \leq .52) = P(0.48 \leq B/2000 \leq .52)$$
$$= P(960 \leq B \leq 1040)$$

= Binom.Dist.Range(2000, 0.5, 960, 1040)

$= 0.93$

□

Quick Review

After reading this section, you should be able to:

- Identify Bernoulli and Binomial variables associated with experiments with independent trials.
- Calculate expectation and variance of Bernoulli and Binomial variables.
- Use Excel to calculate probabilities related to Binomial random variables.

EXERCISE 5.6

1. A coin is tossed 100 times. Use Excel to find the probability that the number of heads will be between 35 and 65.
2. A die is rolled 100 times. The random variable X counts the number of times the result is 1 or 6.
   (a) What is the expectation and variance of X?
   (b) Use Excel to find the probability that the value of X will be between 30 and 40.
3. What is the probability that a student can get at least 60 percent marks by just guessing in a True-False test with 10 questions? What is the probability if there are 20 questions?
4. Four independent trials are conducted of a Bernoulli variable. X counts the number of times a success is immediately followed by a failure. Find the pdf of X.
5. X is a Binomial variable with $n = 25$ and $p = 0.2$. Find $P(\mu_X - \sigma_X \leq X \leq \mu_X + \sigma_X)$.

[3] The Binom.Dist.Range function is not present in Calc.

6. X is a Binomial variable. Keeping n fixed, which value of p leads to maximum variance?
7. X is a Binomial variable with parameters n and p. What can you say about the random variable $Y = n - X$?

5.7 Sampling

Imagine that some candidates contest an election in a constituency with several thousand voters. A newspaper wishes to run a weekly survey on their relative popularity. It is impractical for the newspaper to find out the opinion of each voter, and so it only approaches a few. The question is, of course, how many to approach in order for the result to have a reasonable accuracy? Now, once you approach a small fraction of the population, nothing can be said for certain about accuracy. It is always possible that by chance you talk mainly to people whose opinion is completely contrary to the generally held one. But probabilistic statements may be made. One may be able to make a statement along the lines of there being a 95 percent probability that the measurement error is within 2 percent.

The act of picking a random member of a population and observing or measuring a particular property of that member can be represented by a random variable X. For example, we select a voter at random and ask her if she would vote for a particular candidate or not. A positive answer can be allotted the value 1 and a negative answer can be allotted 0. This creates a Bernoulli random variable X. If 60 percent of all the voters do intend to vote for the candidate, and each voter is equally likely to be randomly selected, then X has parameter $p = 0.6$.

The act of picking a random member and observing the corresponding value of X is called **sampling.** When we sample n members of a population, we create random variables $X_1, \dots, X_n$, one for each act of sampling. We assume that the sampling is *with replacement* so that the sample space is the same for each X_i. We also assume that the act of sampling is not intrusive and does not cause any changes in the population. This ensures that each X_i has the same pdf. And finally, we assume that each observation is independent of the others, so that the X_i are independent random variables. When all these conditions are met, we call $X_1, \dots, X_n$ a **random sample.**

We will only consider the simplest kind of random sampling where the observed property is represented by a Bernoulli random variable. Then the random sample consists of independent Bernoulli random variables, which have a common parameter p. Then their sum $B = X_1 + \cdots + X_n$ is a Binomial random variable with parameters n and p, and counts the number of recorded successes. The random variable $\overline{X} = B/n$ is called the **sample mean** and counts the relative frequency of successes. When the random experiment of observing the random sample is actually carried out, the observed value of X_i is denoted x_i, and the observed value of the sample mean is $\overline{x} = \dfrac{1}{n}(x_1 + \cdots + x_n)$.

Example 5.7.1 An opinion poll collects 300 responses from voters and records that 56 percent of them favour candidate C and 44 percent favour candidate T. Can we conclude from this that C is likely to win the election? We know that it is possible that only 50 percent of actual voters favour C and that we just happened to talk to a group among whom C was more popular. But how likely is it that a group of 300 would be 6 percent off from the overall support of 50 percent? If it is very unlikely, we could reject the possibility of C's popularity being only 50 percent and decide that she is indeed the likely winner.

What we need to decide is how unlikely the poll result has to be before we reject the hypothesis of 50 percent support. For example, suppose we had decided before running the poll that we would reject the hypothesis of 50 percent support for C only if, assuming the hypothesis, the observed support $\overline{x}$ satisfies $P(\overline{X} \geq \overline{x}) < 0.05$. We now calculate

$$\begin{aligned} P(\overline{X} \geq \overline{x}) &= P(B/300 \geq 0.56) \\ &= P(B \geq 168) \\ &= 1 - P(B \leq 167) \\ &= \text{Binom.Dist(167, 300, 0.5, 1)} \\ &= 0.022 \end{aligned}$$

We see that the probability of the random sample reporting a 6 percent or greater *excess* support for C was only 0.022, well below our threshold of 0.05. Therefore, we would reject the hypothesis of only 50 percent support for C. □

We can approach the problem of accepting or rejecting a candidate value p_0 of p in another way. First, we decide on a probability α which sets the threshold for events to be considered unlikely, and is called the **level of significance.** The typically used values of α are 0.1, 0.05 and 0.01. Next, we look for a value c such that $P(B \leq c) = 1 - \alpha$ when B has parameter $p = p_0$. If we can find this c, then we have a simple test: Reject the hypothesis of $p = p_0$ if the observed value of B is greater than c. For this reason, we call c the **critical value** for our testing procedure. The values of B which are greater than c constitute the **rejection region,** while those which are less than or equal to c constitute the **acceptance region** (Figure 5.7.1).

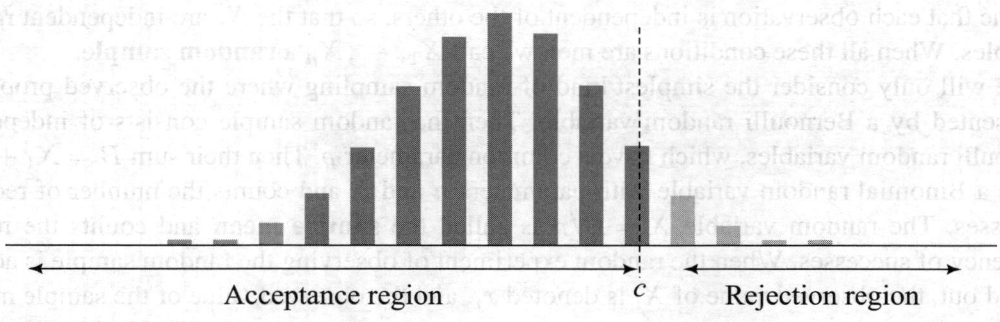

Figure 5.7.1 Suppose α is our significance level and $P(B \leq c) = 1 - \alpha$. Then, observing a value $d > c$ would lead to rejecting our assumed value of p.

However, there is a small complication. Since the probabilities of B change in discrete steps, the probability $P(B \leq k)$ may never exactly equal $1 - \alpha$. Instead, at some location, it will jump from just below $1 - \alpha$ to just above $1 - \alpha$. This location, can be found using the Binom.Inv(n, p, $1 - \alpha$) function, which gives the least value c such that the Binomial random variable B with parameters n and p satisfies $P(B \leq c) \geq 1 - \alpha$. The value c can be used as the critical value.

Example 5.7.2 Consider the opinion poll of Example 5.7.1. We wished to test the assumption $p = 0.5$. We had $n = 300$ and $\alpha = 0.05$. So we calculate

$$c = \text{Binom.Inv(300, 0.5, 0.95)} = 164$$

The observed value of B is $n\overline{x} = 300 \times 0.56 = 168$. Since it is greater than the critical value c, we reject the assumption. □

In the opinion poll example, we already had a suspicion that one candidate was more popular, and we tested specifically to confirm or reject that suspicion. The test we used is called a **one-sided test** because we only considered the extreme values on one side of the probability distribution. If we had not had such a suspicion, we would have tested for A's support being significantly above *or* below 50 %. This would have led to a **two-sided test** in which we consider both ends of the probability distribution.

In a two-sided test with the level of significance α, we distribute the probability α equally on both sides of the distribution. This leads to two critical values c_1 and c_2 such that the assumed value of p is rejected if the observed value of B is less than c_1 or more than c_2 (Figure 5.7.2).

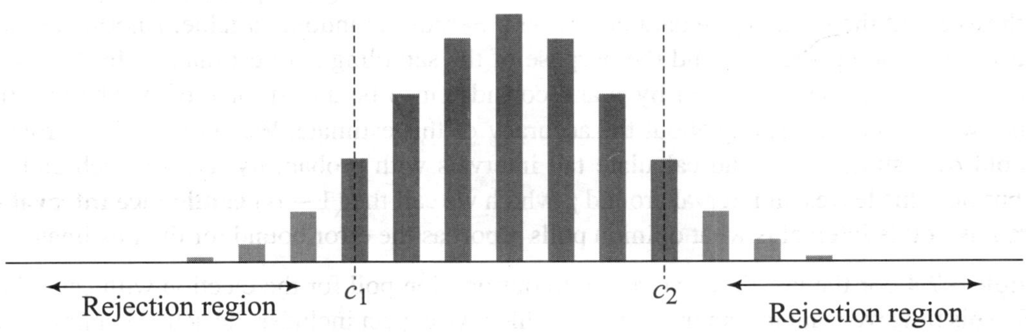

Figure 5.7.2 In a two-sided test, we have two critical values c_1 and c_2. The assumed value of p is rejected if the observed value of B is less than c_1 or more than c_2.

The two-sided critical values can be found using Excel as follows:

$$c_1 = \text{Binom.Inv(n, p, } \alpha/2)$$

$$c_2 = \text{Binom.Inv(n, p, } 1 - \alpha/2)$$

Example 5.7.3 Let us again consider the opinion poll of Example 5.7.1. We wish to test the assumption $p = 0.5$ with a two-sided test to establish whether either A or B has a clear lead over the other. Retaining the $\alpha = 0.05$ level of significance, we find the two critical values:

$$c_1 = \text{Binom.Inv(300, 0.5, 0.025)} = 133$$

$$c_2 = \text{Binom.Inv(300, 0.5, 0.975)} = 167$$

The corresponding values of $\overline{X}$ are $133/300 = 44.33$ percent and $167/300 = 55.67$ percent. If the observed sample mean is between these values, we would not be able to reject the $p = 0.5$ hypothesis.

This acceptance region is quite wide and may not be effective in confirming differences between the popularities of the candidates. The cure, if practical, is to increase the sample size. For example,

$$c_1 = \text{Binom.Inv(1000, 0.5, 0.025)} = 469$$

$$c_2 = \text{Binom.Inv(1000, 0.5, 0.975)} = 531$$

The corresponding values of $\overline{X}$ are 46.9 percent and 53.1 percent. The acceptance region has roughly halved in width, but at the cost of much greater effort in sampling. The other way to narrow the acceptance region is to work with a greater α, that is, to tolerate a greater likelihood of error. With $\alpha = 0.1$, a sample size of $n = 500$ gives roughly the same acceptance region for $\overline{X}$ as $\alpha = 0.05$ with $n = 1000$:

$$c_1 = \text{Binom.Inv(500, 0.5, 0.05)} = 232$$

$$c_2 = \text{Binom.Inv(500, 0.5, 0.95)} = 268$$

The corresponding values of $\overline{X}$ are 46.4 percent and 53.6 percent. □

So far we have been concerned with the problem of using sampling to accept or reject a hypothesis about the value of the parameter p of a Bernoulli random variable. Another possibility is that we have no candidate p and the purpose of the sampling is to estimate p. In this case, we denote the observed sample mean by $\hat{p}$ and consider it to be an estimate of p. The question is whether we can say something about the accuracy of this estimate. We choose a low probability threshold α, assume $p = \hat{p}$ and calculate tail intervals with probability $\alpha/2$ on each end of the distribution. This leaves an interval around $\hat{p}$ which we call the $(1 - \alpha)$ **confidence interval** for p. The radius of this interval is what opinion polls report as the error bound for their estimates.

Example 5.7.4 For the last time, we return to our opinion poll for the election with candidates A and B. This time we want to establish a range which we expect includes the actual support level for A. We assume $p = \hat{p} = 0.56$ and calculate the two end-points of the 95 percent confidence interval for np:

$$d_1 = \text{Binom.Inv(300, 0.56, 0.025)} = 151$$

$$d_2 = \text{Binom.Inv(300, 0.56, 0.975)} = 185$$

This gives the 95 percent confidence interval for p: $(151/300, 185/300) = (50.33, 61.67)$. Typically, this calculation would be reported as 'The margin of error is 5.67 percent at a confidence level of 95 percent'.

In fact, opinion polls typically report a margin of error of 3 percent at a confidence level of 95 percent. This corresponds to a sample size of 1,000. □

Quick Review

After reading this section, you should be able to:

- Use sample mean to accept or reject hypotheses about the value of the parameter of a Bernoulli random variable.
- Use sample mean to calculate confidence intervals for the value of the parameter of a Bernoulli random variable.
- Choose appropriate sample size for the desired significance or confidence level.

EXERCISE 5.7

1. A coin is tossed 1,000 times and heads are observed 530 times. Would this lead to the rejection of the hypothesis that the coin is fair, with a two-sided test at the 5 percent significance level?
2. A coin is tossed 1,000 times and heads are observed 530 times. Would this lead to the rejection of the hypothesis that the coin is fair, as opposed to being biased in favour of heads, with a one-sided test at the 5 percent significance level?
3. How many times should we toss a coin so that there is at least a 95 percent probability that the relative frequency of heads is within 0.01 of the expected value of 0.5?
4. Suppose that 540 heads and 460 tails resulted from 1,000 tosses of a coin. Find a 95 percent confidence interval for the probability of a head. Does this appear to be an unbiased coin?
5. A pollster is organising an opinion poll for an election featuring two candidates. Is there a sample size that will give a 95 percent confidence interval of width 0.03 or less for the proportion supporting a particular candidate, no matter what the popularity levels of the candidates may be?

Review Exercises

1. Urn 1 contains a black ball and a red ball, and Urn 2 contains a white ball and a red ball. Describe the sample space of the following random experiments using set notation:
   (a) First an urn is selected and then a ball is picked from that urn.
   (b) A ball is picked from each urn.
2. Is classical probability a suitable model for the following?
   (a) The probability of picking a boy from a class of 35 students.
   (b) The probability of picking an even number from the set of natural numbers.

(c) The probability that the price of a certain share will rise by more than 5 percent in the next week.

3. Let A and B be events for a random experiment, with $P(A) = 0.5$ and $P(B) = 0.7$. What are the minimum and maximum possible values of $P(A \cap B)$?

4. Let A and B be events from a sample space $\mathcal{S}$, such that $P(A) = 0.4$ and $P(B) = 0.7$. Which of the following statements must be true?

(a) $A \subset B$. (b) $A \cup B = \mathcal{S}$. (c) $A \cap B \neq \emptyset$.

5. Let A, B and C be events for a random experiment. Show that

$$P(A \cup B \cup C) = P(A) + P(B) + P(C) - P(A \cap B) - P(A \cap C) - P(B \cap C) + P(A \cap B \cap C)$$

6. Show that $P(A \cap B) \geq P(A) + P(B) - 1$.

7. The Venn diagram represents the readership of *Hindustan Times* (H), *Times of India* (T) and *The Indian Express* (I) in a locality of Delhi.

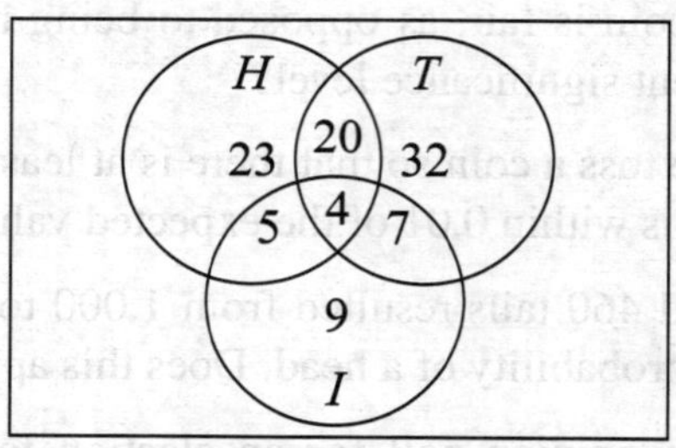

A reader is selected at random. Find the probability that the reader:

(a) Reads *Hindustan Times* but not The *Times of India*.
(b) Reads *Hindustan Times* or The *Times of India* but not The *Indian Express*.
(c) Reads exactly two of these newspapers.
(d) Reads The *Indian Express*.
(e) Reads The *Indian Express*, given that he or she reads *The Times of India*.

8. In the above exercise, let H be the event that the reader reads *Hindustan Times* and T the event that he or she reads *Indian Express*. Are the two events independent?

9. Suppose $P(A) = 0.4$ and $P(A \cup B) = 0.6$. Find $P(B)$ if:

(a) A and B are mutually exclusive. (b) A and B are independent.

10. Three urns are numbered 1–3. Each contains 10 balls, which are coloured as follows:

	Urn 1	Urn 2	Urn 3
Black	9	8	7
White	1	2	3

First an urn is selected at random, and then a random ball is picked from that urn.

(a) What is the probability of picking a black ball?

(b) If a black ball is picked, what is the probability that it came from Urn 2?

11. A random variable X takes values $0, 1, 2, 3, 4, 5$. For $k = 0, 1, \ldots, 5$, the probability $P(X = k)$ is proportional to k. Calculate the pdf and expectation of X. What happens if the range of X is $0, 1, \ldots, n$?

12. Suppose X is symmetric, that is, it satisfies $P(X = x) = P(X = -x)$ for every value x. Show that its expectation is zero.

13. When can a random variable X have zero variance?

14. Let X and Y be jointly distributed. Is it always true that $\sigma^2_{X+Y} \geq \sigma^2_X + \sigma^2_Y$?

15. Suppose $P(X = x) = P(Y = x)$ for every x. Does this mean $X = Y$?

Bibliography

Books and Articles

[1] E. S. Wentzel. 1982. *Probability Theory (First Steps)*. Moscow: Mir Publishers.

[2] L Rastrigin. 1973. *This Chancy, Chancy, Chancy World*. Moscow: Mir Publishers.

[3] Sheldon Ross. 2013. *A First Course in Probability*. 9th edition. London: Pearson.

[4] John Haigh. 2012. *Probability: A Very Short Introduction*. Oxford: Oxford University Press.

[5] Charles M. Grinstead and J. Laurie Snell. 1997. *Introduction to Probability*. 2nd revised edition. American Mathematical Society. `http://www.dartmouth.edu/~chance/teaching_aids/books_articles/probability_book/book.html` (accessed 27 January 2017).

Websites

[W1] David Aldous. Overview of Probability in the Real World Project. `http://www.stat.berkeley.edu/~aldous/Real-World/` (accessed 27 January 2017).

[W2] Census of India. SRS Based Abridged Life Tables 2003–07 to 2006–10. `http://www.censusindia.gov.in/vital_statistics/SRS_Based/SRS_Based.html` (accessed 12 January 2017).

[W3] NRICH. Statistics Curriculum Mapping. `http://nrich.maths.org/9011` (accessed 27 January 2017).

6
CHAPTER
Symmetry

My work has always tried to unite the true with the beautiful and when I had to choose one or the other, I usually chose the beautiful.

Hermann Weyl, a mathematician who contributed greatly to twentieth century mathematics.[1]

What does symmetry have to do with mathematics and Hermann Weyl? Most of us, when asked the question 'what does symmetry mean?' or 'what does it signify?', would try to describe symmetry as signifying regularity or of two parts of an object being the same. Another idea that seems inherent while discussing symmetry is that of beauty. In nature, in the world around us, we are often spellbound by beautiful objects, and more often than not these are also objects that seem more symmetrical. This is not just a phenomenon associated with the present time, for if we consider monuments of the past or ornamental objects from ancient times, we find that they are abundant as examples of symmetry and beauty.

In 1951, Hermann Weyl gave a series of lectures on symmetry at Princeton University, which combined seamlessly the beauty of symmetry and that of the mathematics involved in studying symmetry. His lectures traversed time, space and mathematics to bring alive the connections between symmetry of ornamental objects, monuments, objects in nature, paintings and art and the underlying principles of the mathematics involved. He published a book titled *Symmetry* in 1952 [5] that details the series of lectures from the year before.

Symmetry is a topic that certainly unites truth and beauty in the manner that Herman Weyl wished to in his work. Hopefully, this chapter will convince the readers of the same. The aim of the chapter is to introduce simple ideas which can be used to appreciate and study symmetry. The formal mathematics governing the study of symmetry can be quite sophisticated and is beyond the scope of this book; nevertheless, a flavour or a glimmer of the same will be presented.

Figures 6.0.1–6.0.5 show a series consisting of pairs of pictures. Can you decide which of the pair is more 'symmetrical'?

[1] You can find out more about Weyl and his work in [W12].

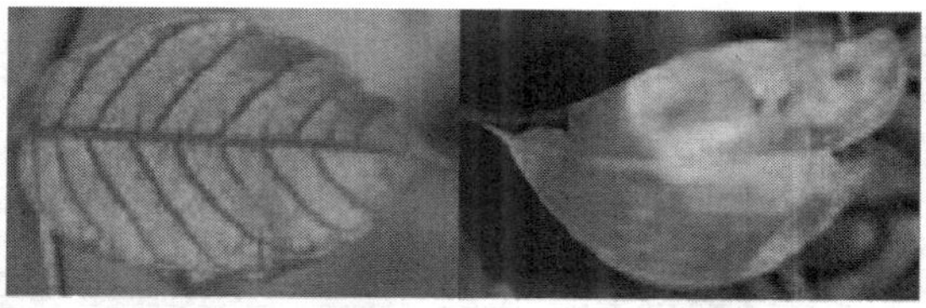

Figure 6.0.1 Leaves. [C6]

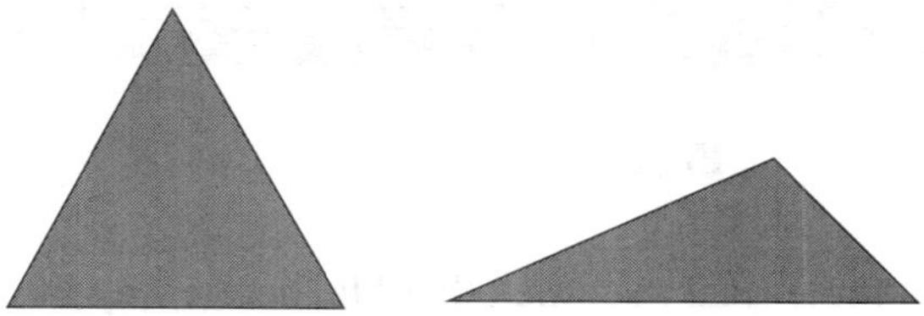

Figure 6.0.2 Triangles.

Figure 6.0.3 Animals. [C6]

Figure 6.0.4 Money signs.

You will notice that in all the figures, the object on the left seems more symmetrical. However, in the first and possibly the third picture, this may not be as obvious. Is there some way of stating

Figure 6.0.5 Flowers. [C6]

when one object is more symmetrical than the other? Is there a way in which we can describe and analyse symmetries of any object? We will attempt to do this in the next few sections.

6.1 Working Definition of Symmetry

Consider the following situation. You have a cut-out of a square made with a slightly thick white sheet of paper. The cut-out is placed on another sheet of paper and its outline drawn. What are the actions that you can perform on the cut-out so that at the end of the action the cut-out comes back and occupies the exact same position described by the outline? The actions allowed cannot involve tearing the cut-out and reconstituting it back to form a square. Further, imagine that you have performed one such action and your friend who is watching you has turned away or closed their eyes while you have performed the action.

If you ask your friend whether anything had been done to the cut-out of the square, the answer would be an emphatic no. This is because the cut-out started in the position of lying within the outline, and after the action was performed, it still occupied the exact same place and position.

An intuitive or working definition of symmetry is the following. A **symmetry** of an object is an action that can be performed on the object so that at the end of the action the object is left looking exactly the same as before, occupying the exact same place and position as before. In other words, an observer who has closed their eyes while the action is performed will have no idea that any action has been performed on the object. Also, note that the action cannot involve tearing or breaking of the object to reconstitute it.

Now that we are armed with an intuitive notion of symmetry, we face two basic questions. One is, can we actually list all the symmetries (actions performed) of a given object or at least describe the different types of symmetries possible? This in fact brings us to the second pertinent question: if an observer has closed their eyes while the symmetry is performed, they cannot know that an action has been executed. Given this, we can ask how can one describe a symmetry? Indeed, it is only when we answer the second that we can make some progress on the first.

Let us go back to the example we started out with. We notice that we can keep track of the entire square cut-out if we know where the vertices of the cut-out are located. Therefore, it makes sense to mark the vertices of the square cut-out as A, B, C, D, in say a counterclockwise direction. See Figure 6.1.1.

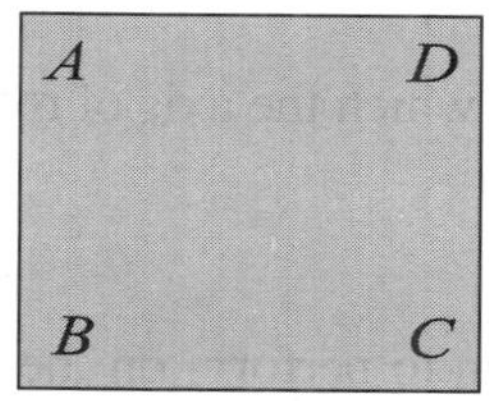

Figure 6.1.1 Square with vertices marked.

We draw the outline of the cut-out on a sheet of paper and once again mark the vertices of the outline as shown in Figure 6.1.2.

Figure 6.1.2 Outline of square cut-out on a sheet of paper.

We begin by placing the cut-out of the square within the outline as shown in Figure 6.1.3.

Now, if a symmetry is performed on the square cut-out, it has to keep the cut-out occupying the space within the outline. It is a fairly simple matter to see where the vertices marked on the cut-out lie in comparison with the fixed outline. Thus, even though the symmetry leaves the square cut-out occupying the same position and space, the marking of the vertices allows us to track the symmetry. Note that if the action is not a symmetry, then there will be some part of the square which is lying outside the outline.

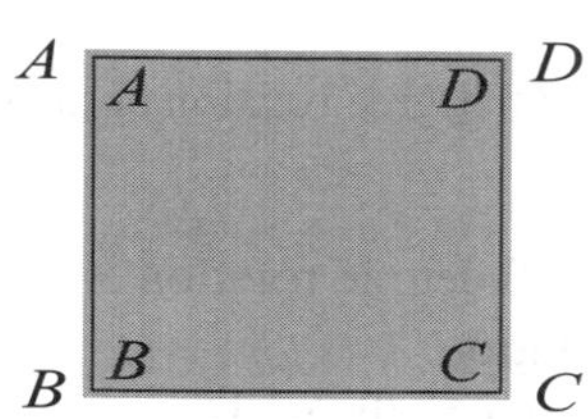

Figure 6.1.3 Square cut-out placed within outline.

We present two examples, one that is a symmetry of a square and one which is not. The figures accompanying show the final positions after the action has been performed on the square cut-out.

Both these actions involve rotation of the cut-out. In general, to describe a rotation of a planar or two-dimensional object, we need to specify certain details. These are:

1. The axis of rotation.
2. The rotocenter or the point through which the axis of rotation cuts the plane of the object.
3. The direction of rotation.
4. The angle of rotation.

The details for the rotations we plan to perform on the square cut-out are as follows: the axis of rotation is a line perpendicular to the plane of the paper on which the outline is placed; the rotocenter is the point at which the diagonals of the square meet; the direction of rotation will be anticlockwise or counterclockwise; the angle of rotation in the first case will be 60 degrees and in the second case will be 90 degrees. To perform these rotations by yourselves, you can use a pencil or a pen to act as the axis of rotation. Place the cut-out on the outline and poke or pin the square cut-out at the rotocenter with the pencil. The required rotation can then be performed by rotating the cut-out to the desired angle.

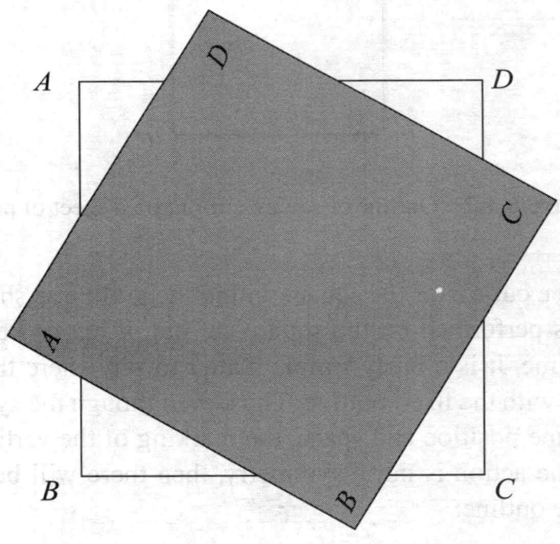

Figure 6.1.4 60 degree rotation of the square cut-out.

It is clear from Figure 6.1.4 that a 60 degree rotation is not a symmetry of a square. Now, let us examine the case of the 90 degree rotation.

In Figure 6.1.5, we note that the square cut-out is occupying the space within the outline. So if there had been no markings on the square, it would not have been possible for an observer who has shut their eyes while the 90 degree rotation is performed to know that an action has taken place. Thus, the 90 degree rotation is the symmetry of a square. We can describe the 90 degree rotation by tracking the start and end position of the four vertices.

Figure 6.1.5 90 degree rotation of the square cut-out.

We notice that after the 90 degree rotation, the vertex A has moved to position B of the outline and so on. Thus, we can write:

$$\begin{aligned} R_{90} : A &\rightarrow B \\ B &\rightarrow C \\ C &\rightarrow D \\ D &\rightarrow A. \end{aligned}$$

Can you now guess which angles will give us symmetries of a square? The answer should be easy to work out with the help of the cut-out. One answer is of course obvious, the 0 degree rotation or the Do-Nothing-Rotation (DNR). Figure 6.1.3 describes the end position of the DNR. In terms of the vertices, it can be described as:

$$\begin{aligned} R_{0} : A &\rightarrow A \\ B &\rightarrow B \\ C &\rightarrow C \\ D &\rightarrow D. \end{aligned}$$

Figures 6.1.6 and 6.1.7 show the other rotational symmetries of a square, namely the 180 degree rotation and the 270 degree rotation, respectively. Their descriptions in terms of the vertices follow the respective figures.

Description of the 180 degree rotation in terms of the vertices

$$\begin{aligned} R_{180} : A &\rightarrow C \\ B &\rightarrow D \\ C &\rightarrow A \\ D &\rightarrow B. \end{aligned}$$

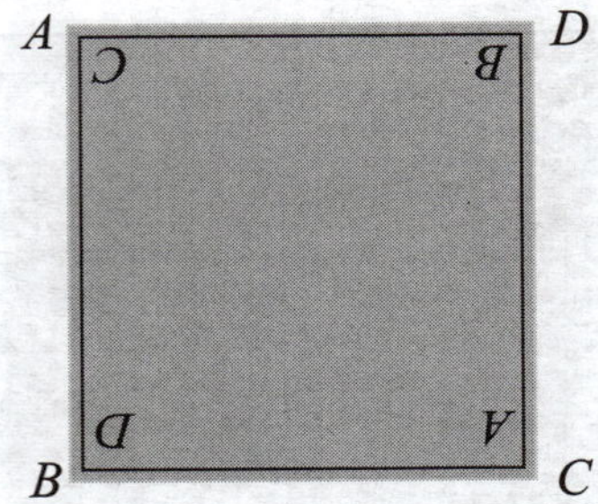

Figure 6.1.6 180 degree rotation of the square cut-out.

Figure 6.1.7 270 degree rotation of the square cut-out.

Description of the 270 degree rotation in terms of the vertices

$$\begin{aligned} R_{270} : A &\to D \\ B &\to A \\ C &\to B \\ D &\to C. \end{aligned}$$

By rotating the cut-out about the rotocenter with a pencil as an axis, one can conclude that the four rotations by angles of 0, 90, 180 and 270 are the only rotational symmetries of a square. A square does not have any other rotational symmetries. Does the square have any other symmetries? Let us explore this further.

Once again we make use of the square cut-out. This time, let us also mark the same vertices at the back of the square cut-out as we did in the front. Thus, at the back of vertex A, we will mark vertex A and so on. Now, place the square cut-out within the outline as shown in Figure 6.1.3. This will be our starting position. Now, lift the cut-out above the outline by holding it at the mid-points of sides AD and BC, flip the cut-out and place it back within the outline. Since there are vertex markings at the back of the cut-off, we can 'track' this symmetry as well. This symmetry is called a flip or a reflection about the line L_V joining the mid-point of the sides AD and BC. The final position is shown in Figure 6.1.8 along with L_V as a dotted line.

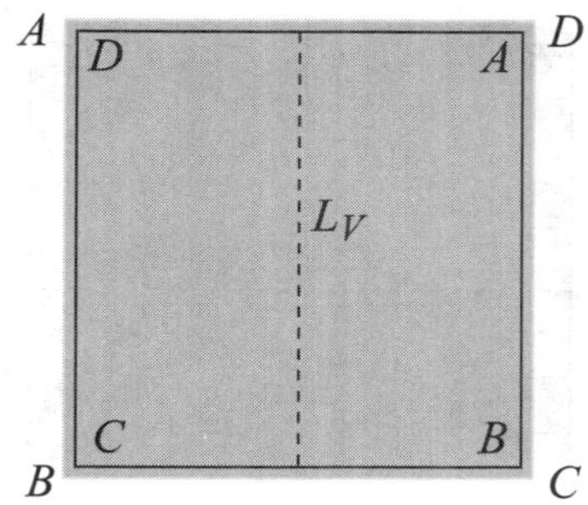

Figure 6.1.8 Vertical flip of the square cut-out.

The symmetry described in Figure 6.1.8 is also called a mirror reflection as when we place a mirror on the dotted line L_V, the reflection of the left half of the square in the mirror completes the square. The line L_V is called the **line of reflection or the line of mirror reflection**. Another experiment that can be conducted is to fold the square cut-out along the dotted line L_V. Then we see that the two parts of the square on either side of the line match perfectly when we fold along L_V. Further, vertices A and D will superimpose on each other as will B and C. This is another way of discovering 'mirror symmetries or reflections or flips'. All we have to do is to find lines that we can draw on our square cut-out so that when we fold along the line, the two-parts on either side of the line will match up exactly. If the two-parts do not match up, then the line about which we are folding is not a line of reflection and the corresponding flip will not be a symmetry. Let L_H and L be the lines marked on the square as shown in Figure 6.1.9. Which of these is a line of mirror symmetry and why?

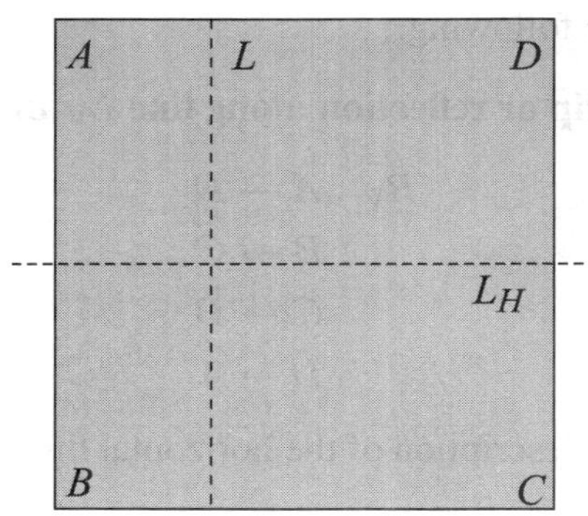

Figure 6.1.9 L and L_H marked on the square cut-out.

Note that L_H is the line joining the mid-points of the sides AB and DC. If we fold along L_H, the two-parts on the either side of L_H will match up perfectly. However, this will not be the case if we fold along L. Thus, we get a symmetry when we reflect about the line L_H and Figure 6.1.10 shows the final position of the cut-out after the horizontal flip or reflection.

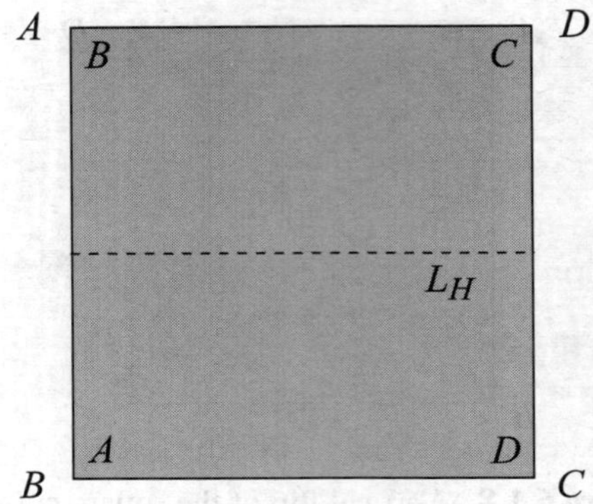

Figure 6.1.10 Horizontal flip of the square cut-out.

Before we locate other mirror symmetries of a square, it will be worth our while to give the details required to describe a reflection of a planar or a two-dimensional object as we did in the case of a rotation. For a mirror reflection, all we need to specify is the **line of reflection.** Once this is known, we will be able to track the mirror symmetry in a very similar fashion to that of a rotational symmetry.

For objects like a square or a triangle, we use vertices to track symmetries. This is because the position of the vertices completely determine the position of the square or the triangle. For other objects, like a leaf or an ornamental design, we will have to find special points on the object which we can refer to as 'defining points'. These will be points which will help us identify the position of the object and should largely be like vertices are for a square or triangle. We can label them with letters of the alphabet, just like with vertices.

From Figure 6.1.8, we get the following.

Description of the vertical flip or reflection along line L_V in terms of the vertices

$$\begin{aligned} R_V : A &\to D \\ B &\to C \\ C &\to B \\ D &\to A. \end{aligned}$$

From Figure 6.1.10, we get a description of the horizontal flip.

Description of the horizontal flip or reflection along line L_H in terms of the vertices

$$\begin{aligned} R_H : A &\to B \\ B &\to A \\ C &\to D \\ D &\to A. \end{aligned}$$

There are just two more mirror symmetries of a square. These are achieved when we reflect about the diagonals. A description of each is given below.

Description of the diagonal flip or reflection along the diagonal AC in terms of the vertices

$$\begin{aligned} R_{AC} : A &\to A \\ B &\to C \\ C &\to B \\ D &\to D. \end{aligned}$$

Description of the diagonal flip or reflection along the diagonal BD in terms of the vertices

$$\begin{aligned} R_{BD} : A &\to D \\ B &\to B \\ C &\to C \\ D &\to A. \end{aligned}$$

It is fairly easy to check that no other line of mirror symmetry exists for a square. Thus, so far we have discovered eight symmetries of a square, four rotations and four reflections. We know that there are no further rotations and no further reflections, but are there any more symmetries of a square which may possibly not be reflections or rotations? We will answer this in the next section.

Quick Review

After reading this section, you should be able to:

- Identify defining points for tracking rotations and reflections.
- Describe rotational symmetries.
- Decide reflection or mirror symmetries.
- Find rotations and reflections for simple planar shapes.
- Decide whether an action performed is a symmetry or not.

EXERCISE 6.1

1. Which of the following describes a symmetry of the given figure? Note that in the following pictures, the position of the figure after an action is performed is shown with respect to the outline of the figure.

(a)

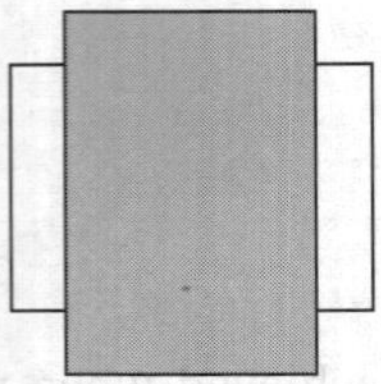

(d)

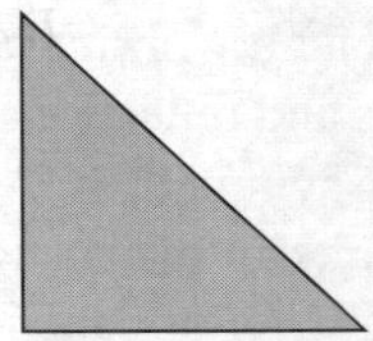

(b)

(e)

(c)

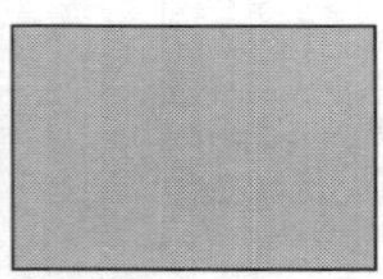

(f)

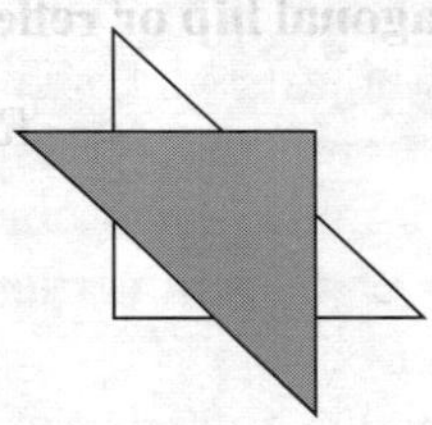

2. For the given figures, namely a rectangle, a query balloon and a triangle, describe all the rotational symmetries and reflection symmetries by identifying defining points on the figure.
3. Make a cut-out of an equilateral triangle and draw its outline on a sheet of paper. Mark the vertices on the cut-out and on the outline as A, B and C in an anticlockwise direction. On the cut-out, draw dotted lines joining each vertex to the mid-point of the opposite side. Name the line joining vertex A to the mid-point of side BC as L_A. Similarly, name the other two lines L_B and L_C. Denote the point of intersection of the three lines as O. By pinning the triangle to the outline with a pencil, verify that an equilateral triangle has exactly three rotational symmetries, namely those of 0, 120 and 240 degrees about the rotocenter O. Verify by folding along the dotted lines that the equilateral triangle has three reflection symmetries. Convince yourself that the equilateral triangle has no other rotational or reflection symmetries. Describe the three rotational symmetries and three mirror symmetries of the equilateral triangle.
4. Repeat the preceding exercise with an isoceles triangle which is not equilateral. Does the isoceles triangle have fewer rotational symmetries or reflection symmetries as compared to an equilateral triangle? If so, can you give a reason for this?
5. Repeat the preceding exercise for a rectangle which is not a square. Does the rectangle have fewer rotational symmetries or reflection symmetries as compared to a square? If so, can you give a reason for this?

6.2 Symmetry of Finite Planar Figures

In the previous section, we had a working definition of symmetry and learnt about two types of symmetries, namely rotations and reflections. We saw that a square has four rotational symmetries and four reflection or mirror symmetries. A square is an example of a **finite planar figure.** We have already encountered the term planar figure which just means a two-dimensional figure or a figure that can be drawn on a sheet of paper. A finite planar figure is a planar figure around which we can draw a rectangle. We will encounter planar objects that are not finite in the next section.

It turns out that a finite planar figure has at most two kinds of symmetries: rotations and reflections. It is easy to see that all objects will always possess a 0 degree rotation or a *DNR*. Thus, a finite planar figure will always have at least one rotational symmetry, but it may or may not have a mirror symmetry. For example, a scalene triangle has only one symmetry which is the *DNR*. We shall encounter other examples a little later.

An important result that can be proved mathematically but which is beyond the scope of this book is the following.

If a finite planar object has only finitely many symmetries, then it has either an equal number of rotational symmetries and reflection symmetries or it has only rotational symmetries.

Are there finite planar objects which have infinitely many symmetries or do finite planar objects only have finitely many symmetries? Think about this. The answer shall be provided later. The square is an example of a finite planar object which has an equal number of rotational and mirror symmetries. It has only eight symmetries of which four are rotations and four are reflections. The scalene triangle referred to earlier is an example of a finite planar figure which has only one symmetry, which is a rotation, namely the *DNR*.

We can classify finite planar figures with finitely many symmetries into two types based on the aforementioned result.

If a finite planar figure has an equal number n, of rotational symmetries and reflection symmetries, then we say that the figure is of type D_n and we say that the figure is of **Dihedral type of degree** n. So, such figures have exactly $2n$ symmetries of which n are rotations and n are reflections. The equilateral triangle and the square are examples of the Dihedral type. The equilateral triangle has six symmetries of which three are rotations and three are reflections. Therefore, it is of type D_3, that is, Dihedral of degree 3. The square is of type D_4, that is, Dihedral of degree 4.

This fact about the equilateral triangle and a square can indeed be generalised. After all, the equilateral triangle is the regular three-sided figure, that, is a figure in which all sides are equal and all internal angles are equal. The square is the regular four-sided figure. If we take a regular n-sided figure or polygon, then it turns out that such a figure has exactly $2n$ symmetries consisting of n rotations and n reflections.

If a finite planar figure has exactly n symmetries consisting only of rotations, then we say that the figure is of type C_n or that the figure is of **Cyclic type of degree** n. The scalene triangle has only one symmetry which is the *DNR*. So, it is of type C_1 or is said to be of Cyclic of degree 1.

Figures 6.2.1 and 6.2.2 provide some pictures which are of type C_n for various n.

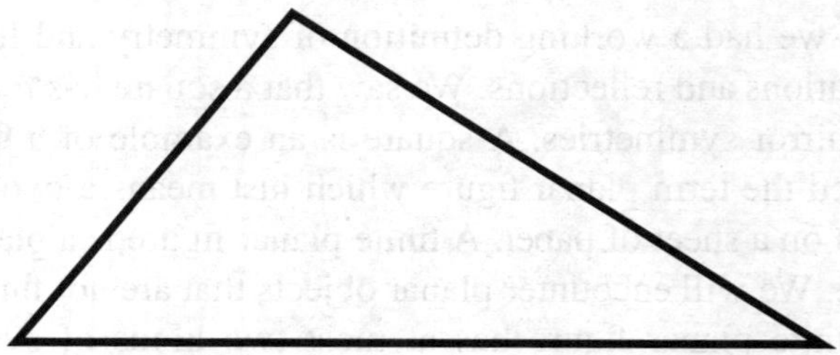

Figure 6.2.1 Scalene triangle of type C_1.

As discussed previously, the scalene triangle only has one symmetry, namely the zero degree rotation or *DNR*.

Figure 6.2.2 Figure of type C_2.

Figure 6.2.2 has exactly two symmetries, both rotations. One is of course the *DNR* and the other is a 180 degree rotation. Can you figure out the rotocenter? If you examine the figure carefully, it is a rectangle with two arrows added. A rectangle which is not a square is of type D_2. Such a rectangle has four symmetries consisting of two reflections and two rotations. The reflections are about lines joining the mid-points of opposite sides.

The effect of adding arrows as in Figure 6.2.2 destroys the reflection symmetries of a rectangle but preserves the rotational symmetries. One must take care though while adding the two arrows so that the 180 degree rotation of the rectangle is preserved. Make a cut-out of Figure 6.2.2 and mark its outline on a sheet of paper. The rotocenter is the point of intersection of the diagonals of the rectangle. If you turn the cut-out through an angle of 180 degrees in the counterclockwise direction about an axis perpendicular to the sheet of paper and passing through the rotocenter, then the arrow which was pointing upward will occupy the position of the downward-pointing arrow of the outline and vice-versa.

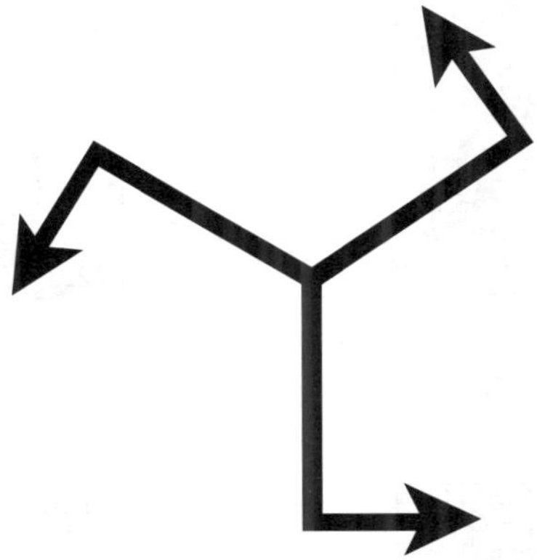

Figure 6.2.3 Figure of type C_3.

Figure 6.2.3 has exactly three symmetries. These are rotations of degrees 0, 120 and 240, respectively. Try to analyse an equilateral triangle to find out how the figure was constructed. Figure 6.2.4 is of type C_4 and has only four symmetries consisting of rotations of degrees 0, 90, 180 and 270.

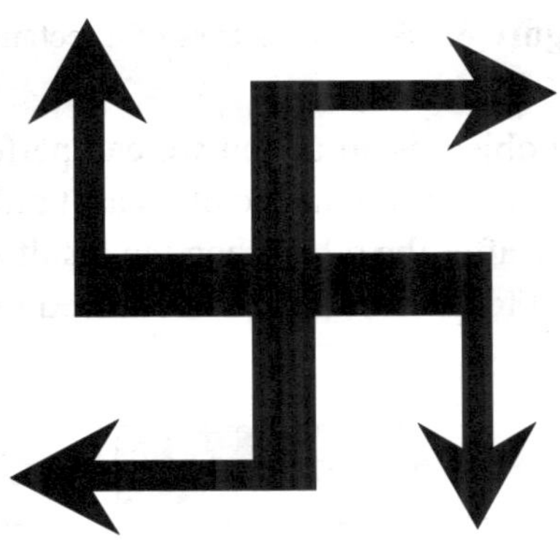

Figure 6.2.4 Figure of type C4.

Given an object, the collection of all its symmetries has some very interesting properties. Let us denote the object under study as X and let the collection of all symmetries of the object X be $\text{Sym}(X)$. So, for example, X could be a square and then $\text{Sym}(X)$ would consist of the eight symmetries of the square discussed earlier, or X might be an equilateral triangle, then $\text{Sym}(X)$ would consist of the three rotations of degrees 0, 120 and 240 and the three reflections about the respective lines joining the vertices and the mid-point of the opposite side. We are now going to investigate some properties that the collection $\text{Sym}(X)$ satisfies.

When we want to illustrate the properties, we shall take X as a rectangle which is not a square. Figure 6.2.5 shows the four symmetries of X. Thus, $\text{Sym}(X) = \{R_0, R_{180}, R_V, R_H\}$.

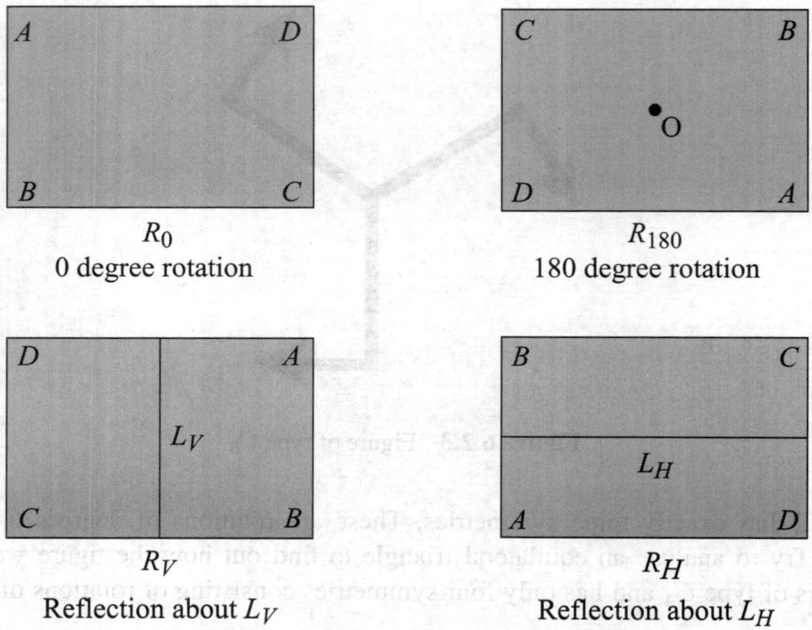

Figure 6.2.5 Symmetries of a rectangle.

Recall that a symmetry of an object is an action we can perform on the object that leaves the object looking exactly like it did and in the same position as it did in the beginning. So, if we apply two symmetries on the object one after the other, then the result will be a symmetry. For example, if we apply a 180 degree rotation to the rectangle X followed by a horizontal reflection, then we will again get a symmetry of X.

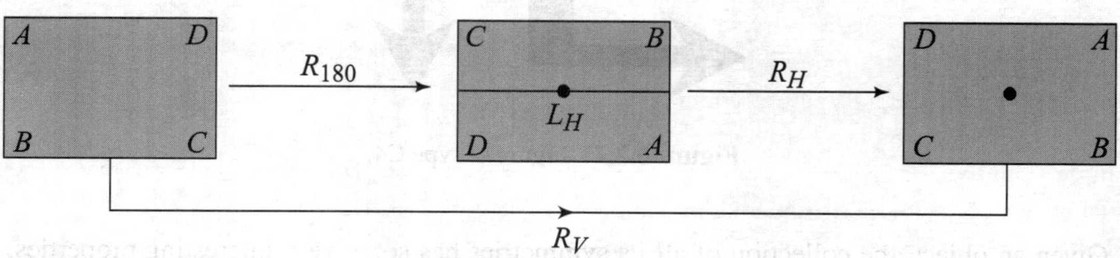

Figure 6.2.6 Symmetry followed by a symmetry is again a symmetry.

Figure 6.2.6 shows that the result of R_{180} being followed by R_H is R_V. We write this as $R_H \circ R_{180} = R_V$. The property that a symmetry followed by a symmetry is again a symmetry is called the property of **closure.** Table 6.2.1 shows all the possibilities for a symmetry followed by a symmetry in the case of a rectangle. Such a table is called a **multiplication table** for symmetries.

Table 6.2.1 Multiplication Table for the Symmetries of a Rectangle

Row ∘ Col	R_0	R_{180}	R_H	R_V
R_0	R_0	R_{180}	R_H	R_V
R_{180}	R_{180}	R_0	R_V	R_H
R_H	R_H	R_V	R_0	R_{180}
R_V	R_V	R_H	R_{180}	R_0

It turns out that if R, S and T are symmetries of the same object, then $R \circ (S \circ T) = (R \circ S) \circ T$. This property is called **associativity.** For example, if we take R_{180}, R_H and R_V and use the information in Table 6.2.1, then we get

$$R_{180} \circ (R_H \circ R_V) = R_{180} \circ R_{180} = R_0 \,.$$

Also,

$$(R_{180} \circ R_H) \circ R_V = R_V \circ R_V = R_0 \,.$$

Thus, $R_{180} \circ (R_H \circ R_V) = (R_{180} \circ R_H) \circ R_V$.

The collection of all symmetries of an object always has a special symmetry called the **Identity,** usually denoted by I. For any symmetry R of the object, we have that $R \circ I = R = I \circ R$. If we refer to the example of a symmetries of a rectangle, Table 6.2.1 shows that the identity symmetry here is R_0 or *DNR*. In fact, for any object X, there is always a symmetry, namely *DNR*, and it can be verified that *DNR* is the identity symmetry denoted by I.

Another property is that for any symmetry R of an object X, there is always a symmetry R' of X such that $R \circ R' = I = R' \circ R$. In other words, for every symmetry R, there is a symmetry R' such that R followed by R' or vice-versa results in the identity symmetry. We say that R' is the **inverse** of R. If we refer back to our example where X is a rectangle, we find that the inverse of R_0 is itself, as is the case for R_{180}, R_H and R_V. This is however not usually the case. If we take X as the equilateral triangle and R as the 120 degree rotation, then its inverse is the 240 degree rotation. Another special feature when X is a rectangle is that $R \circ S = S \circ R$ for all symmetries R, S of X. This too is not usually the case. Try to find two symmetries R and S of an equilateral triangle such that $R \circ S \neq S \circ R$.

Thus, we have learned that for any object X and its collection of symmetries $\text{Sym}(X)$, the following holds:

S1 Closure A symmetry followed by a symmetry is again a symmetry or equivalently for all R, S in $\text{Sym}(X)$, we have $R \circ S$, which is again a symmetry belonging to $\text{Sym}(X)$.

S2 Associativity For any R, S and T in $\text{Sym}(X)$, it is always the case that $R \circ (S \circ T) = (R \circ S) \circ T$.

S3 Existence of Identity There is a special symmetry in $\text{Sym}(X)$, namely *DNR* $= I$ such that for all R in $\text{Sym}(X)$, we have $R \circ I = R = I \circ R$.

S4 Existence of Inverses For each R in Sym(X), there is an R' in Sym(X) such that $R \circ R' = I = R' \circ R$.

These aforementioned four properties make Sym(X) a **group** with respect to $\circ$. Recall that $\circ$ is just the way in which we can combine two symmetries. A group is an object of study in the area of abstract algebra and is beyond the scope of this book.[2]

Now, we come to the answer to a question posed earlier. A circle is a finite planar figure which has infinitely many symmetries, both rotations and reflections. A rotation by any angle about the centre will be a symmetry as will any reflection about a diameter.

Quick Review

After reading this section, you should be able to:

- Identify finite planar objects.
- Decide whether the collection of symmetries of a finite planar object is finite or not.
- Classify the collection of symmetries of a finite planar object into D_n or C_n type, provided the collection of symmetries is finite.
- Find the result of a symmetry followed by a symmetry.
- Find the inverse of a symmetry.

EXERCISE 6.2

1. For each of the following figures, draw/write down the following.
   (a) The lines of reflection or mirror symmetry, if any.
   (b) A description of the axis of rotation and the rotocenter.
   (c) The reflections, with their effect on suitably marked vertices.
   (d) The rotations, with their effect on the marked vertices.
   (e) The symmetry type of the figure.

---

[2] If it turns out that for all R and S in Sym(X), $R \circ S = S \circ R$, that is, the order in which you apply symmetries does not matter, then we say that Sym(X) is an **abelian** group.

The term abelian comes from the name of the famous eighteenth century Norwegian mathematician Abel. You can find out more about Abel and his work at the MacTutor History of Mathematics archive [W10]. If Sym(X) is not abelian, that is, there exist R and S such that $R \circ S \neq S \circ R$, then we say that Sym(X) is non-abelian.

It can be shown that if Sym(X) is of D_n type and $n \geq 3$, then it is a non-abelian group, and when Sym(X) is of C_n type, then it is an abelian group, for all n. Another example of an abelian group is a set that one is familiar with. The set of integers is an abelian group with respect to addition.

(i)

(ii)

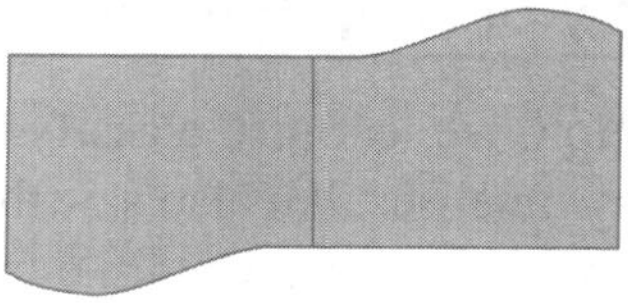

(iii)

(iv)

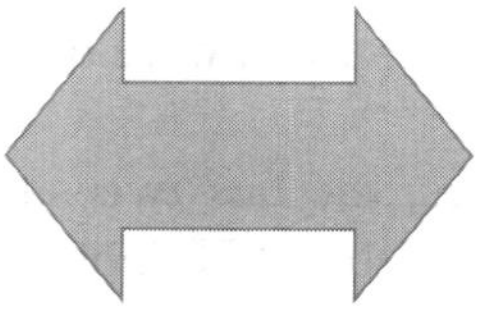

(v)

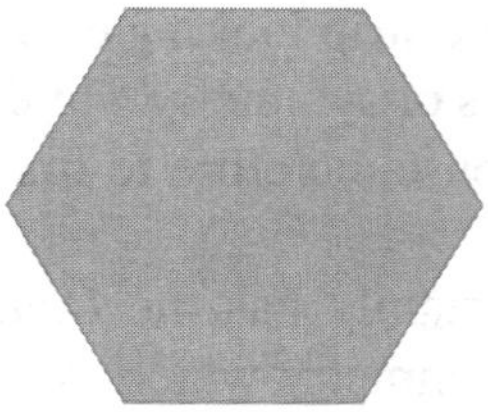

2. State whether true or false. Give reasons.
   (a) There is a finite planar figure with exactly one rotation and two reflections.
   (b) There are finite planar figures with no rotations except the do-nothing rotation.
   (c) If a finite planar figure has only finitely many symmetries and it has a reflection, then it must have symmetry type D_n for some n.
   (d) Every finite planar figure has a reflection symmetry.
   (e) If a finite planar figure has only finitely many symmetries, then it has an equal number of rotations and reflections.
3. Find the following symmetries. All rotations take place in the anticlockwise direction unless stated otherwise.
   (a) Reflection about line joining vertex A to mid-point of opposite side (BC) followed by a 120 degree rotation in an equilateral triangle.
   (b) Inverse of the reflection about a diagonal of a square.
   (c) Rotation of 180 degrees applied twice in a rectangle.
   (d) Inverse of a 90 degree rotation in a square.
   (e) 180 degree rotation followed by a vertical reflection in a square.

6.3 Symmetry of Strip Patterns

In the previous section, we learnt to classify the symmetry type of a finite planar figure that had only finitely many symmetries. We also saw that it is possible for a finite figure like a circle to have infinitely many symmetries. Now, we turn our attention to certain types of 'infinite' figures. In this section, we will study strip patterns, also called frieze patterns. So what is a strip pattern?

A strip pattern is a pattern that is created by taking a basic motif or design and placing this motif at equal intervals along a horizontal line. So the basic motif or design is repeated at equal distances to the left and to the right. You have to use your imagination to think of this as repeating endlessly to the right as well as the left along the line with equal spacing between any two successive motifs. The pictures in Figure 6.3.1 are examples of strip patterns. We are of course constrained by the finiteness of the page, so you will have to visualise the pattern continuing infinitely to the left and to the right.

In the given examples, the last two strips are actually decorations that appear in the Lodhi tombs in Delhi and the Qutabshahi tombs in Hyderabad. Ancient monuments and modern buildings abound with examples that could be considered as strip patterns, albeit with a bit of visioning on the part of the viewer in order to continue the pattern infinitely to the right and left.

A symmetry of a strip pattern is any action that leaves the strip looking exactly the same in the exact same position. We have already encountered two types of symmetry while dealing with finite planar figure, namely rotations and reflections. It makes sense therefore to first look for rotations and rotocenters, reflections and lines of reflection.

As in the case of finite figures, every strip pattern certainly has the *DNR* symmetry. Now, suppose that a given strip pattern also has a rotational symmetry which is not *DNR*. Mentally rotate a strip pattern, remember that it has to come back and occupy the same position and look exactly as it did when you started. Now, can you guess what the angle of rotation could possibly be? The answer is that if at all there is a non-*DNR* rotation, then it has to be a 180 degree rotation. Let us now consider the strip patterns listed in Figure 6.3.1 and see which of them has a 180 degree rotation as a symmetry and also find the rotocenters.

(A)

(B)

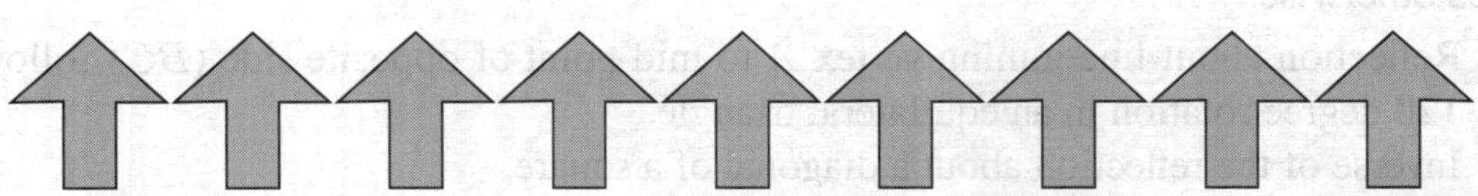

(C)

(Figure 6.3.1 contd.)

(Figure 6.3.1 contd.)

(D)

(E)

(F)

Figure 6.3.1 Strip patterns. [C6]

Figure 6.3.2 shows the strip patterns of Figure 6.3.1 having undergone a 180 degree rotation. We can see that in Figure 6.3.2, the strips (B), (C) and (D) do not look the same after the 180 degree rotation. In these three cases, the 180 degree rotation is not a symmetry of the original strips.

Strip (A) in fact looks exactly as it did before the 180 degree rotation. In this case, we do have a rotational symmetry of 180 degrees. We have marked dots and small diamonds at various points on Strip (A) to indicate possible rotocenters. Since a strip pattern is an infinite pattern repeating

(A)

(B)

(C)

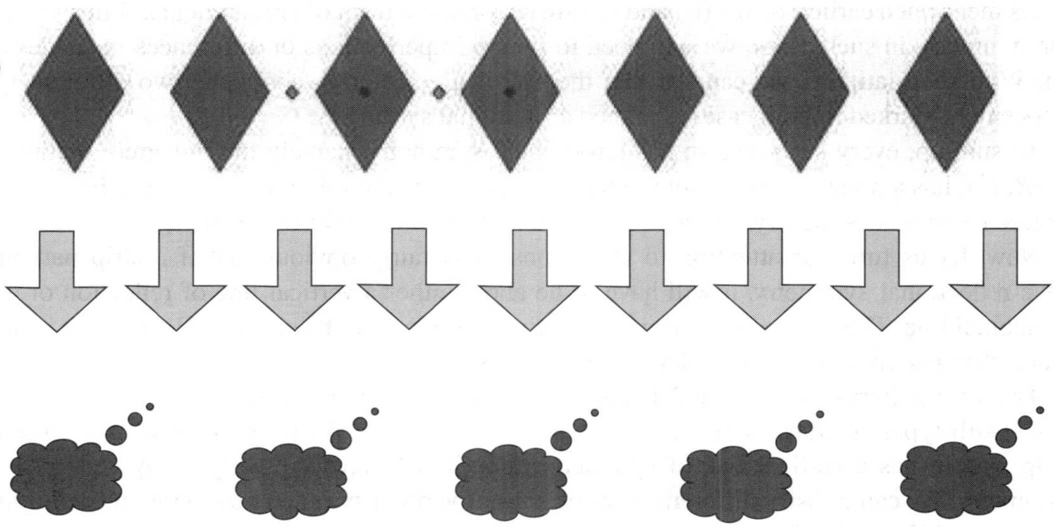

(Figure 6.3.2 contd.)

(Figure 6.3.2 contd.)

(D)

(E)

(F)

Figure 6.3.2 180 degree rotations on strip patterns of Figure 6.3.1. [C6]

endlessly to the left and to the right, if it has a rotocenter it will have infinitely many rotocenters. In the case of Strip (A) too, the centre of every basic motif (marked by the dot) will be a rotocenter. The mid-point between two basic motifs (marked by the small diamond) will also be a rotocenter. However, there are only **two types of rotocenters,** namely the dot type and the small diamond type. We say that two **rotocenters are of the same type** if there is a symmetry of the strip which can take one rotocenter to the other.

As mentioned earlier, Strips (E) and (F) are decorations from old monuments. When we analyse for symmetry in such cases, we will need to ignore imperfections or differences in shades and so on. With such latitude, we can see that the 180 degree rotation about the two different type of rotocenters marked in both cases is indeed a rotational symmetry.

To sum up, every strip pattern has a rotational symmetry, namely the 0 degree rotation or the *DNR*. If it has a nonzero degree rotational symmetry, then the rotational symmetry has to be a 180 degree rotational symmetry. But a strip pattern may or may not have a 180 degree rotation.

Now, let us turn our attention to reflections. It is fairly obvious that if a strip pattern does have reflectional symmetry, it will have to be about either a vertical line of reflection or along a horizontal line. The lines of reflection for a strip pattern cannot be oblique. However, it is possible that a strip pattern does not have any reflectional symmetry.

The strip patterns of Figure 6.3.1 that possess reflectional symmetries are presented in Figure 6.3.3 with types of lines of reflection marked on the strips. As in the case of rotocenters, if a strip pattern has a vertical line of symmetry, then it will have infinitely many vertical lines of symmetry. We can classify different types of lines of vertical symmetry by using the definition that two vertical **lines of reflection are of the same type** if there is a symmetry that takes one to the other. Note that there can only be one horizontal line of reflection. In the case of Strips (E) and (F), we have to ignore imperfections and colours as before.

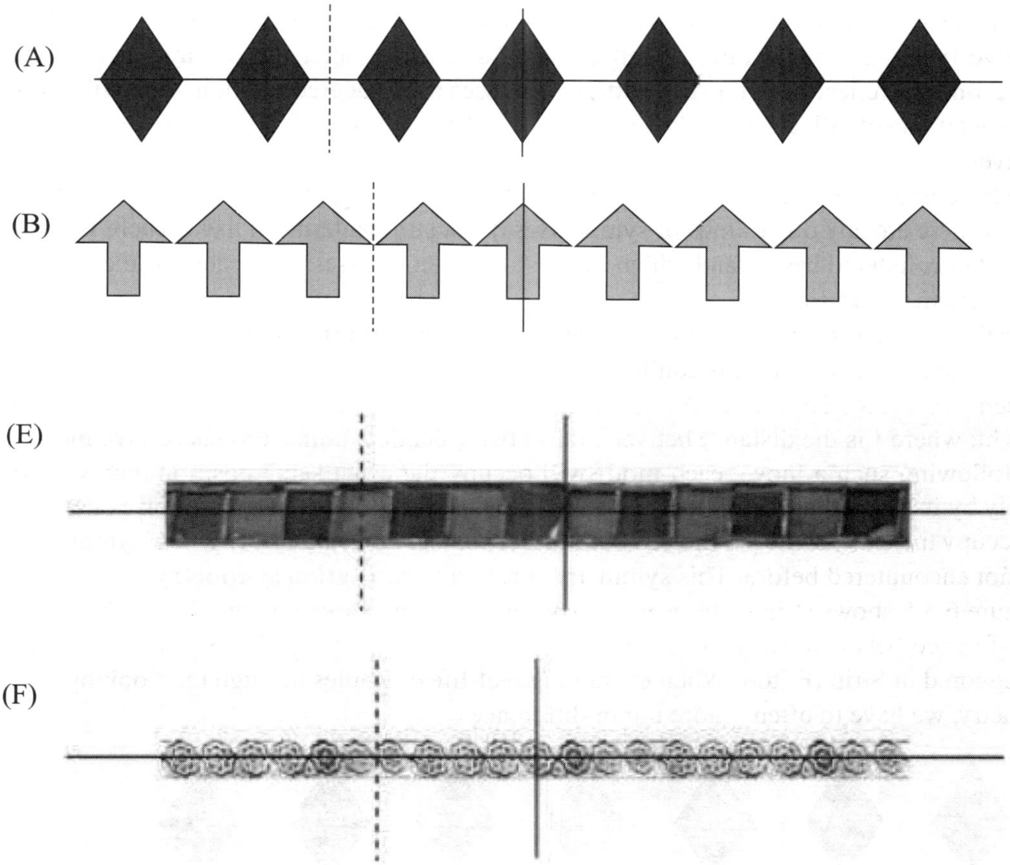

Figure 6.3.3 Strip patterns with reflection symmetries. [C6]

We see from Figure 6.3.3 that Strips (A), (B), (E) and (F) have reflection symmetries whereas Strips (C) and (D) do not have any reflection symmetries. Strips (A), (B), (E) and (F) each possesses two types of vertical lines of reflection (denoted by the dotted line and the normal line). Note that there will be infinitely many vertical lines of reflection of each type. Strips (A), (E) and (F) also have a horizontal line of reflection, but Strip (B) does not have a horizontal line of reflection. You can also have strips with a horizontal line of reflection and no vertical line of reflection. See Figure 6.3.4.

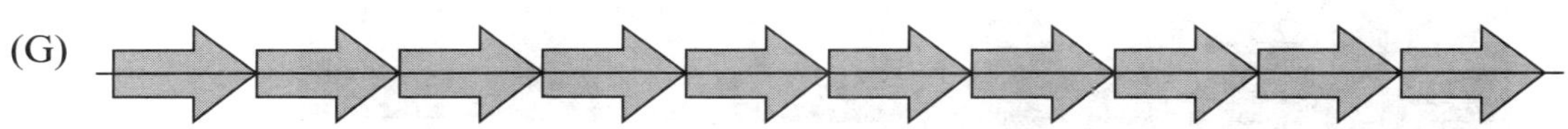

Figure 6.3.4 Strip pattern with a horizontal reflection symmetry and no vertical line of symmetry.

Some questions that arise naturally are: Will a strip that has a vertical line of reflection always have two types of vertical lines of reflection? Can a Strip pattern have three or more types of vertical lines of reflection? If a strip pattern possesses a 180 degree rotation, then will it also have a horizontal line of reflection? Is the vice-versa true? Try to think of the answers to these questions intuitively.

We have now investigated rotations and reflections of strip patterns. The question remains as to whether there are any other kinds of symmetries that a strip pattern can have. There is one that we have not encountered before and which every strip pattern possesses because of the way in which a strip pattern is constructed.

Recall that to construct a strip pattern, we choose a basic motif and repeat it at equal intervals on a horizontal line. The pattern is continued infinitely to the left and to the right with equal spacing between two successive motifs. Now, imagine ordering every motif to move a fixed distance **t** to the right, where **t** is the distance between (any) two (identical points on) successive motifs. Then, after following such a move, each motif will occupy the exact same position that was occupied initially by its immediate neighbour to the right. Thus, the strip pattern will look exactly as before and occupy the exact same position as before. Hence, this move describes a new symmetry that we have not encountered before. This symmetry is called a **translation symmetry.**

Figure 6.3.5 shows strip patterns with their respective **t**s marked on the figure. We are ignoring the difference between two consecutive tiles due to colour in Strip (E). Similar differences have been ignored in Strip (F) too. When examining real-life examples through the 'looking glasses' of symmetry, we have to often ignore minor differences.

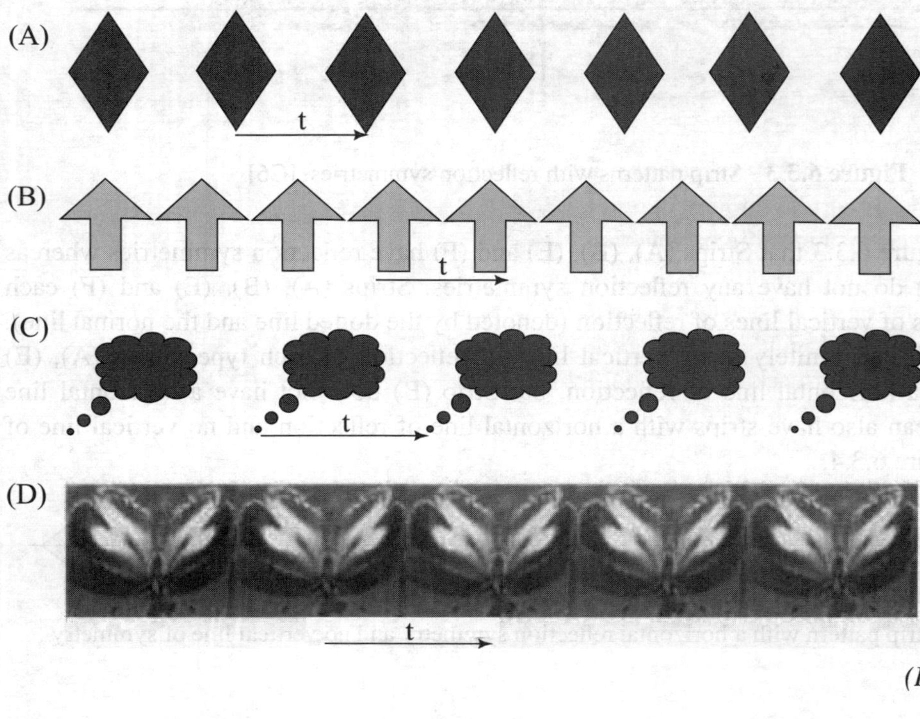

(Figure 6.3.5 contd.)

(Figure 6.3.5 contd.)

(E)

(F)

Figure 6.3.5 Strip patterns with t's shown. [C6]

If we denote by T the symmetry describing the move to the right by a distance **t,** then T^2 will denote the symmetry that involves moving the strip pattern to the right by a distance of **2t.** We can similarly have a translation T^n for any natural number n. So, it should be fairly obvious that T^0 will denote the *DNR* symmetry and T^{-1} will denote the symmetry which involves moving the strip pattern to the left by a distance **t.** Thus, for each natural number n, the translation symmetry T^{-n} will denote the movement of the strip pattern by a distance **nt** to the left. Consequently, a strip pattern always possesses infinitely many translation symmetries, namely T^m for each integer m.

The aforementioned discussion makes clear that every strip pattern has infinitely many translation symmetries. Recall that towards the end of Section 6.2, there was a discussion about the collection of all symmetries of an object X forming a group under the operation of a symmetry followed by a symmetry. The discussion there holds true for any object X and in particular also when X is a strip pattern.

We have seen earlier too that a symmetry followed by a symmetry is always a symmetry. This property was referred to as closure. In a finite planar figure, a rotation followed by a rotation is aways a rotation.[3] Several questions arise that are similar. What is the symmetry that occurs when a rotation is followed by a reflection or a reflection is followed by a rotation? What happens if a translation is followed by a rotation or vice-versa?

It turns out that when a reflection is followed by a translation, the resultant symmetry is not a reflection or a rotation or a translation but a new symmetry called a **glide reflection.** Let us examine the figures below to understand a glide reflection.

Figure 6.3.6 shows the effect on the capital letter R of a reflection followed by a translation on a planar surface. Figure 6.3.7 shows the same net effect as in Figure 6.3.6 but as a glide reflection. A **glide reflection** is described as a single symmetry which is the result of a reflection followed by a translation in a direction parallel to that of the line of reflection.

[3] In group theoretic terms, this means that the subcollection of rotations of a finite planar figure itself forms a group and is referred to as a subgroup. However, in a strip pattern when we apply two distinct rotations, that are 180 degree rotations one after the other, the result can be a translation and not a rotation. So, we may not have the rotations of a strip pattern forming a subgroup. The collection of all translations of a strip pattern does form a subgroup of the full symmetry group.

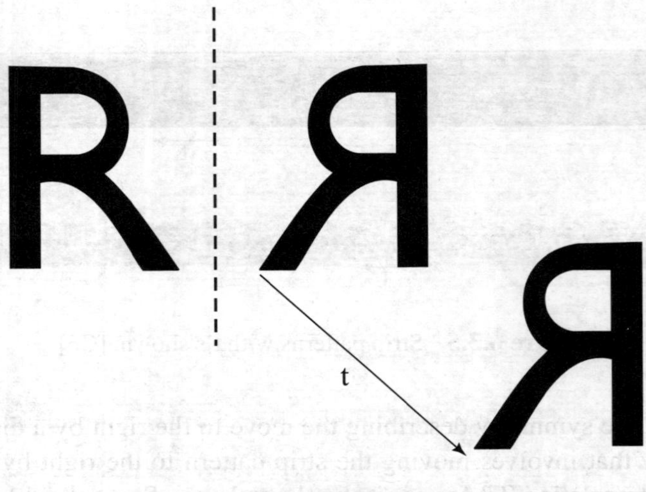

Figure 6.3.6 Glide reflection in a plane.

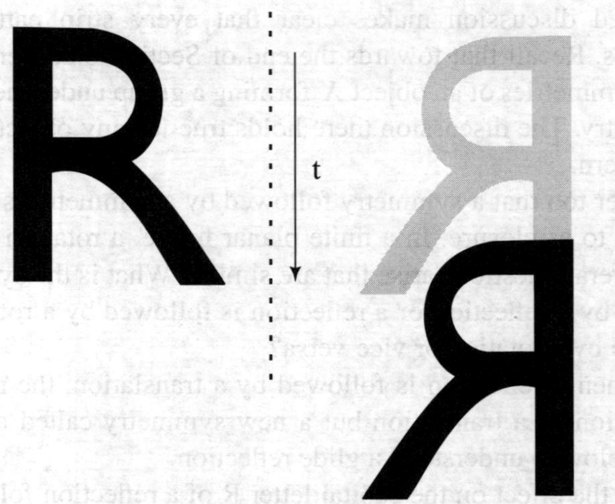

Figure 6.3.7 Reflection followed by a translation in a plane.

Figures 6.3.6 and 6.3.7 show a glide reflection in a plane. Let us turn our attention to strip patterns in which glide reflections can occur. One way that a glide reflection can occur in a strip pattern is if it possesses a reflection symmetry about a horizontal line. Figure 6.3.8 illustrates how a horizontal reflection followed by a translation in a strip pattern gives rise to a glide reflection. The basic motifs in the strip patterns have been numbered so that we are able to track the effect of the symmetry applied.

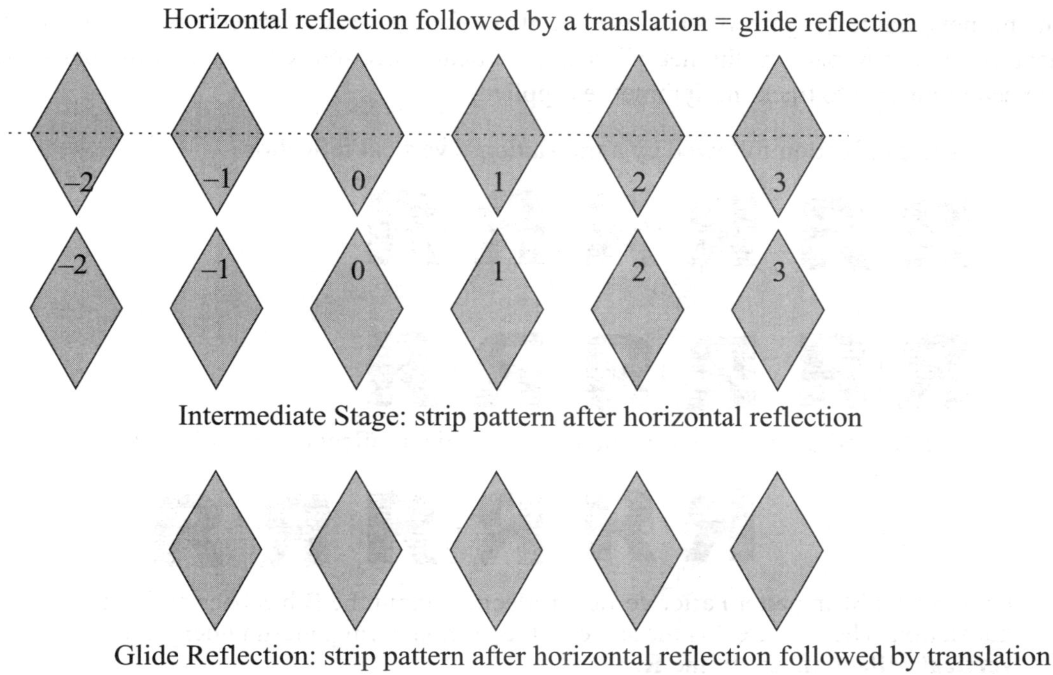

Figure 6.3.8 Glide reflection in a strip pattern.

However, it is possible for a glide reflection to occur in a strip pattern without the strip pattern possessing a horizontal reflection. How do we construct such a strip pattern? You can also try this method as a task. Choose a motif. Then perform a glide reflection in the plane, on the motif. The line of reflection and translation should both be in the horizontal direction. The result will be the motif and its image after the glide reflection. This 'double motif' will form our basic motif for creating a strip pattern. You can check that the resulting strip pattern does not have reflection symmetry about a horizontal line but has glide reflections.

The next figure illustrates this (Figure 6.3.9). The basic glide reflection will take an R to the upside-down R to the right of it and the upside-down R to the normal R on its right. The dotted line shows the line of reflection in the glide reflection and the arrow shows the amount of translation for the glide reflection.

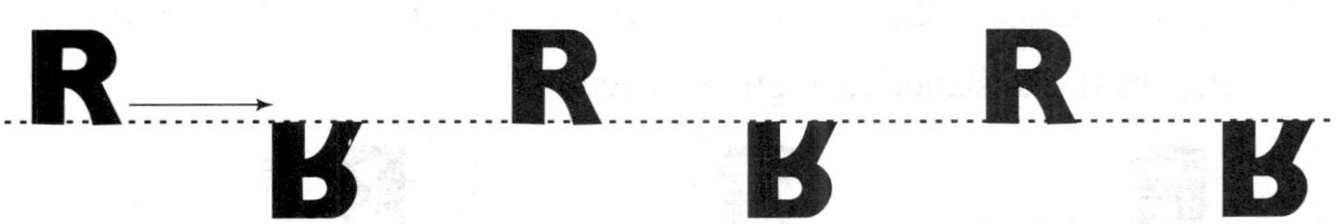

Figure 6.3.9 Glide reflection in a strip pattern that has no horizontal reflection.

In the next figure (Figure 6.3.10), we see that when a vertical reflection is followed by a translation in a strip pattern, the net effect is a vertical reflection. Once again, the basic motifs have been numbered to track the symmetries applied.

Vertical reflection followed by a translation = vertical reflection

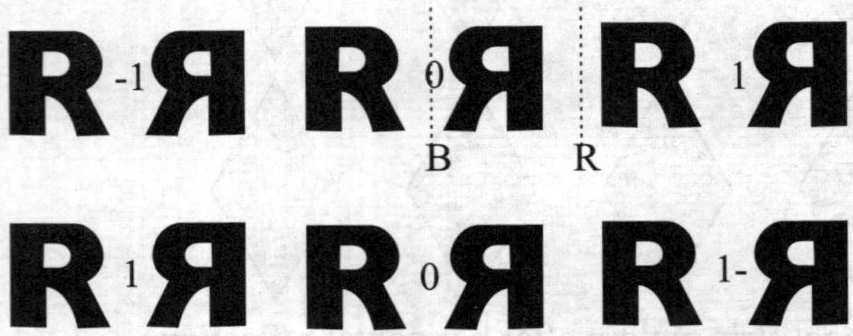

Intermediate Stage: strip pattern after vertical reflection about line B

Final Stage: Strip pattern after vertical reflection about the **B** has been followed by a translation. The net effect is the same as the original strip pattern undergoing a **vertical reflection** about line **R**.

Figure 6.3.10 Vertical reflection followed by a translation in a strip pattern.

To summarise, every strip pattern will have infinitely many translation symmetries and *DNR*. It may or may not possess a 180 degree rotation, reflections about a vertical line, reflections about a horizontal line and glide reflections.

It is possible to classify strip patterns as we did for finite planar figures that have only finitely many symmetries. This is done on the basis of studying the full symmetry groups of the strip pattern and would require a reasonable knowledge of group theory. This is beyond the scope of this book. However, we can certainly state that there are **seven** different types of strip patterns. Examples of the seven types are in Figures 6.3.11 and 6.3.12.

Pattern I: only translation

Pattern II: translations and glide reflection

(Figure 6.3.11 contd.)

(Figure 6.3.11 contd.)

Pattern III: translations and vertical reflections

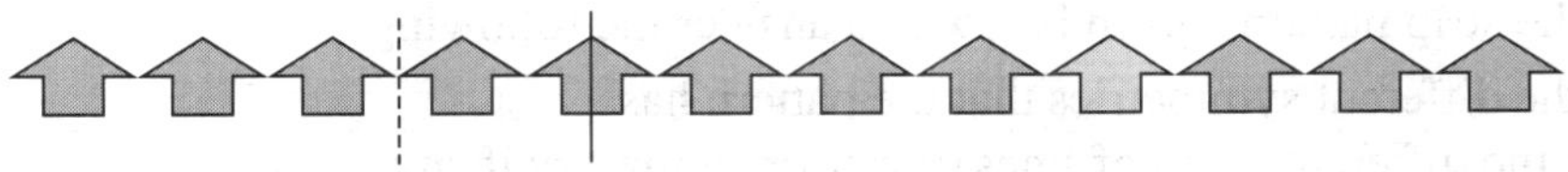

Pattern IV: translations and 180 degree rotations

Figure 6.3.11 Seven types of strip patterns: Patterns I–IV.

Pattern V: translations, glide reflections, vertical reflections and 180 degree rotations

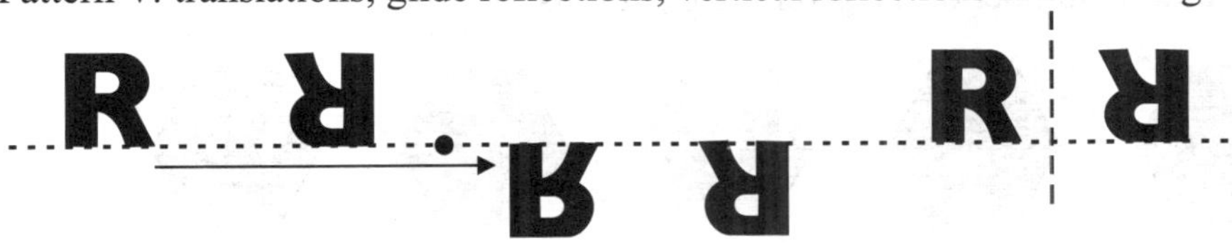

Pattern VI: translations, horizontal reflection and glide reflections

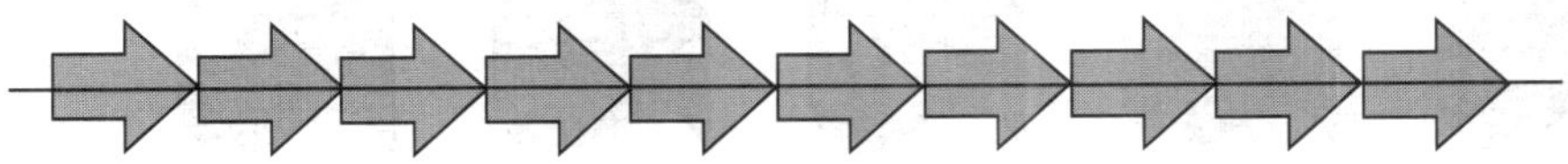

PatternVII: translations, horizontal reflection, vertical reflections, 180 degree rotations and glide reflections

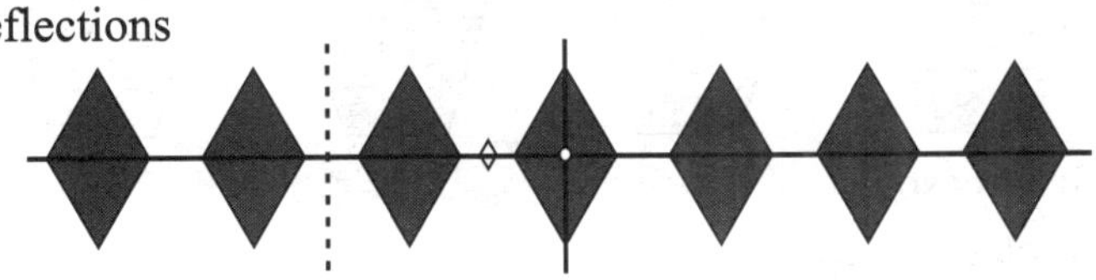

Figure 6.3.12 Seven types of strip patterns: Patterns V–VII.

Quick Review

After reading this section, you should be able to:

- Identify strip or frieze patterns.
- Identify the symmetries that a strip pattern possesses.
- Identify types of rotocenters and lines of reflection if any.
- Find the result of a symmetry followed by a symmetry for a strip pattern.
- Find the inverse of a symmetry for a strip pattern.
- Identify the 'Pattern Number' (I–VII) to which the given strip pattern belongs.

EXERCISE 6.3

1. Examine the strip patterns given below and answer the following.
   (a) List the different symmetries that the pattern has.
   (b) Mark the different types of lines of mirror symmetry if any.
   (c) If there is a non-do-nothing rotation, then mark the different types of rotocenters and describe the rotations.
   (d) If there is a glide reflection, then mark the line of reflection and translation to depict the glide reflection.
   (e) Will there always be a translation symmetry? If so, draw an arrow to mark the direction of the translation.
   (f) Identify the Pattern Number of the given strip patterns.

   (i) 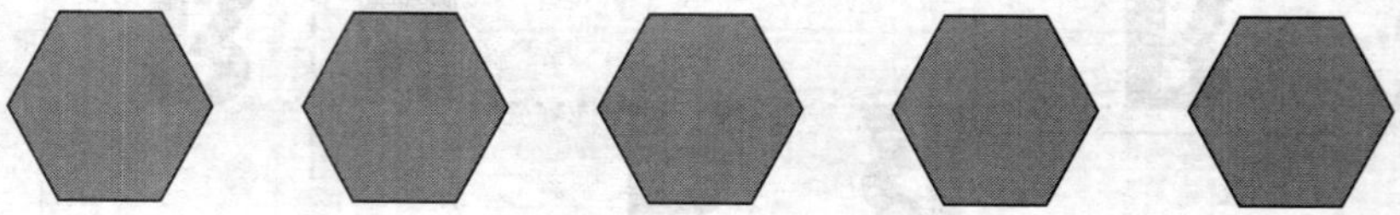

   (ii) 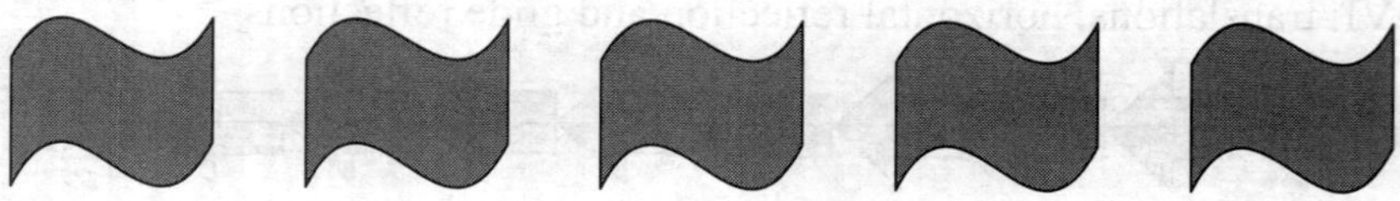

   (iii) 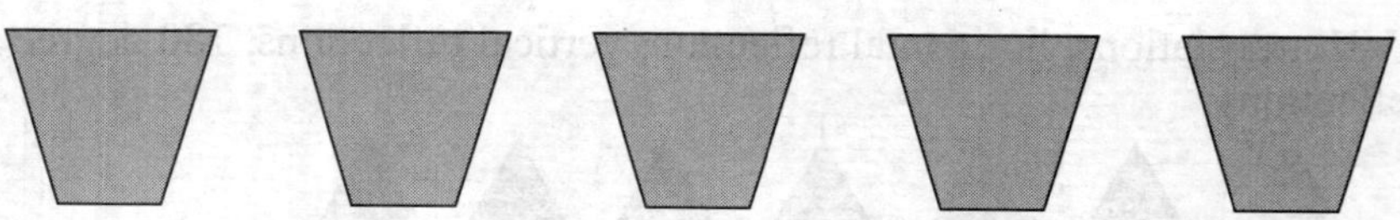

2. State whether true or false. Give reasons.
   (a) Every strip pattern has a rotational symmetry.
   (b) There are strip patterns with only finitely many symmetries.
   (c) Every strip pattern has a translation symmetry.
   (d) If a strip pattern has reflection symmetries about vertical lines, then it must have at least two types of vertical lines of reflection.
   (e) If a strip pattern has 180 degree rotational symmetries, then it must have at least two types of rotocenters.
   (f) A strip pattern can only have one horizontal line of reflection symmetry.
   (g) A strip pattern with a reflection symmetry about a horizontal line has to possess a 180 degree rotational symmetry.
3. Find the following symmetries. All rotations take place in the anticlockwise direction unless stated otherwise.
   (a) $\rho \circ \rho$, where ρ is a 180 degree rotational symmetry of a strip pattern.
   (b) $\rho \circ \sigma$, where ρ and σ are 180 degree rotational symmetries of a strip pattern about two distinct rotocenters of the same type.

(c) Inverse a reflection in a strip pattern.
(d) Inverse of a translation in a strip pattern.
(e) Inverse of a 180 degree rotation in a strip pattern.

6.4 Symmetry of Wallpaper Patterns

In this chapter, so far we have learnt to identify symmetries of finite planar figures and of strip or frieze patterns. In this section, we concentrate on **wallpaper patterns.** As the name indicates, intuitively, it is a pattern which can cover a wall.

We defined strip patterns through the process of creating them. A strip pattern is created by choosing a basic motif and repeating the motif in a single direction (left–right) along a horizontal line at equal intervals indefinitely. A wallpaper pattern is created by choosing a strip pattern and stacking it one on top of the other in the up–down direction at equal intervals indefinitely. A symmetry of a wallpaper pattern is any move that leaves the wallpaper pattern as a whole occupying the same position and looking exactly as it did in the beginning.

Figure 6.4.1 provides examples of wallpaper patterns. By virtue of their construction, wallpaper patterns will certainly have translation symmetries in two different directions. One would be the symmetry that we get by translating the wallpaper pattern by a distance of **t** to the right in the horizontal direction, where **t** is the distance between two identical points in successive motifs in the left-right direction. The other is the symmetry that we get by translating the wallpaper pattern by a distance of **s** in the downward direction, where **s** is the distance between two identical points in successive motifs in the up-down direction. Please note that wallpaper patterns extend infinitely in the left-right and the up-down direction and what is shown here is only a very small part of the wallpaper pattern.

(A)

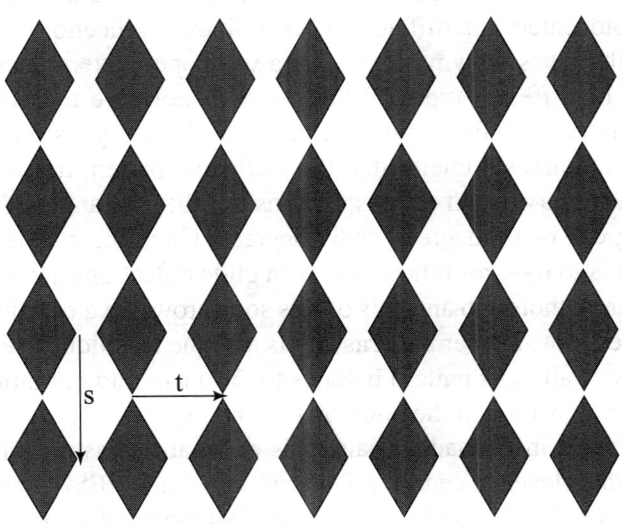

(Figure 6.4.1 contd.)

Figure 6.4.1 contd.)

(B)

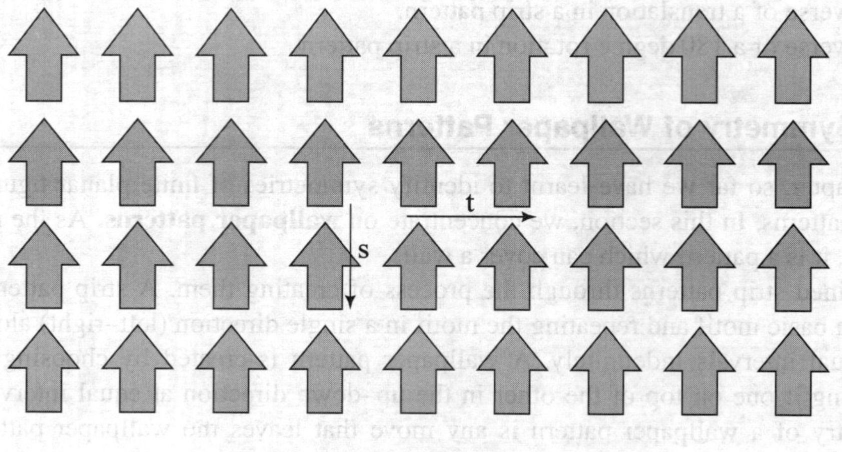

Figure 6.4.1 Wallpaper patterns.

Stacking strip patterns as described in Figure 6.4.1 gives us wallpaper patterns. Every wallpaper pattern has translation symmetries, it may or may not have nonzero degree rotational symmetries, it may or may not have reflection symmetries and it may or may not have symmetries that are glide reflections. Unlike strip patterns, wallpaper patterns can have rotational symmetries which involve rotation other than 180 degrees. They can also have reflections and glide reflections about oblique lines of reflection. As in the case of strip patterns, two rotocenters (lines of reflection) are of the same type (also called equivalent) if there is a symmetry of the wallpaper pattern, that can take one to the other. A basic task, given a wallpaper pattern, is to find the different type of symmetries, different types of rotocenters and different types of lines of reflection.

On the basis of the types of symmetries that a wallpaper pattern possesses, wallpaper patterns are classified into 17 different types. While we will not give the complete classification, we can mention that the method for arriving at the 17 different types first rests on discovering the smallest nonzero rotational symmetry that the wallpaper pattern has. Of course it may not have such a rotational symmetry, but if it does, it turns out that the angle of rotation can only be 180 degrees or 120 degrees or 90 degrees or 60 degrees. Once the smallest rotational symmetry is discovered, the next is to look for reflections, then glide reflections, position of rotocenters, etc., in a systematic manner. A thorough analysis of this sort throws up a definite method for ascertaining the 17 different types of wallpaper patterns and is also the technique used to find out which of the 17 patterns the given wallpaper pattern belongs to. You can find out more about the classification method including an algorithm in the book by Gallian [3].

The Alhambra Palace in Granada, Spain, was originally constructed in the ninth century and then renovated in the eleventh century. This palace is a UNESCO world heritage site and is considered a fine example of Islamic architecture. Examples of all the 17 wallpaper patterns are present in the Alhambra Palace. One of such example is given in Figure 6.4.2. Such wallpaper patterns at Alhambra were a source of inspiration for the artist M. C. Escher.[4] Escher's work lends itself wonderfully to the study of symmetry. Indeed, he used principles of symmetry to create many of his wonderful works. We will look at the connection between Escher like art and symmetry in a latter section.

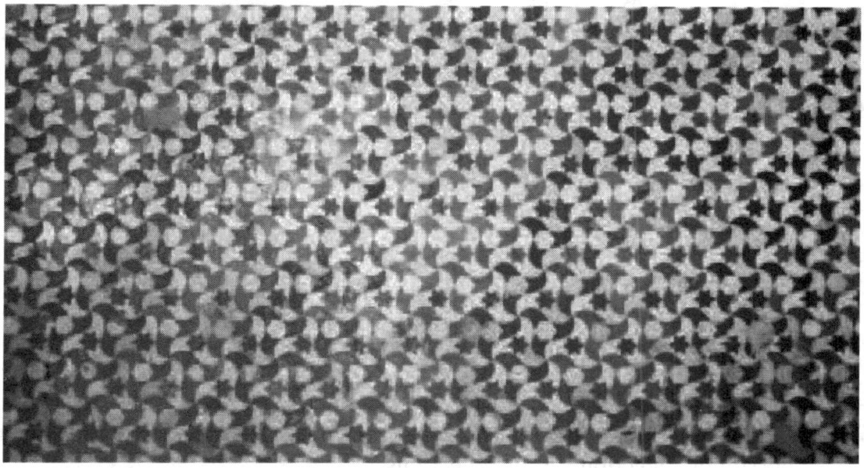

Figure 6.4.2 Wallpaper pattern from the Alhambra Palace. [C4]

Some examples of wallpaper patterns with the different type of symmetries that they possess, rotocenters, lines of reflection are given in Figures 6.4.3 and 6.4.4.

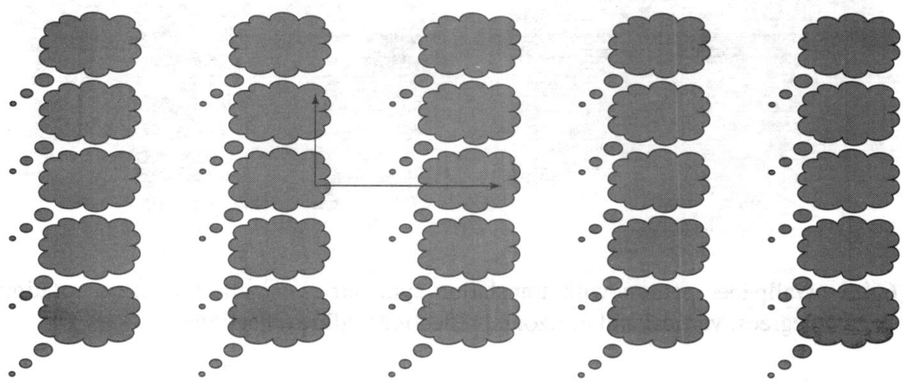

Figure 6.4.3 Wallpaper pattern with only translation symmetries.

[4] You can read about Escher's biography at the MacTutor History of Mathematics archive [W11] and also visit the official website [W8] to see examples of his work.

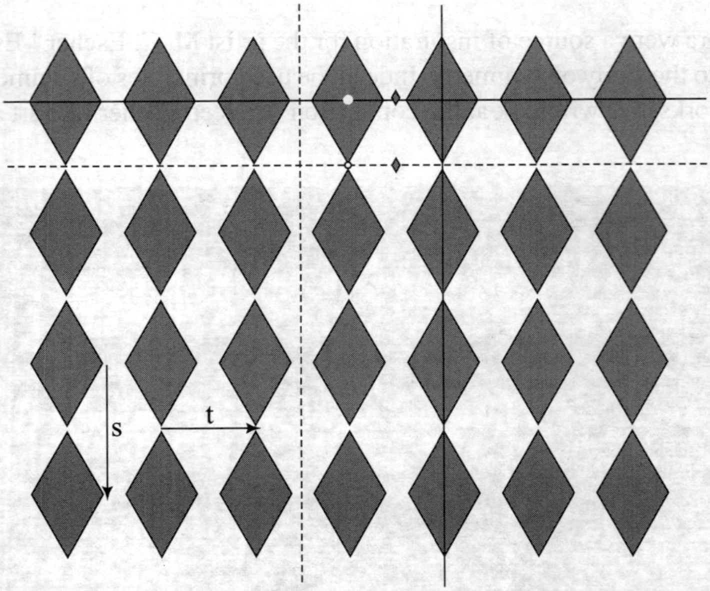

Figure 6.4.4 Wallpaper pattern with translation symmetries, 180 degree rotations, vertical and horizontal reflections, glide reflections.

Try to find the lines of reflection for the glide reflections. Are the lines of reflection for the glide reflections the same as the lines of reflection or different?

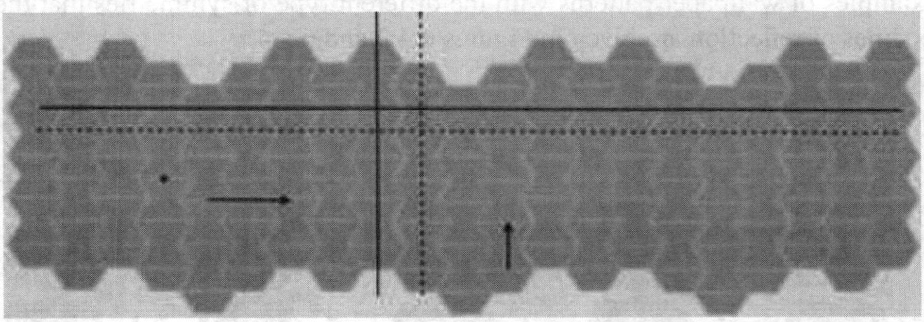

Figure 6.4.5 Wallpaper pattern with translation symmetries, smallest nonzero rotation of 60 degrees, vertical and horizontal reflections, glide reflections.

The wallpaper pattern in Figure 6.4.5 also has nonzero rotational symmetries of degrees 120, 180, 240 and 300. Further, it has oblique lines of reflection, translation along oblique lines and glide reflection along oblique lines. Try to find these. Are the oblique lines of reflection new types of lines of reflection or are they the same as either the horizontal line or the vertical line of reflection? One type of rotocenter is marked on the pattern, are there any more types of rotocenters? If so, find them.

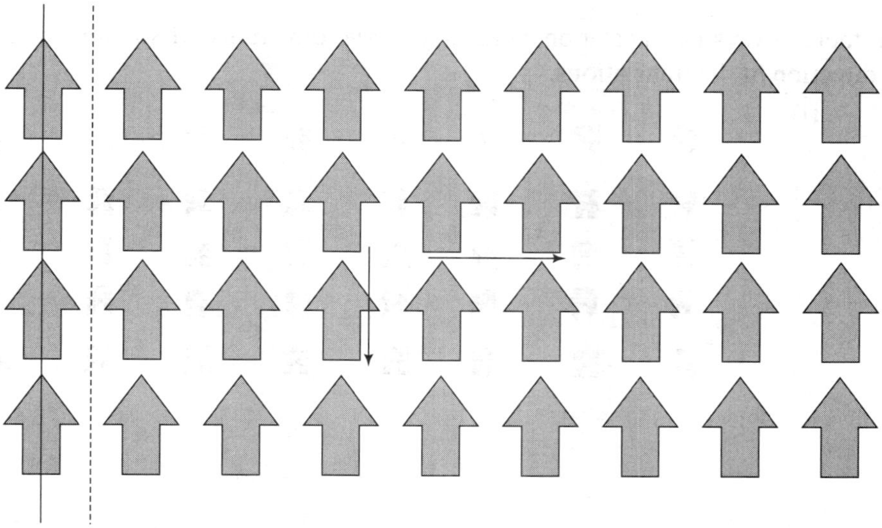

Figure 6.4.6 Wallpaper pattern with translation symmetries, no nonzero rotational symmetries, vertical reflections, glide reflections.

Try to find the lines of reflection for the glide reflections. Are the lines of reflection for the glide reflections the same as the lines of reflection or different?

Quick Review

After reading this section, you should be able to:

- Identify wallpaper patterns.
- Identify the symmetries that the wallpaper pattern possesses.
- Identify translations in two different directions.
- Identify types of rotocenters and lines of reflection, if any.
- Identify the angle of the smallest nonzero rotational symmetry, if any.

EXERCISE 6.4

1. Examine the given wallpaper patterns and answer the following:
   (a) List the different symmetries that the pattern has.
   (b) Mark the different types of lines of mirror symmetry if any.
   (c) If there is a non-do-nothing rotation, then mark the different types of rotocenters and describe the rotations.

(d) Will there always be translation symmetry in two directions? If so, draw arrows to mark the direction of the translations.

(i)

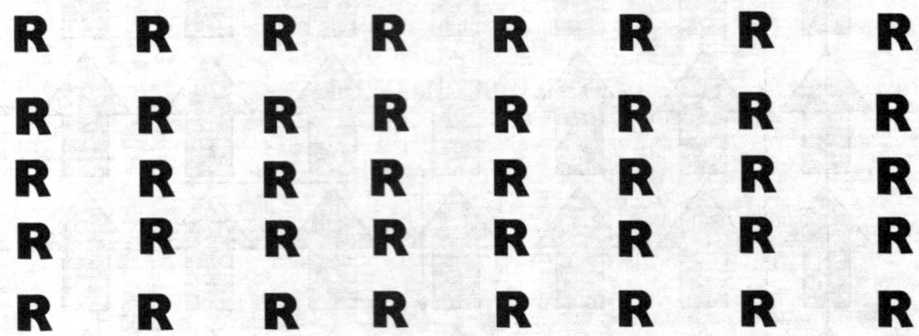

(ii)

(iii)

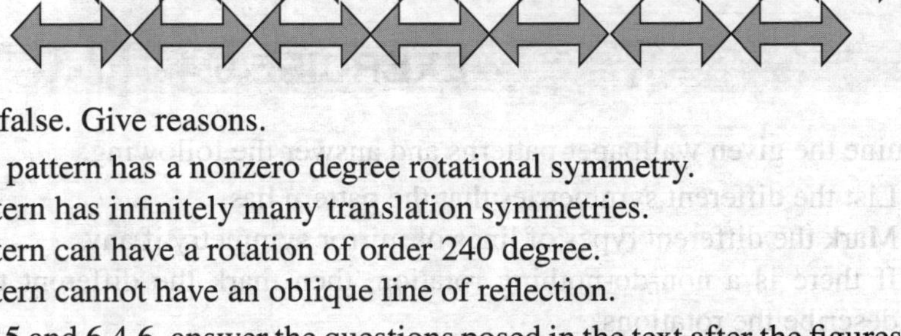

2. State whether true or false. Give reasons.
   (a) Every wallpaper pattern has a nonzero degree rotational symmetry.
   (b) A wallpaper pattern has infinitely many translation symmetries.
   (c) A wallpaper pattern can have a rotation of order 240 degree.
   (d) A wallpaper pattern cannot have an oblique line of reflection.
3. For Figures 6.4.4, 6.4.5 and 6.4.6, answer the questions posed in the text after the figures appear.

6.5 Case Study: Humayun's Tomb

Humayun was the second ruler of the Mughal dynasty, after Babar. He succeeded to the Mughal throne in 1530 but was defeated and driven out of Delhi in 1540 by Sher Shah Suri. After a period of 15 years in exile, Humayun claimed back the throne by defeating Sher Shah's successor and son Sikander Shah Suri in 1555. Towards the end of January 1556, Humayun fell while climbing the steps of the library in Sher Mandal, Purana Qila, Delhi and died. Nine years later, in 1565, Humayun's senior widow Haji Begum commissioned a tomb for him. The tomb reflects Indo-Islamic architectural style and is set in a garden. Though built entirely of red sandstone, Humayun's tomb is considered a prototype for the world-famous Taj Mahal at Agra. In this section, we shall explore designs and decorative patterns found in the tomb using the lens of symmetry.

Figure 6.5.1 shows a water colour painting of Humayun's Tomb in 1820 by an unknown artist. Though we have only concentrated on the notions of symmetry for two-dimensional or planar objects, the intuitive notions extend to three dimensions or spatial objects. The painting however is two-dimensional and we can examine it from the point of view of planar symmetry. An important fact that one has to keep in mind while exploring symmetry in real life is that one may have to ignore some objects like trees or a slight misalignment or a bit of the design so that one may be able to analyse the main object under study in a fulfilling manner.

Figure 6.5.1 Water colour painting of Humayun's Tomb, c. 1820 [C1].

If we ignore the trees and bushes and concentrate on the main building, then the Tomb is a fine example of bilateral symmetry with the line of mirror reflection passing through the middle of the central dome.

Figure 6.5.2 Photograph of Humayun's Tomb. [C6]

Figure 6.5.2 shows a recent photograph of the Tomb showing the line of reflection. In the photograph, we ignore the tree on the side.

The next picture we shall analyse is an artist's rendition of the layout of Humayun's Tomb (Figure 6.5.3). In the layout, consider the octagonal area marked 2 which has Isa Khan's Tomb and Mosque. Isa Khan was an Afghan nobleman in Sher Shah Suri's court and his tomb, set in an octagonal garden, was built during the lifetime of Isa Khan in 1547-48. Ignoring the building in the centre and the building on one of the sides, the outer and inner octagonal walls are examples of finite objects possessing symmetry type D_8 consisting of eight rotations and eight reflections. On the other hand, when we consider the square area at the top depicting the char-bagh (or quadrilateral garden) with Humayun's Tomb located in the centre, we find that here the symmetry type is like that of a square, namely, D_4 symmetry consisting of four rotations of degrees 0, 90, 180 and 270 and four reflections along the diagonals and the lines joining the mid-points of opposite sides.

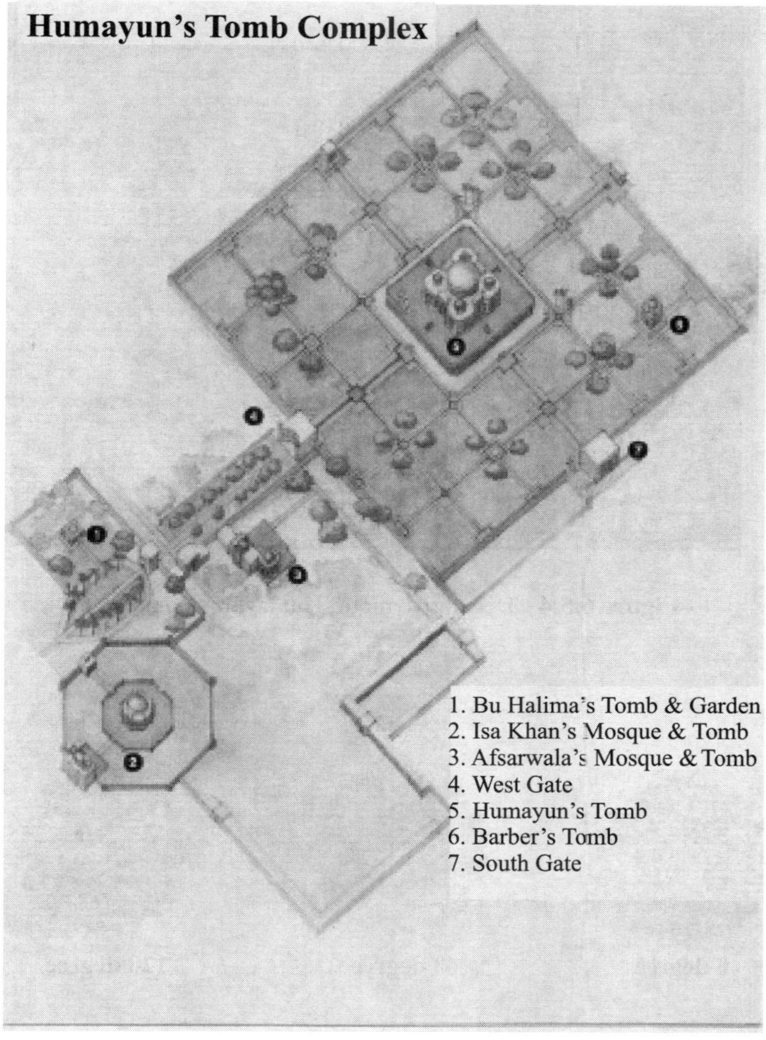

Figure 6.5.3 Layout of Humayun's Tomb. [C6]

The picture in Figure 6.5.4 is of a decorative motif found in Humayun's Tomb.

Studying the flower design carefully while ignoring the line design on the outer part, we find that the basic symmetry pattern is that of a hexagon and that the flower design possesses symmetry type D_6 consisting of six rotational symmetries and six reflection symmetries. The picture below (Figure 6.5.5) shows the result of the six rotational symmetries on the figure. The motif or figure has been rotated by the respective degree mentioned below the figure. The flower design in the centre appears the same in each of the six figures in the picture, only the changes in the outer line design indicate that the figure has been rotated.

Figure 6.5.4 Decorative motif, Humayun's Tomb. [C6]

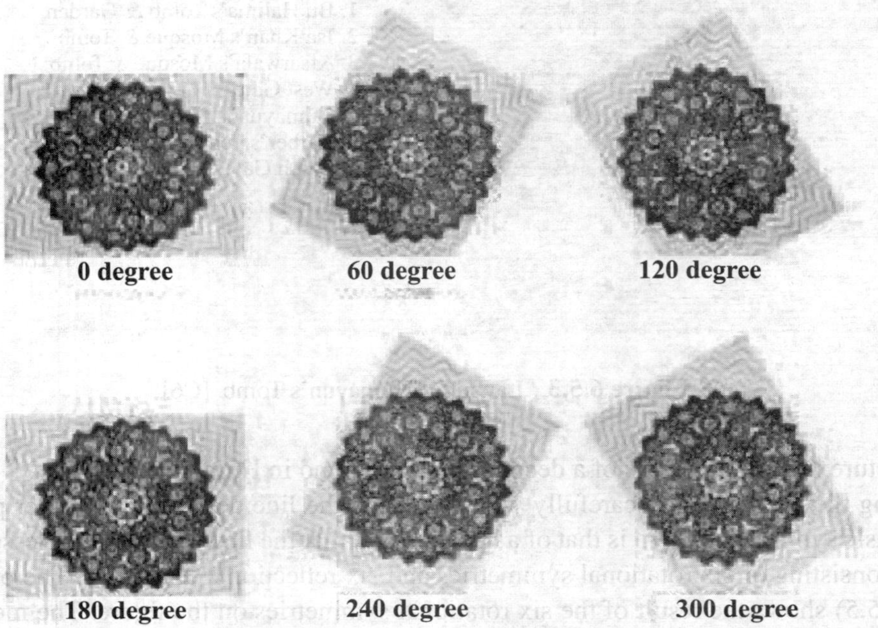

Figure 6.5.5 Rotational symmetries of a decorative motif, Humayun's Tomb. [C6]

Figure 6.5.6 shows the six lines of reflection on the decorative motif.

Figure 6.5.6 Lines of reflection of a decorative motif, Humayun's Tomb. [C6]

In monuments, borders that run along walls, patterns on the edges of roofs, staircase or balcony railings, designs on fence walls, similar objects can provide an ample source for strip patterns. One can also use decorative motifs found in the monuments to create one's own strip pattern to analyse.

Figure 6.5.7 shows a decoration found in the Humayun's Tomb complex which can be regarded as a strip pattern. When we analyse the strip pattern below for symmetries, it is not difficult to see that the strip pattern possesses the following symmetries: translations, 180 degree rotations, reflection along a horizontal line, reflections along vertical lines and glide reflections along the horizontal line of reflection. Thus, it falls into pattern VII in the classification of strip patterns. The two different types of rotocenters, two different types of lines of vertical reflection and the horizontal line of reflection are all marked on Figure 6.5.7.

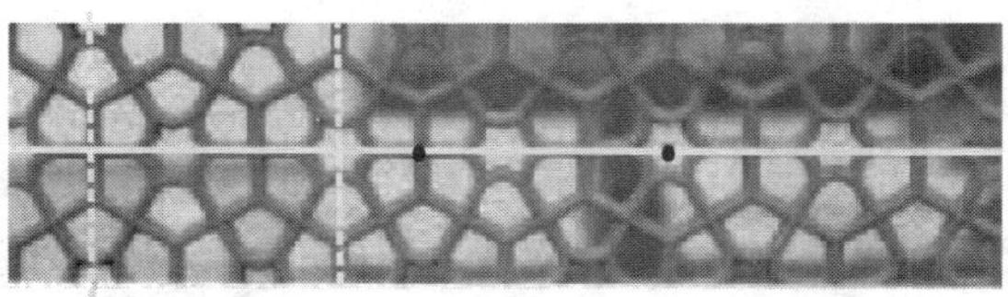

Figure 6.5.7 Strip pattern I, Humayun's Tomb. [C6]

Analyse the strip pattern shown in Figure 6.5.8 and figure out its classification.

Figure 6.5.8 Strip pattern II, Humayun's Tomb. [C6]

Now, let us turn our attention to possible wallpaper patterns. Good sources of wallpaper patterns in Mughal monuments are the 'window jalis', the grill type of decoration or lattice work found in window spaces. Decorative floor tiles can also provide wallpaper patterns.

The picture in Figure 6.5.9 is of a window jali from Humayun's Tomb. The wallpaper pattern in the figure has translations, 180 degree rotations, vertical reflections, horizontal reflections and glide reflections.

Figure 6.5.9 Wallpaper pattern I, Humayun's Tomb. [C6]

The picture shown in Figure 6.5.10 is of decorative designs on the floor in a part of the Humayun's Tomb complex.

Figure 6.5.10 Decorative floor design, Humayun's Tomb. [C6]

Using the decorative floor design in the figure, we can construct a wallpaper pattern. This is shown in Figure 6.5.11. This wallpaper pattern has translations, 180 degree rotations, reflections in two different directions and glide reflections. The translations, different types of rotocenters and the different types of lines of reflection are marked on the figure.

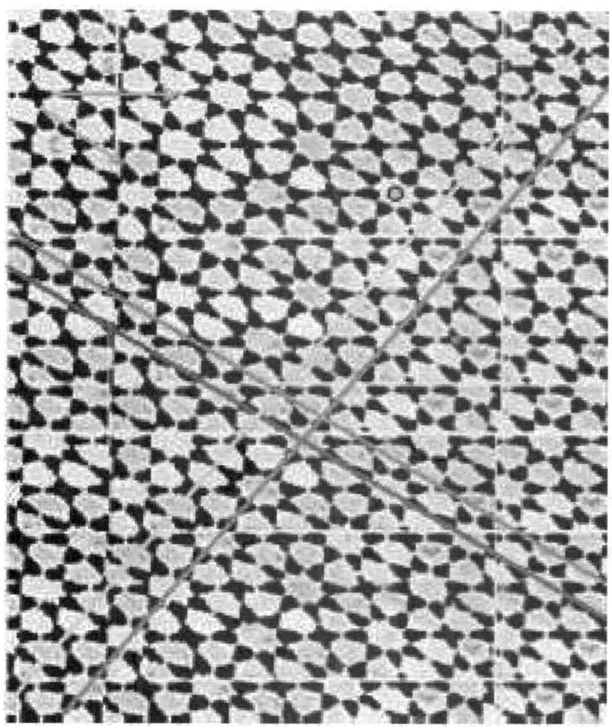

Figure 6.5.11 Wallpaper pattern II, Humayun's Tomb. [C6]

Quick Review

After reading this section, you should be able to:

- Analyse and classify motifs found in buildings and monuments.
- Identify the decorative patterns that can be analysed as strip patterns.
- Classify such strip patterns.
- Identify decorations that can be analysed as wallpaper patterns.
- Identify symmetries of such wallpaper patterns.

EXERCISE 6.5

1. Do a similar case study for an old monument in your area.
2. Analyse a modern building in the same manner using the lens of symmetry.
3. Analyse and classify the strip pattern in Figure 6.5.8. For the wallpaper patterns in Figures 6.5.9 and 6.5.11, find the lines of glide reflection.
4. The following are floor plans, photos, decorative motifs, strip patterns and wallpaper patterns from the Purana Qila in Delhi.
   (a) Analyse the floor plans and photos for symmetry.
   (b) Analyse and classify the decorative motifs.
   (c) Analyse and classify the strip patterns.
   (d) For the wallpaper patterns, mark the translations, the different types of lines of reflection if any, the different types of rotocenters if any and the smallest non-do-nothing rotation. If there are glide reflections, then mark the line of reflection and translation to depict the glide reflection.

(i)

(ii)

(iii)

(iv)

(v) 

Source: [C6].

6.6 Case Study: Abstract Art

At first I had no idea at all of the possibility of systematically building up my figures. I did not know . . . this was possible for someone untrained in mathematics.

— M. C. Escher

Maurits Escher was a Dutch artist who created wonderful artwork that brings to the fore the beauty of symmetry and mathematics. His foray into creating motifs which could fill the plane began after a visit in 1922 to the Alhambra Palace. However, it was only after a second visit in 1936 that he immersed himself almost obsessively into creating his plane-filling artwork inspired by the wallpaper patterns in Alhambra. His brother Berend was a professor of geology at the University of Leiden. On seeing Escher's work on plane-filling designs, Berend sent Maurits a list of mathematical articles that he felt would be of assistance. This was Escher's first encounter with that level of mathematics.

The following is an extract from the biography of M. C. Escher in the MacTutor History of Mathematics archive [W11]:

> Escher read Pólya's 1924 paper on plane symmetry groups. Although he did not understand the abstract concept of groups discussed in Pólya's paper he did understand the 17 plane symmetry groups described there. He subsequently taught himself the principles by which each of the 17 groups operated.

Over the next few years, Escher used his new mathematical insights to produce a series of intricate coloured drawings. In this section, we will analyse some abstract art which is similar to that of Escher's wonderful artwork. Due to copyright issues, we are unable to use Escher's original artwork in this section. However, the reader can find the entire collection on the official website

[W8]. The aim will be to examine the examples of art either as a finite planar object or a strip pattern or a wallpaper pattern. We however begin with two examples of Escher's own work. These are wall sculptures that Escher constructed.

The first is a photograph taken by Bouwe Brouwer which has been converted into monochrome (Figure 6.6.1). The description given by the photographer is as follows: 'Wall tableau of a tessellation by local resident M. C. Escher on the Princessehof ceramics museum, Leeuwarden. The tableau appears to be based on the ink and watercolour Regular Division of the Plane Drawing #47, a study for Verbum, 1942'.

Figure 6.6.1 Wall sculpture by Escher, Leeuwarden. [C2]

As a wallpaper pattern, this has translation in two directions. It has no nonzero degree rotation, no reflections and no glide reflections.

The second example of a wall sculpture by M. C. Escher is mounted on the walls of the water purification plant in the Hague (Figure 6.6.2). The photograph has been taken by Reinier De Jong and has been converted to monochrome for use here with permission from the photographer.

This wallpaper pattern too has translation in two directions. It has no nonzero degree rotation, no reflections and no glide reflections.

Next, we have a floor puzzle inspired by Escher's work. This was one of the puzzles which was part of an exhibition at the International Congress on Mathematical Education held in Seoul, South Korea in 2012.

Figure 6.6.2 Wall sculpture by Escher, the Hague. [C3]

Figure 6.6.3 Floor puzzle inspired by Escher's work. [C6]

If one analyses the photograph in Figure 6.6.3 carefully, one will notice that the plane-filling design of lizards is using the principle of filling the plane with equilateral triangles. (The circle encloses three lines in which each adjacent pair encloses an angle of 120 degrees.) This also shows that if we were to concentrate only on the three lizard whose faces meet in the circle and ignore the colour differences, then that finite planar motif will be of C_3 symmetry type. The three symmetries will be the 0 degree, 120 degree and 240 degree rotations. As a wallpaper pattern, this will have symmetries consisting of translations in two directions, nonzero degree rotations of 120 degree and 240 degree, no reflections or glide reflections.

The next artwork we analyse is also a plane-filling template of four lizards by Kathy Barbo. The photograph of the artwork is by Kathy Barbo and is being used here with permission from the artist. More examples of her Escher-inspired work can be seen on her website *Art Projects for Kids* [W1].

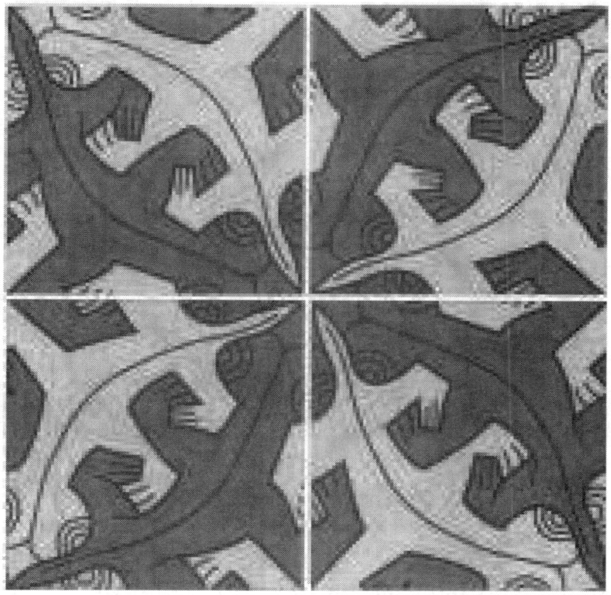

Figure 6.6.4 The four lizard template. [C7]

As a finite planar object, the motif in Figure 6.6.4 has four symmetries consisting of the 0 degree, 90 degree, 180 degree and 270 degree rotations and hence will be of type C_4. Figures 6.6.5 and 6.6.6 show respectively a strip pattern and a wallpaper pattern created using this template repeatedly.

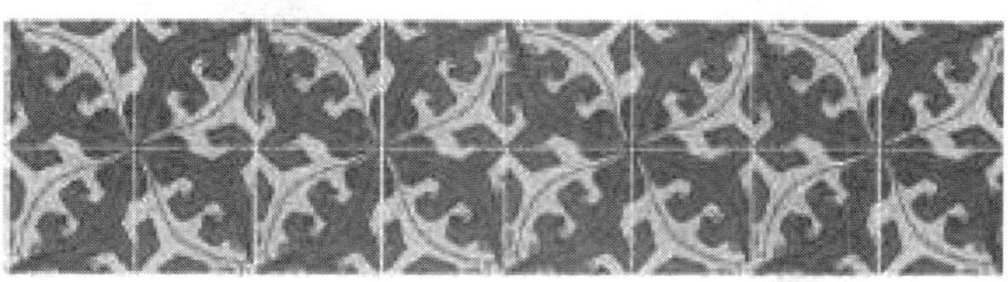

Figure 6.6.5 The four lizard strip pattern.

The strip pattern in the figure is of the Pattern IV type. The symmetries it has are translations and nonzero degree rotations of 180 degree.

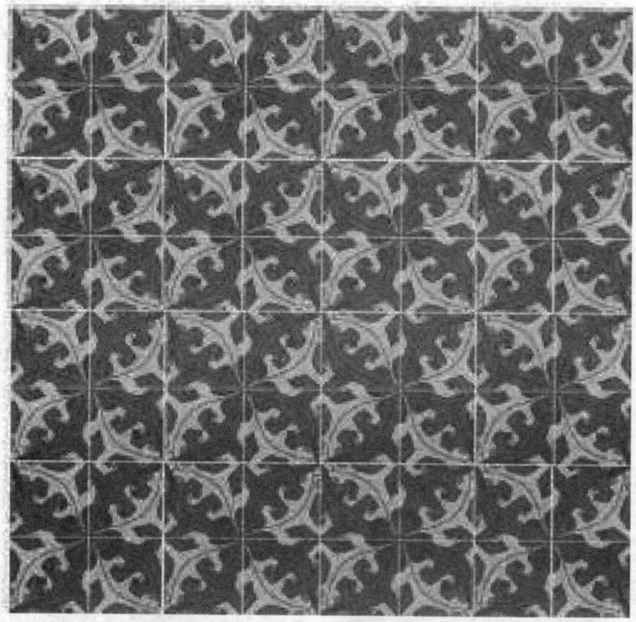

Figure 6.6.6 The four-lizard wallpaper pattern.

The wallpaper pattern in Figure 6.6.6 has the following symmetries: translation in two directions and nonzero degree rotations of 90 degree, 180 degree and 270 degree. It does not have reflections or glide reflections.

In 1941, Escher created a woodcut titled *Plane-Filling Motif with Reptiles* [W4]. Figure 6.6.7 has been inspired by this woodcut. It shows both the original motif and the motif after having undergone a 180 degree counterclockwise rotation.

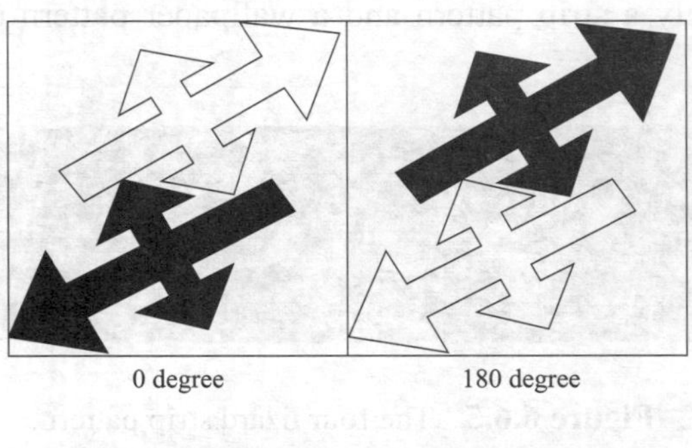

Figure 6.6.7 The Yin-Yang arrows.

If we consider the original motif without ignoring any of the facets in the figure, then we find that the only symmetry possible is the zero degree rotation and so the figure will be of C_1 type. If, however, we ignore the colours, then we see that there is an additional rotational symmetry of 180 degrees. There are still no reflection symmetries and so in this case the figure will have C_2 type.

The following wallpaper pattern has been created using the Yin-Yang arrows motif. It has translation in two directions. It also has nonzero degree rotations of 90 degree, 180 degree and 270 degree. It does not have reflections or glide reflections. Figure 6.6.8 shows the wallpaper pattern undergoing the aforementioned rotations. A small black dot has been added to the wallpaper pattern to track the rotations.

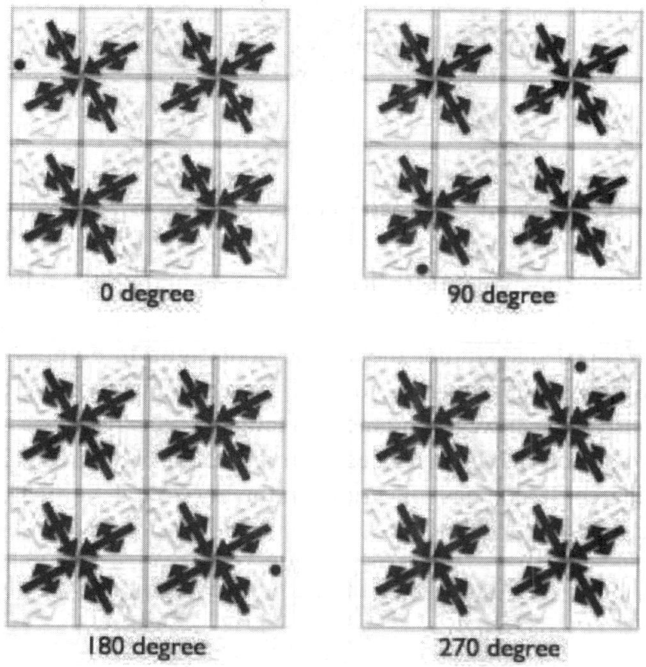

Figure 6.6.8 The Yin-Yang arrows wallpaper pattern.

Escher produced a large body of work, many of which lend themselves to analysis using the lens of symmetry. Indeed, given what we know about Escher, he blended his genius as an artist with that of a nascent but untrained mathematical eye for symmetry to bring forth his creations.

We end this section with two examples of artwork by Regolo Bizzi.[5] The photographs of the artwork are also by the artist and monochrome versions are being used with his permission.

The first artwork has been inspired by regular pentagons (Figure 6.6.9). If variations in colour are ignored, it has symmetry type C_5. The rotational symmetries will be of 0 degree, 72 degree, 144 degree, 216 degree and 288 degree. It has no reflection symmetries.

[5] More examples can be seen at the artist's Instagram account [W2].

Figure 6.6.9 Pentagonal motif by Regolo Bizzi. [W2]

The second piece is a wallpaper pattern inspired by a regular dodecagon. A dodecagon is a 12-sided figure. A regular dodecagon has 24 symmetries consisting of 12 rotations and 12 reflections. Its symmetry type is D_{12}. If one studies the central dodecagon in Figure 6.6.10, one notices that adjacent edges are slightly different. Due to this, the symmetry type of the central dodecagon becomes like that of a regular hexagon, namely symmetry type D_6. As a wallpaper pattern, this has translations in two directions, nonzero degree rotations of 60 degree, 120 degree, 180 degree, 240 degree and 300 degree. It also has reflections in the horizontal, vertical and oblique directions as well as glide reflections.

Figure 6.6.10 Dodecagon-inspired wallpaper pattern by Regolo Bizzi. [W2]

Quick Review

After reading this section, you should be able to:

- Analyse and classify motifs found in works of art.
- Identify artwork that can be analysed as strip patterns.
- Classify such strip patterns.
- Identify artwork that can be analysed as wallpaper patterns.
- Identify symmetries of such wallpaper patterns.

EXERCISE 6.6

1. Do a similar case study using artwork. Escher's work itself will provide many more examples. Another website where such examples can be found is the collection of Mathematical Art Galleries of the Bridges Foundation [W3].
2. The following are some more examples of artwork done by Escher.
   (a) Analyse the finite motifs for symmetry.
   (b) Recognise possible strip patterns and classify them.
   (c) Recognise possible wallpaper patterns. Analyse and identify the different symmetries of the wallpaper patterns.
   (d) For the wallpaper patterns, mark the translations, the different types of lines of reflection if any, the different types of rotocenters if any and the smallest non-do-nothing rotation. If there are glide reflections, then mark the line of reflection and translation to depict the glide reflection.
      (i) Plane-filling motif with reptiles [W4]
      (ii) Fish, Fish/Bird [W5]
      (iii) Development II [W6]
      (iv) Crab [W7]
3. Choose a Warli painting. Analyse the Warli painting's symmetry through the following: symmetry of finite motifs and symmetry of strip patterns.

6.7 Symmetry Around Us

The techniques we have developed so far allow us to see symmetry in objects all around us, whether it is the tiles in a pavement that look like wallpaper patterns or the iron grill of railings that seems like a strip pattern or a wild flower or the patterns on a ladybug or the logo of the local Metro. In this section, we provide a non-exhaustive list placed in four categories that can give you a chance to explore symmetry. We also provide some visual examples. We have broadly divided our choices into four categories with the last being 'others'.

Architecture related:
Buildings, bridges, electricity grids, window shapes, decorative motifs, fences, balcony and staircase railings, window and door grills, tiles on walls, floors, border tiles, bathroom tiles, pavement tiles.

Textile related:
Carpets, rugs, cushion covers, table cloths, bedsheets or bedcovers, curtains, runners, garments, sarees, dupattas.

Nature related:
Insects, flowers, leaves, birds, animals, snowflakes, crystals, geographical formations, nests, hives.

Others:
Vases, pottery, paintings, artwork, book covers, logos, road signs, kolam designs, rangoli, lampshades, shadows, jewellery.

Figure 6.7.1 illustrates some examples from the aforementioned lists:

Figure 6.7.1 Architecture related. [C6]

The pictures in the collage in Figure 6.7.1 are as follows in a clockwise direction from the top left: window jali or grillwork at Amer Fort, Jaipur; flower motif at one end of a pole used in the roof of Gyeongbok Palace in Seoul; roof decoration from Gyeongbok Palace, Seoul; bridge in Singapore, window jali or grillwork from Gyeongbok Palace; decorative mosaic in a mirror at Amer Fort, Jaipur and door from a Buddhist temple in Seoul.

Figure 6.7.2 Textile related. [C6]

Figure 6.7.2 shows examples of motifs, strip and wallpaper patterns that abound in textiles of various hues and colours. The collection in the picture consists of examples from curtains, saris, shirts, and tunics or kurtas.

The central picture in Figure 6.7.3 shows reflection symmetry involving natural formations. The other pictures in a clockwise direction are: flower exhibiting five-fold rotation symmetry; dragonfly; fern, leaf, spotted deer, all exhibiting bilateral or reflection symmetry; snowflake designs [C5]; spot-billed ducks and their reflections creating mirror symmetry; flower exhibiting D_4 symmetry type.

Figure 6.7.3 Nature related. [C5] and [C6]

Figure 6.7.4 Others. [C6]

Starting clockwise from the top left corner, we have a road sign; strips formed by rangoli designs; logo of the Delhi metro; a digit in the ancient Babylonian script; a symbol of the 33rd Infantry Division of the United States Army; ornamental pottery from the Gyeongbok Palace in Seoul; a rangoli motif; wallpaper created by a shadow (Figure 6.7.4).

We end this section with a few more illustrations.

The collage in Figure 6.7.5 shows decorative motifs on pottery discovered in the Indus Valley Civilization, which existed around 5,000 years ago. These are displayed in the National Museum in Delhi, India. The decorations can be studied as finite motifs or as strip patterns or wallpaper patterns.

Fine examples of decorative motifs, wallpaper patterns and strip patterns can be found at the Amer Fort near Jaipur which is over 400 years old (Figure 6.7.6).

The murals from Nahargarh Fort (Figure 6.7.7), built a little less than 300 years ago, give us illustrations of decorative motif and strip patterns as does the decoration in an archway in Figure 6.7.8 from an old church in Goa over 400 years old.

The journey we began in our quest to understand symmetry stops here as far as this book is concerned. We have merely touched the surface of a topic that is immense in terms of beauty

Figure 6.7.5 Pottery from the Indus Valley Civilization. [C8]

Figure 6.7.6 Amer Fort, Jaipur. [C6]

Figure 6.7.7 Nahargarh Fort, Jaipur. [C6]

Figure 6.7.8 Decoration in an archway. [C6]

and depth. One pathway that you can explore on your own is symmetry of three-dimensional objects. This would lead to the question as to what the regular three-dimensional structures are? (The counterparts to the two-dimensional regular n-gons.) For each n is there a regular n-faced three-dimensional figure? The answer is no. Indeed there are only five regular three-dimensional figures. Another topic that connects itself naturally to wallpaper patterns is the study of tilings. The bibliography at the end of this chapter will also give you a list of books and articles that you could consult should you wish to explore symmetry further.

Quick Review

After reading this section, you should be able to:

- Analyse and classify motifs.
- Identify and analyse strip patterns.
- Classify such strip patterns.
- Identify and analyse wallpaper patterns.
- Identify symmetries of such wallpaper patterns.

EXERCISE 6.7

1. Find the symmetry type of the following finite planar figures in Figure 6.7.4.
   (a) Road Sign
   (b) Delhi Metro Sign
   (c) Babylonian Digits
   (d) Plus Sign
2. Do the following for the collages and illustrations given in this section.
   (a) Analyse the finite motifs for symmetry and classify them by type.
   (b) Recognise possible strip patterns and classify them.
   (c) Recognise possible wallpaper patterns. Analyse and identify the different symmetries of the wallpaper patterns.
   (d) For the wallpaper patterns, mark the translations, the different types of lines of reflection if any, the different types of rotocenters if any and the smallest non-do-nothing rotation. If there are glide reflections, then mark the line of reflection and translation to depict the glide reflection.

Review Exercises

1. For the figures given below, describe all the rotational symmetries and reflection symmetries by identifying defining points on the figure.

(i)  (ii)  (iii) 

Source: [C6].

2. For each of the following figures, draw or write down the following:
   (a) The lines of reflection or mirror symmetry if any.
   (b) A description of the axis of rotation and the rotocenter.
   (c) The reflections, with their effect on suitably marked vertices.
   (d) The rotations, with their effect on the marked vertices.
   (e) The symmetry type of the figure.

(i)

(ii)

(iii)

(iv)

(v)

3. State whether true or false. Give reasons.
   (a) It is not possible for a finite planar figure to have infinitely many symmetries.
   (b) Given any n, there is a finite planar figure with exactly n reflections and n rotations as symmetries.
   (c) Every finite planar figure has a reflection symmetry.
   (d) Every planar figure has a rotational symmetry.
   (e) There is a figure with exactly two rotational symmetries and no reflection symmetry.
4. Find the following symmetries. All rotations take place in the anticlockwise direction unless stated otherwise.
   (a) Two distinct reflections of an equilateral triangle applied one after the other.
   (b) Two distinct reflections of a non-square rectangle applied one after the other.
   (c) Vertical reflection followed by a 180 degree rotation in a square.
   (d) Inverse of the 90 degree rotation for a square.
   (e) Two distinct reflections of a regular hexagon applied one after the other.
5. State whether true or false. Give reasons.
   (a) A strip pattern can only have one horizontal line of reflection symmetry.
   (b) A strip pattern with a reflection symmetry about a horizontal line has to possess a 180 degree rotational symmetry.
   (c) There are strip patterns with no nonzero degree rotational symmetries.
   (d) Every strip pattern has infinitely many symmetries.
   (e) There exist strip patterns with only finitely many translation symmetries.
6. Examine the strip patterns given below and answer the following.
   (a) List the different symmetries that the pattern has.
   (b) Mark the different types of lines of mirror symmetry if any.
   (c) If there is a non-do-nothing rotation, then mark the different types of rotocenters and describe the rotations.
   (d) If there is a glide reflection, then mark the line of reflection and translation to depict the glide reflection.
   (e) Will there always be a translation symmetry? If so, draw an arrow to mark the direction of the translation.
   (f) Identify the Pattern Number of the given strip pattern.

(i)

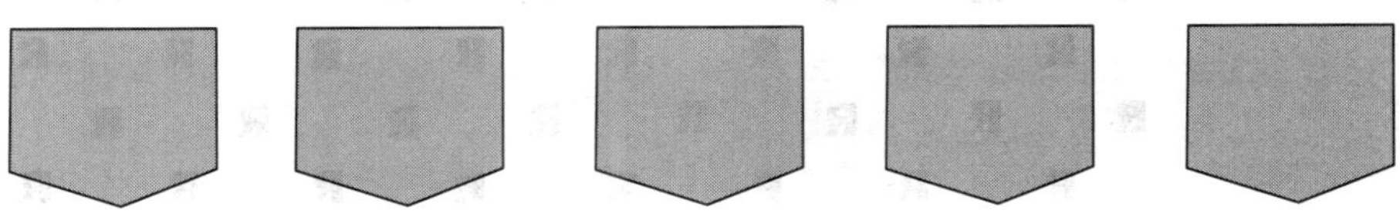

(ii)

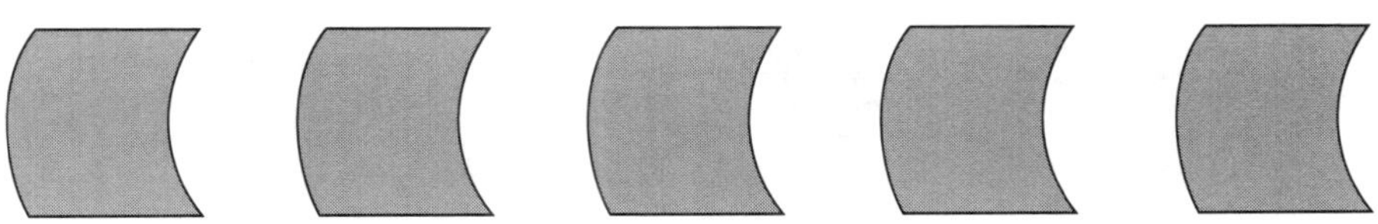

(iii) 

(iv) 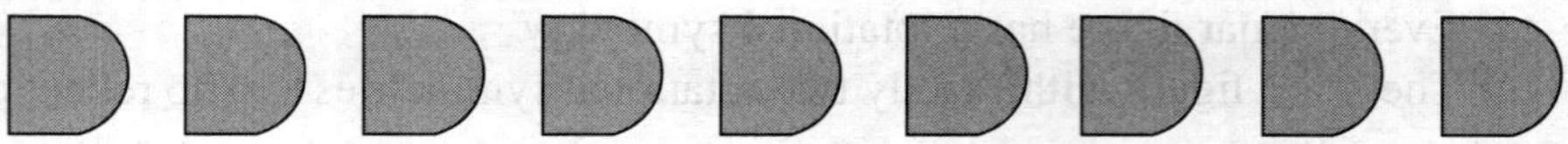

7. Find the following symmetries. All rotations take place in the anticlockwise direction unless stated otherwise.
   (a) $\rho \circ \sigma$, where ρ and σ are 180 degree rotational symmetries of a strip pattern about two rotocenters of different types.
   (b) $\rho \circ \sigma$, where ρ is a reflection of the strip pattern about a horizontal line of reflection and σ is a 180 degree rotational symmetry of a strip pattern.
   (c) Inverse of a glide reflection in a strip pattern.
   (d) Inverse of a reflection about a vertical line of reflection in a strip pattern.
8. Examine the wallpaper patterns given below and answer the following.
   (a) List the different symmetries that the pattern has.
   (b) Mark the different types of lines of mirror symmetry if any.
   (c) If there is a non-do-nothing rotation, then mark the different types of rotocenters and describe the rotations.
   (d) Will there always be translation symmetry in two directions? If so, draw arrows to mark the direction of the translations.

(i) 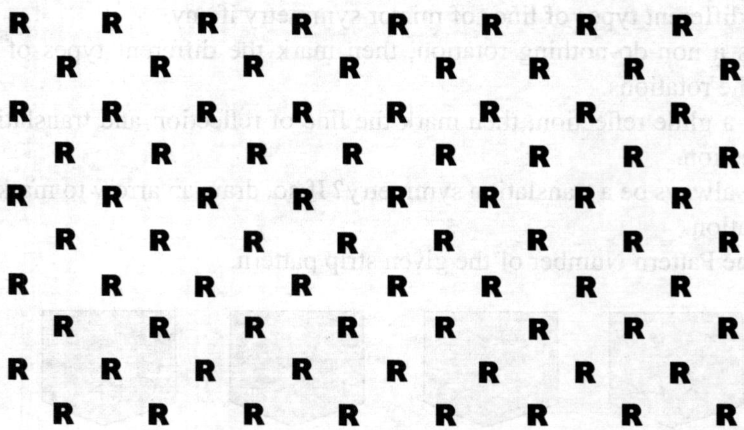

(ii)

(iii)

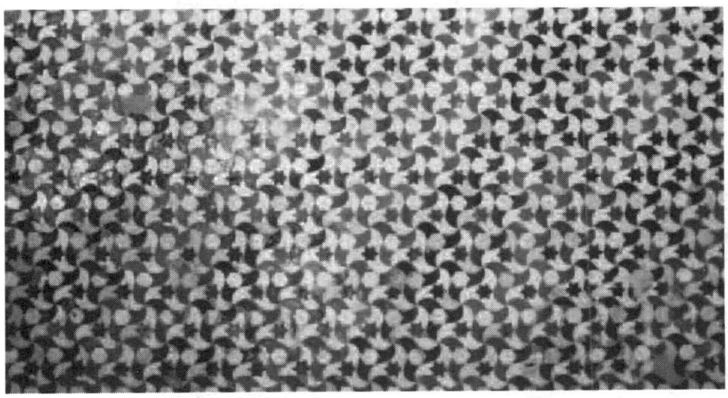

Source: [C4].

9. State whether true or false. Give reasons.
   (a) The rotocenters in a wallpaper pattern always lie on lines of reflection.
   (b) A wallpaper pattern can only have one horizontal line of reflection symmetry.
   (c) A wallpaper pattern always has a vertical line of mirror symmetry.
   (d) Every wallpaper pattern has a glide reflection.
10. The following are floor plans, photos, decorative motifs, strip patterns and wallpaper patterns from the Purana Qila in Delhi.
    (a) Analyse the floor plans and photos for symmetry.
    (b) Analyse and classify the decorative motifs.
    (c) Analyse and classify the strip patterns.
    (d) For the wallpaper patterns, mark the translations, the different types of lines of reflection if any, the different types of rotocenters if any and the smallest non do-nothing rotation. If there are glide reflections, then mark the line of reflection and translation to depict the glide reflection.

(i)

(ii)

(iii)

(v)

(iv)

Source: [C6].

Bibliography

Books and Articles

[1] Marcus Du Sautoy. 2008. *Symmetry: A Journey into the Patterns of Nature*. USA: Harper Collins.

[2] David W. Farmer. 1998. *Groups and Symmetry*. India: Universities Press.

[3] Joseph A. Gallian. 1999. *Contemporary Abstract Algebra*. 4th edition. New Delhi: Narosa Publishing House.

[4] Roger C. Lyndon. 1985. *Groups and Geometry*. UK: Cambridge University Press.

[5] Hermann Weyl. 1952. *Symmetry*. USA: Princeton University Press.

Websites

[W1] Kathy Barbro. *Art Projects for Kids*. `http://artprojectsforkids.org` (accessed 27 January 2017).

[W2] Regolo Bizzi. *Official Website*. `https://www.instagram.com/regolo54/` (accessed 27 January 2017).

[W3] The Bridges Organization. *Mathematical Art Galleries*. `http://gallery.bridgesmathart.org/exhibitions` (accessed 27 January 2017).

[W4] M. C. Escher. *Plane Filling Motif with Reptiles*. `http://www.mcescher.com/gallery/back-in-holland/plane-filling-motif-with-reptiles/` (Accessed: 28 Jan. 2017).

[W5] M. C. Escher. *Fish-Fish Bird*. `http://www.mcescher.com/gallery/recognition-success/no-109-ii-fish-fishbird/` (accessed 28 January 2017).

[W6] M. C. Escher. Development. http://www.mcescher.com/gallery/switzerland-belgium/development-iii/ (accessed 28 January 2017).

[W7] M. C. Escher. *Crab*. http://www.mcescher.com/gallery/symmetry/no-40-crab/ (accessed 28 January 2017).

[W8] The M. C. Escher Company. *Official Website of M. C. Escher*. http://www.mcescher.com/ (accessed 27 January 2017).

[W9] A. Nelson, H. Newman and M. Shipley. *Plane Symmetry Groups*. https://caicedoteaching.files.wordpress.com/2012/05/nelson-newman-shipley.pdf (accessed 27 January 2017).

[W10] J J O'Connor and E F Robertson. *Niels Henrik Abel*. http://www-history.mcs.st-and.ac.uk/Biographies/Abel.html (accessed 27 January 2017).

[W11] J J O'Connor and E F Robertson. *Maurits Cornelius Escher*. http://www-history.mcs.st-andrews.ac.uk/Biographies/Escher.html (accessed 27 January 2017).

[W12] J J O'Connor and E F Robertson. *Hermann Klaus Hugo Weyl*. http://www-history.mcs.st-and.ac.uk/Biographies/Weyl.html (accessed 27 January 2017).

[W13] Wikipedia. *Frieze Group*. https://en.wikipedia.org/wiki/Frieze_group (accessed 28 January 2017).

Image Credits

[C1] Anonymous. *Mausoleum of Humayun, Delhi*. Online Gallery of the British Library. http://www.bl.uk/onlinegallery/onlineex/apac/addorimss/m/019addor0001809u00000000.html(accessed 27 January 2017).

[C2] Bouwe Brouwer. *Wall Tableau by Escher*. https://commons.wikimedia.org/wiki/File:Leeuwarden_-_Tegeltableau_Escher.jpg (accessed 27 January 2017).

[C3] Reinier De Jong. *Metamorphosis*. http://www.reinierdejong.com/2012/07/metamorphosis/ (accessed 27 January 2017).

[C4] Patrick Gruban. *Alhambra*. CC BY-SA 2.0 license. https://flic.kr/p/218ij (accessed. 27 January 2017).

[C5] Petr Kratochvil. *Snowflake Pattern*. http://www.publicdomainpictures.net/view-image.php?image=19145&picture=snowflake-pattern (accessed 27 January 2017).

[C6] Photo by Geetha Venkataraman.

[C7] Photo by Kathy Barbo.

[C8] Photo by Amber Habib.

7

CHAPTER

Perspective: Art and Mathematics

There are three aspects to perspective. The first has to do with how the size of objects seems to diminish according to distance: the second, the manner in which colours change the farther away they are from the eye; the third defines how objects ought to be finished less carefully the farther away they are.

—Leonardo da Vinci

Human beings have been expressing themselves through art from time immemorial. The earliest examples date back to about 30,000 or 40,000 years ago. These were cave paintings, often depicting animals and hunting scenes. In India, the Bhimbetka caves in Madhya Pradesh provide wonderful examples of such cave art. Over the ages and across the continents, examples abound but what seems common to all until a few centuries ago is that, even though the paintings were depicting three-dimensional objects, albeit on two-dimensional surfaces, they did not convey the three dimensionality convincingly. On the other hand, photographs manage to capture what the eye sees on a piece of paper, namely, the three dimensionality of the scene. However, the paintings of yesteryear did not do so. The space occupied by an object in a painting was often according to the importance given to it by the artist rather than a true reflection of its size and position. It was only during the Renaissance that artists began to fully capture three dimensionality in their work, much like a photograph. The key to this is the art of 'perspective' and the tool of 'projection'.

Modified versions of perspective are currently being applied to computer games, CAD software and computer graphics. The set of points in the scene are projected to the plane of the computer screen, and the problem of perspective is to identify the coordinates of these points in the viewing plane. Although developers are currently deploying principles of linear algebra rather than descriptive geometry, it is interesting to learn how geometry helped to achieve perspective. In this chapter, we shall learn some of the mathematical principles that underlie the concept of perspective and make it possible to realistically portray three-dimensional scenes on paper or screen.

7.1 Introduction to Perspective

The word 'perspective' is composed of *per* that means *through* and *specere* which means *to look at*. Perspective in art, in a broader sense, means representing a three-dimensional scene on a two-dimensional surface, such that the feeling of depth is maintained. One can devise a set of rules based on geometry to achieve perspective. Shading and use of tones in colours can also enhance the feeling of varying distance and thus of three dimensionality. In this chapter, we shall mainly focus on the use of geometry to create the feeling of depth in paintings. In order to do so, we first consider some simple examples before we look at the 'rules of perspective' that have evolved over the centuries.

First, we see an example of how one portrayal of an object can convey its three dimensionality, whereas another fails to do so (Figure 7.1.1).

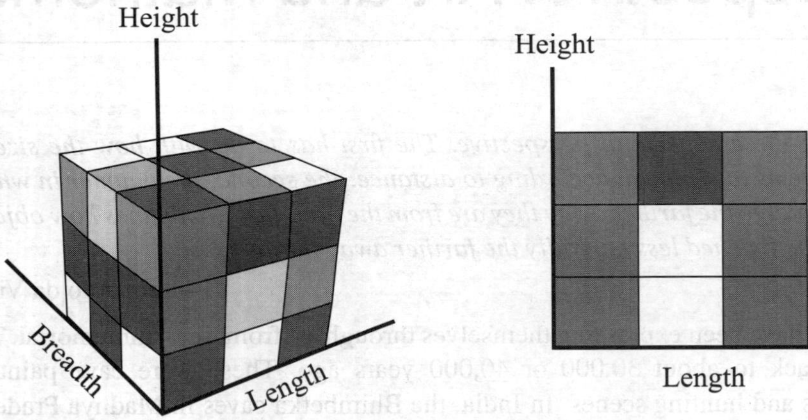

Figure 7.1.1 Rubik's cube and one of its square faces. The image on the left manages to convey the three dimensionality of the cube, for example, it gives the feeling that the top is nearer to us than the base. The image on the right shows one face and there is no feeling of depth.

For another example, consider the two photographs of a glass in Figure 7.1.2.

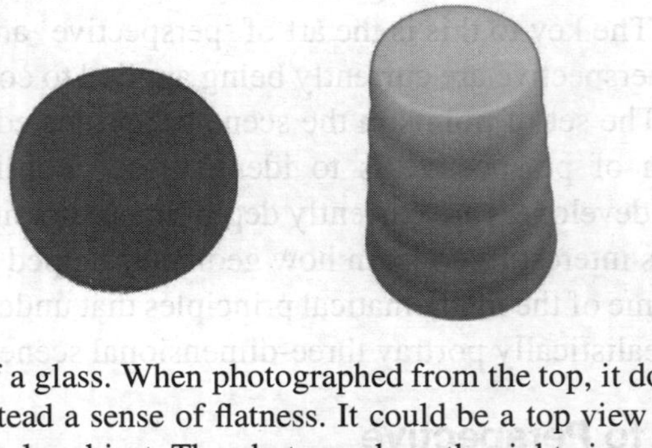

Figure 7.1.2 Two views of a glass. When photographed from the top, it does not give any idea of depth but instead a sense of flatness. It could be a top view of a glass or a bowl or any other circular object. The photograph on the right gives a better impression and reveals that the object has a nearly cylindrical shape. [C28]

Now, let us take a very brief look at how artists have dealt with depth over time. To begin with, Figure 7.1.3 shows two paintings that are examples of prehistoric 'cave art'. While we can easily identify the animals depicted, the paintings do not convey their three dimensionality.

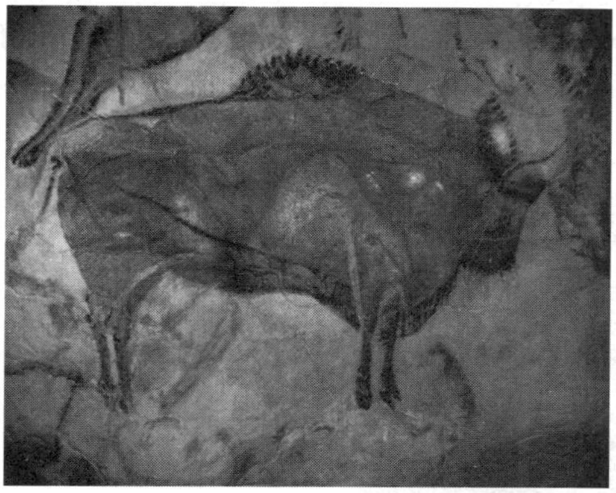

(a) Painting of a bison in the cave of Altamira. [C1]

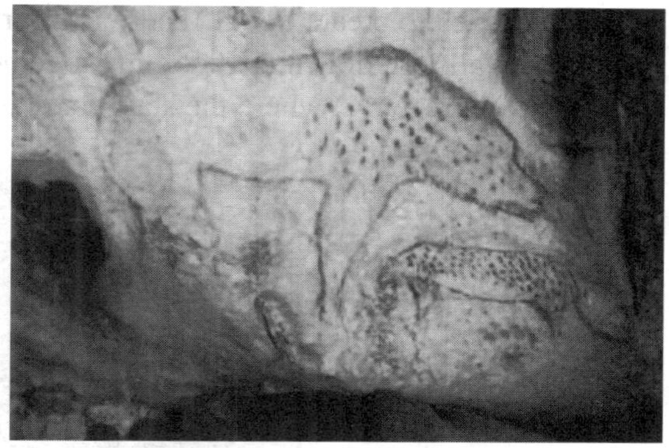

(b) Cave hyena (Crocuta crocuta spelaea) painting found in the Chauvet cave. [C2]

Figure 7.1.3 Examples of cave art.

Figure 7.1.4 Babur receives a courtier, by Farrukh Beg (1589). [C3]

The use of perspective in art became prevalent mainly after 1400 AD. In the prior period, variations of size in a painting indicated relative importance rather than distance or actual size. Important people were deliberately drawn larger in comparison to others, to convey the primacy of their position. However, with the emerging desire of a more physically realistic depiction, artists began experimenting with ideas like drawing distant objects smaller. Another technique was to draw faraway objects higher than nearby ones.

An illustrative example is presented next. Figure 7.1.4 is a sixteenth century painting by the Mughal artist Farrukh Beg and shows a partially successful attempt at conveying depth. The courtiers sitting at the back are depicted higher up in the painting but are of the same size. Slanted lines are used to bring out the depth of the canopy on which emperor Babur is seated. Around this time, the artists were becoming aware of the fact that distant objects should be drawn smaller as compared to the objects that are nearer for better illusion. The building in the left hand top corner of the painting is an outcome of this observation. But other objects retain a flat appearance, and to our eyes—accustomed to photographs—the overall impression is of inconsistency.

The Italian painter Duccio's *The Last Supper* (Figure 7.1.5), painted in the early fourteenth century, also has elements of perspective. Looking at the roof, or the table, we again see the use of slanted lines to create depth. There is a further improvement—the edges along which the roof meets the walls are drawn as if coming together in the distance. Compare this with Farrukh Beg's painting (Figure 7.1.4) where the slant lines are parallel.

Figure 7.1.5 Duccio's *The Last Supper*. [C5]

Figure 7.1.6 Sculpture of Brunelleschi looking at his dome on the Cathedral of Florence. [C10]

However, one can spot inconsistencies on taking a closer look. The point where the roof and wall edges meet when extended is different from the point at which the extended roof supports meet. We shall see later that, if this painting had been painted keeping to the rules of perspective, these different meeting points would have coincided. Further, the table cover does not use the insight of meeting lines.

The discussion above of extended parallel lines meeting at a point is the fundamental tenet in what is called **linear perspective.** The originator of this concept is Filippo Brunelleschi (Figure 7.1.6) who lived in Florence, Italy during 1377 to 1446. In early fifteenth century, he created a panel painting using linear perspective. Apart from his paintings, Brunelleschi was also a sculptor and an architect. He was interested in both mathematics and engineering. The huge dome he designed and built for the Cathedral of Florence is considered the largest masonry dome ever built.

While Brunelleschi's panel painting using perspective is lost, there seems to be some information on an experiment that he conducted which showed his audience that his painting and the actual scene matched perfectly. His experiment, also known as 'Brunelleschi's Peepshow', consisted of a small hole drilled at a central point in the panel painting. The viewer placed his or her eye at the hole at the back part of the panel (which was unpainted). The viewer could then look at the actual scene outside which was also the subject matter of the panel painting. There was a mirror perfectly aligned in front of the painting which could be moved in and out. This allowed the viewer to first look at the actual scenery and then, via the mirror, at the reflection of the panel painting, perfectly coinciding with the earlier real scenery (Figure 7.1.7).

Figure 7.1.7 Illustration of Brunelleschi's Peepshow. [C28]

Brunelleschi did not put down the rules pertaining to perspective in writing. It was Leone Battista Alberti (1404–72), who for the first time systematically presented the geometrical methods of linear perspective in his treatise *On Painting*. This treatise consisted of three books with the first book being the most mathematical. In this first book, he presented the details of steps underlying the use of linear perspective. Soon after, many other artists in Italy started using perspective in their work.

We now take up the use of perspective by Pietro Perugino in his fresco *Christ Handing the Keys to St. Peter* at the Sistine Chapel (Figure 7.1.8), painted during 1481–82.

The depth in the painting is most strongly conveyed by the slant lines that cut the horizontal lines and meet at the temple in the distance (Figure 7.1.9(a)). The horizontal lines divide the painting into three distinct part—foreground, middle ground and background (Figure 7.1.9(b)). The important figures occupy the space in the foreground, whereas other figures, smaller in size,

Figure 7.1.8 *Christ Handing the Keys to St. Peter* by Perugino. [C19]

are placed by the painter in the middle ground. The background is dominated by mountains, trees and three architectural structures.

The artist also uses symmetry in his painting. The figures on both sides of Christ and St. Peter are placed symmetrically. The central building is symmetrical about the line of symmetry (the vertical dotted line) and is flanked by two other buildings symmetrically placed on either side of the central building.

(a) (b)

Figure 7.1.9 Use of linear perspective ideas by Perugino.

Later, the use of perspective was widely incorporated by Leonardo da Vinci, Michelangelo and Raphael, the trinity of Renaissance, in their work.

Although there are many rules of perspective, they are based on the assumption that a single eye is looking at the subject being drawn from a fixed point of view. Since cameras operate on the same principles as the human eye, the images they produce already have the rules of perspective built into them. Nevertheless, photographers who understand the concepts of perspective are able to incorporate elements which further emphasise the sense of depth in their photographs.

Figure 7.1.10 Receding lines meeting at a point create a sense of depth. [C29]

Consider the photograph in Figure 7.1.10. The road, the windows, the rooftops—all these present to us multiple straight lines that recede into the distance and appear to meet at a single distant point. This is a typical element of linear perspective and gives a strong sense of depth.

The photograph in Figure 7.1.11 is of the Qutub Minar in Delhi, clicked through one of the arches of the adjoining Qubbat ul Islam mosque. The height of the Qutub Minar is approximately 73 m. The arch is less than half the height of the Qutub Minar. But it was photographed at an angle that gives an impression of the arch being taller than the Qutub Minar. The illusion is possible due to the lack of any horizontal receding lines in the photograph. Their absence makes it impossible to judge the distance between the arch and the tower.

The next photograph (Figure 7.1.12) is clicked from a slightly different position. By allowing the arch to cut in front of the tower, it makes clear the greater height of the Qutub Minar. Nevertheless, the absence of horizontal receding lines still makes it difficult to judge how far away is the Minar, and we cannot sense just how much taller it is than the arch.

The knowledge of perspective is invaluable to any artist, no matter what technique or school of art he or she may prefer. One who knows and understands the basic theories of perspective can produce work with varying degrees of realism or thoughtful distortion. Although the use of perspective had a great influence in the development of art in Western culture, artists today are discarding its use in order to give wings to their creativity and imaginations.

Figure 7.1.11 Qutub framed I. [C29]

Figure 7.1.12 Qutub framed II. [C29]

Quick Review

After reading this section, you should be able to:

- Evaluate the aspects of a painting or photograph that create a sense of depth.
- Appreciate the role of the study of perspective in the development of art.
- Analyse whether an image is drawn with a consistent use of perspective.

EXERCISE 7.1

1. Look at the following photographs. From what viewpoint do you think the photograph has been clicked? Which one best succeeds in depicting three dimensionality?

(a)

(b)

(c)

2. What objects can have the following top view representation?

(a)

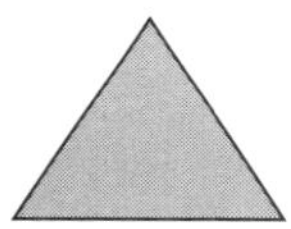

(b)

(c)

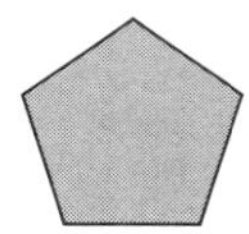

(d)

3. Following is an eighteenth century Rajput painting by the artist Nihal Chand. The artist has tried to depict some objects in linear perspective, while others are not drawn in perspective. Identify the objects that are in perspective and that are not.

Source: [C9].

4. Do a similar analysis as in Exercise 3 for the following paintings.

(a)

Source: [C8].

(b)

Source: [C24].

(c)

Source: [C25].

5. *The School of Athens* fresco by the Italian artist Raphael is one of the finest examples of linear perspective. Describe how the artist has made use of perspective in this piece of art.

Source: [C4].

6. The given painting is titled *Reconstruction of the Temple of Jerusalem* and is from the fifteenth century. Identify the elements that give the illusion of perspective in the painting. Which objects are not in perspective?

Source: [C21].

7. This seventeenth century painting portrays the *Old St. Paul's Cathedral* in London. Describe the geometric errors in the artist's attempt to capture perspective.

Source: [C18].

8. The altarpiece *The Ognissanti Madonna* was created by Giotto for the church of Ognissanti in Florence. Describe how the artist has employed perspective.

Source: [C14].

7.2 Mathematical Principles: Coordinates and Projections

Coordinate Planes and Axes

While studying coordinate geometry in school, you would have become familiar with the use of a coordinate plane to describe the location and shape of objects. A coordinate plane can only handle length and breadth. If we wish to also describe height, we would have to use a third coordinate. To understand the coordinate system with three coordinates, let us take a corner of a room (Figure 7.2.1).

Figure 7.2.1 Coordinate system for three dimensions.

Suppose you stand facing a wall in a room. We will call that wall as the xy-plane. It is between the x-axis and the y-axis. The plane on which you stand is the xz-plane bounded by the x-axis and the z-axis. The third plane is the yz-plane bounded by the y-axis and the z-axis.

Each axis is a number line with its origin at the corner where all three axes meet. Any point in the room can now be represented as a triple (x, y, z) of three numbers. In this triple, the number x gives the location of the foot of the perpendicular dropped from the point to the x-axis, and is called the x-coordinate of the point. Similarly, we have the y and z-coordinates which give the locations of the feet of the perpendiculars dropped on the respective axes.

A point that is on the x-axis will have coordinates (x, 0, 0), a point that is on the y-axis will have coordinates (0, y, 0) and a point that is on the z-axis will have coordinates (0, 0, z). A point that is on the xy-plane has coordinates (x, y, 0). Similarly, a point lying on the yz-plane will have its x-coordinate as 0, and a point that lies on the xz-plane will have y-coordinate as 0.

Example 7.2.1 If a point P in space has coordinates (1, 3, 2), it means that to reach it, you start at the origin and move a distance of 1 unit along the x-axis, then 3 units parallel to the y-axis, followed by 2 units parallel to the z-axis (Figure 7.2.2).

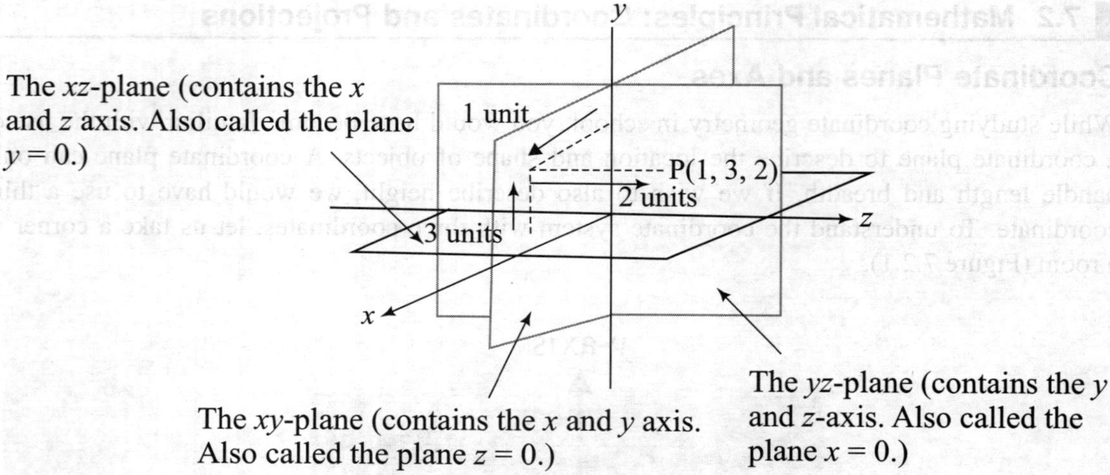

Figure 7.2.2 Locating a point P relative to the coordinate axes and planes.

□

Example 7.2.2 A cuboid is drawn with three of its concurrent edges along the coordinate axes.

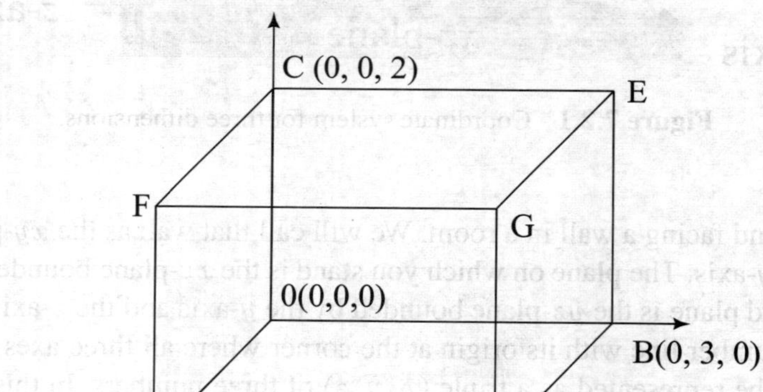

1. Where is the origin located?
2. Label the axes and the coordinate planes.
3. What are the coordinates of the vertices D, E, F and G?

The vertex O that has the coordinates, (0, 0, 0) is the origin. The vertex A has coordinates (1, 0, 0). Only the x-coordinate is nonzero. Therefore, it lies on the x-axis. Hence, the line OA is the x-axis. Similarly, the line OB is the y-axis and the line OC is the z-axis. The coordinate planes are shown as follows.

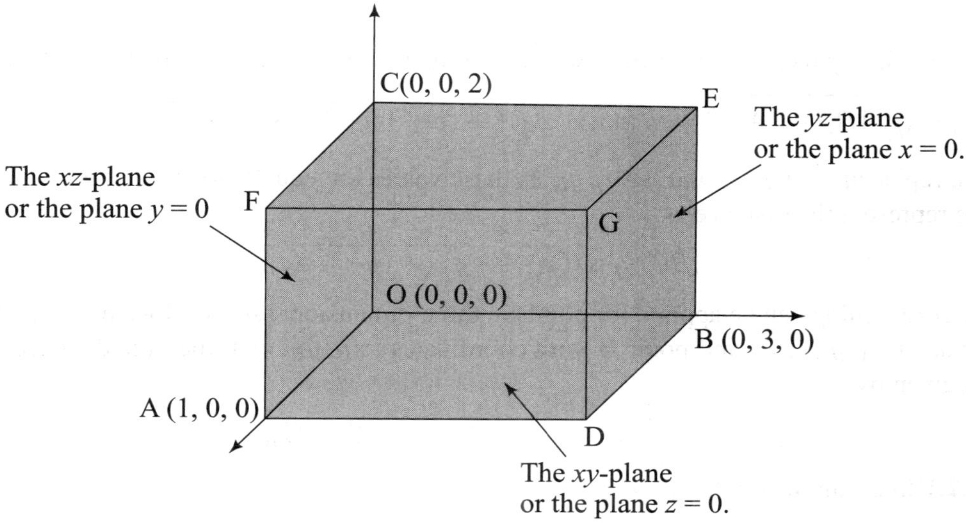

Point D lies on the xy-plane. It is 1 unit along the x-axis and 3 units along the y-axis. Therefore, its coordinates are (1, 3, 0). Point E lies on the yz-plane. It is 3 units along the y-axis and 2 units along the z-axis. Therefore, its coordinates are (0, 3, 2). Point F lies on the xz-plane. It is 1 unit along the x-axis and 2 units along the z-axis. Therefore, its coordinates are (1, 0, 2). Point G is 1 unit along the x-axis, 2 units along the y-axis and 3 units along the z-axis. Therefore, its coordinates are (1, 2, 3). □

Distance Between Two Points

In the figure below, the co-ordinates of point A are (1,1), of point B are (5,1) and of point C are (1,4). The line segment AB is horizontal and the line segment AC is vertical. The distance $|AB|$ between A and B is 4, units whereas the distance $|AC|$ between A and C is 3 units. Since AB and AC are perpendicular to each other, we can apply the Pythagoras theorem to find the distance $|BC|$ between B and C:

$$|BC| = \sqrt{|AB|^2 + |AC|^2} = \sqrt{4^2 + 3^2} = 5$$

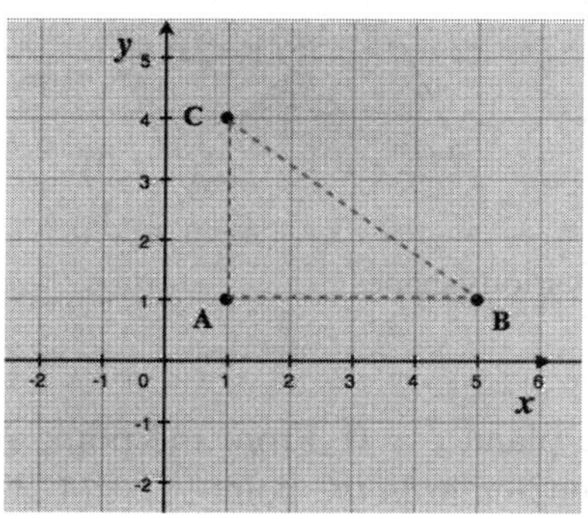

Let us view these calculations from a more general perspective. The distance between A and B is just the difference between their x-coordinates. Similarly, the distance between A and C is the difference between their y-coordinates. We use the notation x_A, x_B and x_C for the x-coordinates of A, B and C, respectively. Similarly, we use the notation y_A, y_B and y_C for their y-coordinates. Then we can write:

$$|AB| = x_B - x_A, \qquad |AC| = y_C - y_A$$

Applying the Pythagoras theorem as we did before, we find the distance between B and C:

$$|BC| = \sqrt{|AB|^2 + |AC|^2} = \sqrt{(x_B - x_A)^2 + (y_C - y_A)^2} = \sqrt{(x_B - x_C)^2 + (y_C - y_B)^2}$$

We have replaced x_A by x_C and y_A by y_B as these values are equal. Noting that $a^2 = (-a)^2$, we can also represent the distance as:

$$|BC| = \sqrt{(x_C - x_B)^2 + (y_C - y_B)^2}$$

Similar reasoning can be applied for points in three-dimensional space. If we have point A with coordinates (x_A, y_A, z_A), and point B with coordinates (x_B, y_B, z_B), then the distance between them is given by

$$|AB| = \sqrt{(x_B - x_A)^2 + (y_B - y_A)^2 + (z_B - z_A)^2}$$

Task 7.2.1 In Example 7.2.2,

1. What is the distance between the origin and the vertex G?
2. What is the length of the diagonal DE?

Perpendicular Projection

Any method of representing three-dimensional objects on a plane involves the projection of points of space to points of the plane. The simplest kind of projection is **perpendicular projection,** in which a perpendicular is dropped from each point of the object to the plane. To understand this kind of projection, we shall first consider perpendicular projection of a point, a line and a planar surface on a plane, and in particular on the plane $z = 0$.

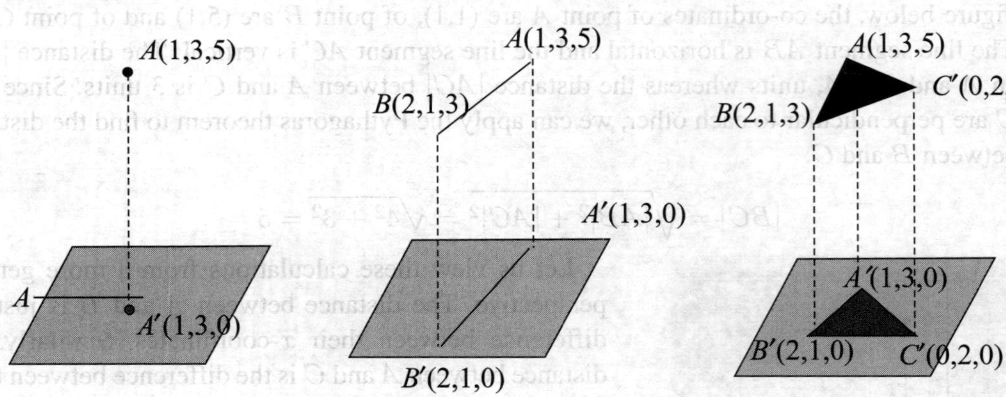

Figure 7.2.3 Perpendicular projection of various shapes.

In Figure 7.2.3, a point $A = (1, 3, 5)$ is projected on the plane $z = 0$. From the point A, a perpendicular is drawn to the plane and the foot of the perpendicular is the point $A' = (1, 3, 0)$.

When a point is projected on the plane $z = 0$, the x and y-coordinates of the point and the projected point stay the same, but the z-coordinate becomes 0.

Similarly, the line segment AB with $A = (1, 3, 5)$ and $B = (2, 1, 3)$ is projected to the line segment $A'B'$, with $A' = (1, 3, 0)$ and $B' = (2, 1, 0)$. We may observe here that the projected line segment is shorter:

$$|AB| = \sqrt{(2-1)^2 + (1-3)^2 + (3-5)^2} = \sqrt{1+4+4} = 3$$
$$|A|'B' = \sqrt{(2-1)^2 + (1-3)^2 + (0-0)^2} = \sqrt{1+4+0} = \sqrt{5} < 3$$

This is a general phenomenon with perpendicular projection. On projection, distances either stay the same or they shrink.

Returning to Figure 7.2.3 once more, we consider the triangle ABC, which is projected to the triangle $A'B'C'$. Although we do not make the calculations here, but it is true that the shrinking of the lengths of its sides causes the angles between the sides to also be distorted. Such distortion is inevitable when we squeeze three dimensions into two.

Perpendicular projections are popular in technical work such as drawings of engineering parts and architectural plans. An example is given (Figure 7.2.4):

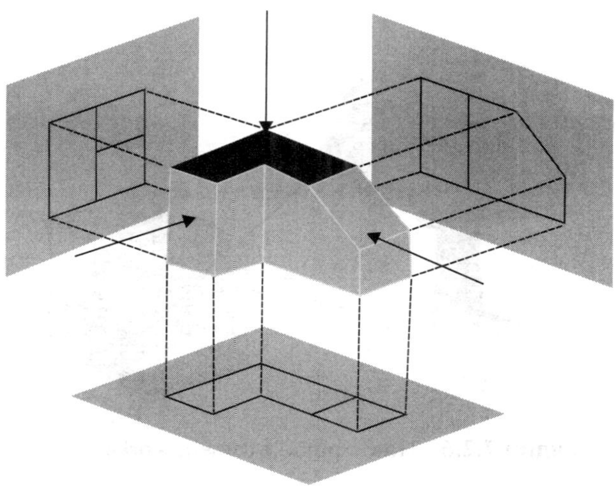

Figure 7.2.4 Different perpendicular projections of the same object. [C15]

Nevertheless, perpendicular projections are not suitable for realistic art, as they do not feature foreshortening. In perpendicular projection, distant objects do not shrink. We do not get the sense of viewing an object from a definite position and distance.

Perspective Projection

Let us now look at a method of projection that more closely mimics what our eye perceives. As before, we have an object whose points are to be mapped to a plane. But instead of dropping perpendiculars from points on the object, we draw lines from them through a specified viewing point, and again mark where these lines cut the plane where the image is to be made. This method is called **perspective projection** or **linear perspective** (Figure 7.2.5).

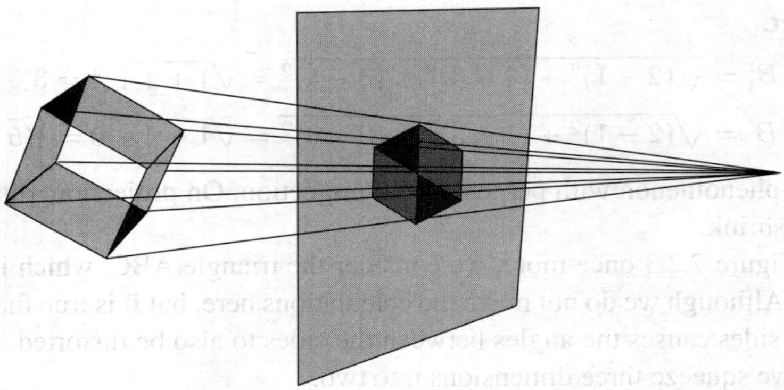

Figure 7.2.5 A perspective projection of a cube. [C16]

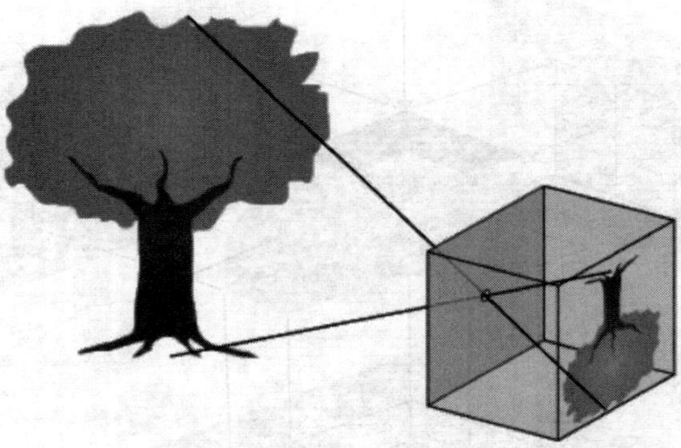

Figure 7.2.6 How a pinhole camera works. [C17]

In linear perspective, the image depends not only on the relative location of the object and the plane, but also on that of the viewing point.

We may note that the images produced by a pinhole camera are also based on linear perspective, with the viewing point (the pinhole) lying *between* the object and the screen on which the image is made (Figure 7.2.6).

Let's take a closer look at the geometry of linear perspective, with the help of three-dimensional coordinates. In Figure 7.2.7, the image of a vase is projected on a plane which is parallel to its axis. The image is perceived by the eye placed at E.

In Figure 7.2.7, the z-axis is drawn as the line passing through the eye at E and perpendicular to the picture plane. The picture plane is designated as the $z = 0$ plane. The vase is placed at

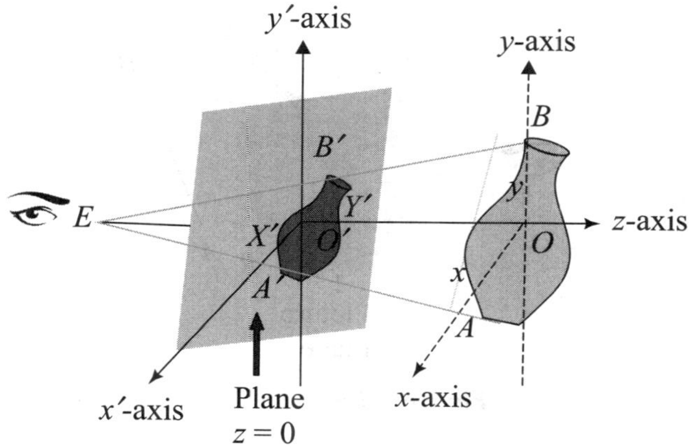

Figure 7.2.7 Projection of a vase on the picture plane, using linear perspective.

a distance z from the eye, whereas the plane $z = 0$ is placed at a distance d from the eye. The triangle EOA is similar to the triangle $EO'A'$, while the triangle EOB is similar to $EO'B'$. Using the properties of similar triangles, we have

$$\frac{d}{z} = \frac{X'}{x} \Rightarrow X' = \frac{xd}{z} \quad \text{and} \quad \frac{d}{z} = \frac{Y'}{y} \Rightarrow Y' = \frac{yd}{z} \tag{7.1}$$

A word of caution: We have assumed that the axis of the vase is parallel to the plane of the picture, and this provides the similar triangles on which we based our calculations. As $z = EO$ is greater than $d = EO'$, the factor $\frac{d}{z}$ will be less than one. Hence, in this case $X' < x$ and $Y' < y$, and the projected distances are shorter than the original ones. But in general, if a line segment is not parallel to the plane of the picture, the similarity rule will not apply and the image may be longer than the line segment.

Rules of Linear Perspective

To draw an object in linear perspective, the artist is supposed to look at the object with one eye only, which we shall take as a point in space. As we did in the example of the vase, we shall assume that the picture plane is situated at $z = 0$ and the eye is situated on the z-axis behind the picture plane. The visual rays from the eye (point) are then drawn. We shall first look at the basic mathematical principles concerning the projections of straight lines.

Rule 2.1: The perspective projection of a straight line is again a straight line.

$PQRS$ is a picture plane (Figure 7.2.8) and AB is the line whose image is sought on the picture plane. The eye is located at E. We connect AE and BE; these lines intersect the picture plane $PQRS$ at A' and B'. Joining A' and B' gives us the perspective image of the line AB.

An important exception is that when the line is perpendicular to the picture plane, A and B will be projected to the same point in the picture plane, and therefore the image of the line will be just a point.

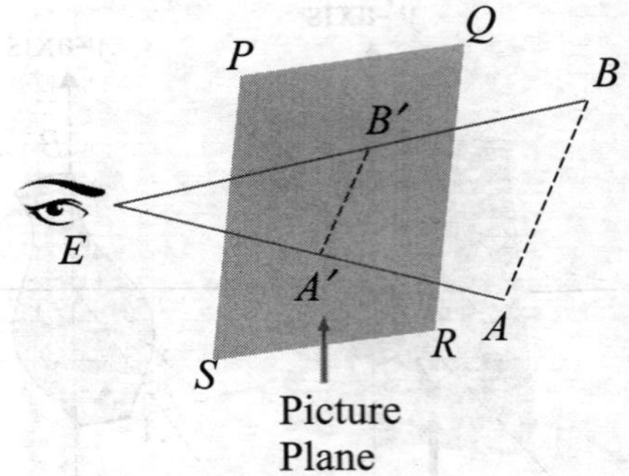

Figure 7.2.8 Rule 2.1.

Rule 2.2: The perspective projection of a vertical line is vertical and of a horizontal line is horizontal.

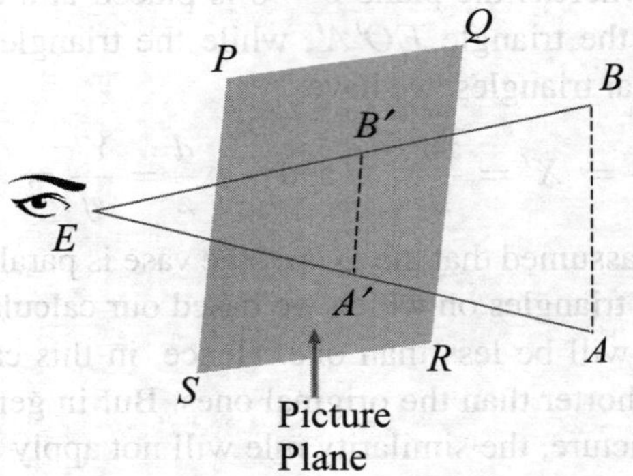

Figure 7.2.9 Rule 2.2.

Recall that we have been taking the z-axis to be perpendicular to the picture plane, while the x-axis and y-axis are in the picture plane. Moreover, the x-axis is horizontal and the y-axis is vertical. Now consider a vertical line AB as in Figure 7.2.9. It is parallel to the y-axis. Let A have coordinates (x_A, y_A, z_A). Then an arbitrary point on the line has coordinates (x_A, y, z_A). When we take the perspective projection of this line in the $z = 0$ plane, the image points have coordinates $\left(\frac{x_A d}{z_A}, \frac{y d}{z_A}, 0\right)$, by Equation 7.1. Thus, the image points also have only the y-coordinate varying and so lie along a vertical line.

By a similar piece of reasoning, horizontal lines also project to horizontal lines.

Since our choices of 'horizontal' and 'vertical' are arbitrary, these arguments also imply that if a line is parallel to the picture plane, it will also be parallel to its projected image. It follows from this that if two lines are parallel to the picture plane and to each other, then their projected images are also parallel in the picture plane.

Rule 2.3: Consider a line segment which is parallel to the picture plane. The length of its perspective projection diminishes in proportion to the distance from the viewer.

The diagram and calculations leading to Equation 7.1 can be modified to show the following. Let the eye be at distance d from the picture plane, and a line segment of length x lie in a parallel plane at a distance z from the eye. Let the perspective projection of this line in the picture plane have length X. Then we find that

$$X = \frac{xd}{z}$$

Rule 2.4: The image of any line L in space that is not parallel to the picture plane will appear to vanish at a point V in the picture plane. V is called the **vanishing point of** L**.** The line through the position of the viewer E and the vanishing point V will be parallel to L.

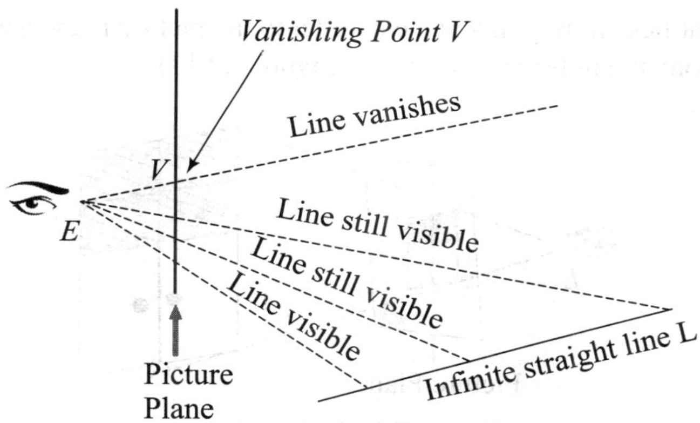

Figure 7.2.10 Rule 2.4.

In Figure 7.2.10, a viewer looks along different lines of sight (dashed lines) at an infinite straight line L in space. As the viewer looks at farther and farther points on the line, he keeps seeing the line. But at a certain moment, his line of sight becomes exactly parallel to L and no longer intersects it. That is the precise moment the line L seems to vanish. The intersection of this special line of sight with the picture plane is the 'vanishing point' of L.

Rule 2.5: Two or more lines that are parallel to each other but not to the picture plane, will have the same vanishing point. The perspective images of these lines will intersect at the vanishing point.

If L and L' are two parallel lines in space and EV is parallel to L (Figure 7.2.11), then it is also parallel to L'. So V serves as the vanishing point for both the lines.

Rule 2.6: If a line L is perpendicular to the picture plane, then the vanishing point of L is the origin, that is, the point on the picture plane directly opposite to the eye location.

If L is perpendicular to the picture plane then it is parallel to the z-axis. The vanishing point of the z-axis is the origin.

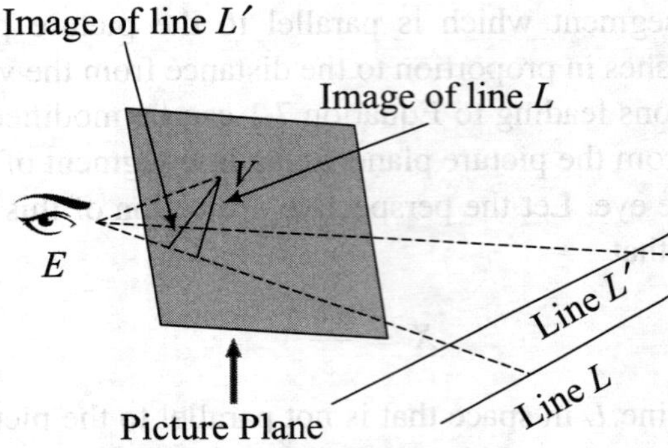

Figure 7.2.11 Rule 2.5.

Rule 2.7: A shape that lies entirely in a plane parallel to the picture plane has a perspective image that is free of distortion: all angles stay the same (Figure 7.2.12).

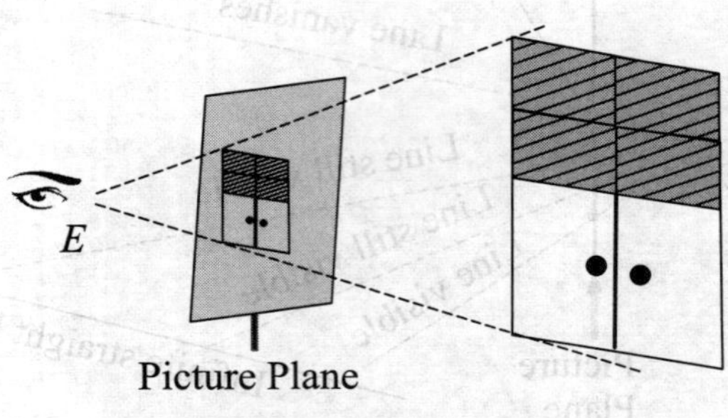

Figure 7.2.12 Rule 2.7.

From Equation 7.1, we have that all lengths in the shape and its image are proportional. Therefore, every triangle in the shape is projected to a similar triangle. Hence, all angles are preserved, and there is no distortion.

Rule 2.8: The perspective image of the centre of a rectangle is the intersection of the images of the diagonals.

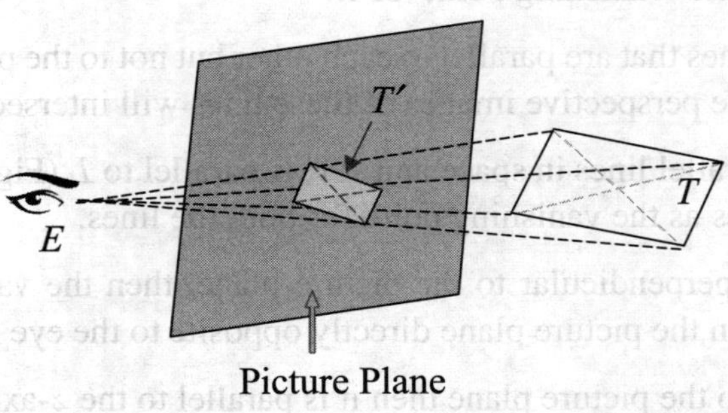

Figure 7.2.13 Rule 2.8.

The centre of the rectangle T in Figure 7.2.13 is the point where the diagonals of T intersect. Therefore, in the perspective image too the centre must be the intersection of the images of the diagonals of T'.

Example 7.2.3 Let us illustrate these rules by looking at the photograph in Figure 7.2.14.

Figure 7.2.14 Photograph illustrating perspective I. [C29]

We assume that the photograph is in the plane $z = 0$ with the y-axis vertical, x-axis horizontal and z-axis projecting out from the plane of the page.

1. The sides of the road are straight lines that remain straight in the image also.
2. The tree trunks are vertical and remain so in the image. They remain parallel in the image plane, but diminish as they become distant.
3. The sides of the road and sidewalk are parallel but not parallel to the plane (in fact, they are perpendicular to the plane). They have the same vanishing point, the origin (Figure 7.2.15).
4. The points on the ground, the tree trunks and a line joining the position where the tree foliage starts form rectangles. The diagonals of these rectangles meet at the centre. The line

Figure 7.2.15 Photograph illustrating perspective II. [C29]

joining these centres would be parallel to the road and therefore has the same vanishing point (Figure 7.2.16).

Figure 7.2.16 Photograph illustrating perspective III. [C29]

□

Quick Review

After reading this section, you should be able to:

- Locate points in a plane or space using coordinate axes and planes.
- Calculate the distance between any two points whose coordinates are known.
- Find the perpendicular projection of an object on a plane.
- Find the vanishing point for parallel lines during perspective projection.
- Use the stated rules for perspective to analyse the use of perspective in an image.

EXERCISE 7.2

1. In Figure 7.2.2, what will be the coordinates of a point

   (a) 5 units directly above P?
   (b) 3 units to the right of P?
   (c) 2 units in front of P?
   (d) 4 units behind P?
   (e) 1 unit to the left of P?
   (f) 7 units directly below P?

2. A pyramid has a square base such that the base is parallel to the xy-plane and the edges of the base are parallel to the x-axis and the y-axis. The coordinates of the vertex A are (2, 1, 2).

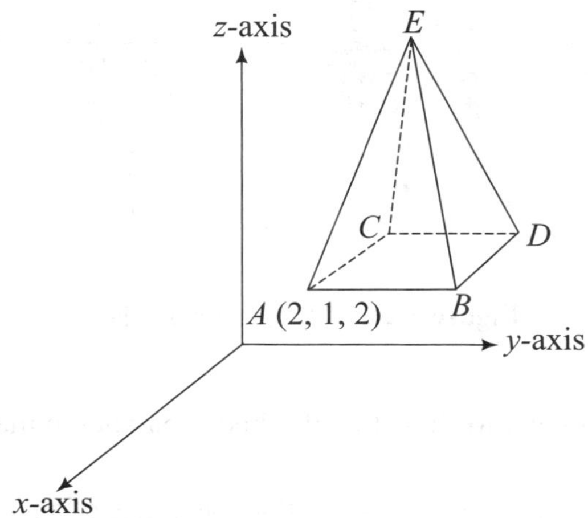

   (a) Find the coordinates of the vertices B, C and D, if it is given that the square base has a side of 2 units.
   (b) Find the coordinates of the vertex E given that the height of the pyramid is 5 units and E is vertically above the centre of the base.
   (c) If A, B, C, D and E are projected on the xy-plane, what will be the coordinates of the projected points?

(d) If A, B, C, D and E are projected on the yz-plane, what will be the coordinates of the projected points?
(e) If A, B, C, D and E are projected on the xz-plane, what will be the coordinates of the projected points?
(f) What is the distance between B and E?

3. In Figure 7.2.17, A and B are the centre of the bottom and top surfaces of a jar. The coordinates of the point A are (5, –2, 3) and of the point B are (5, 2, 3). The eye E is placed at a distance of 7 units from the plane $z = 0$.
(a) Find the coordinates of the points A′ and B′.
(b) What is the height of the jar? What is the height of the image of the jar on the plane $z = 0$?
(c) If the jar is moved 90 units to the right so that the coordinates of the points A and B are (5, –2, 93) and (5, 2, 93), respectively, what will be the coordinates of the points A′ and B′ ? What is the height of the image?
(d) If the coordinates of the points A and B are (5, –2, 993) and (5, 2, 993), respectively, what will be the coordinates of the points A′ and B′ ? What is the height of the image?
(e) Explain how the results obtained in (b)–(d) is consistent with our everyday experience.

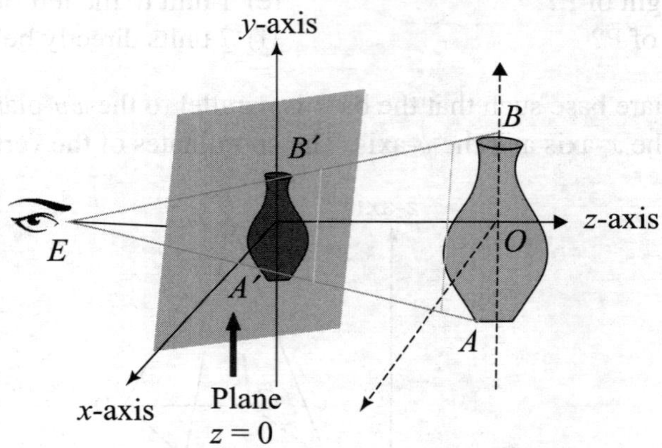

Figure 7.2.17 Projection of a jar.

4. Give a mathematical argument to show that if a line is parallel to the picture plane, it will also be parallel to its image.
5. Use the properties of similar triangles to show that the image of an object parallel to the picture plane is proportional to the object.
6. A box has a shape of a cuboid with dimensions 3 units, 4 units and 2 units with its faces parallel to the coordinate planes and centre C at the point (5, 5, 5).
Draw the image of the box using linear perspective
(a) in the $x = 0$ plane, if the eye is placed at (–5, 5, 5)
(b) in the $y = 0$ plane, if the eye is placed at (5, –5, 5)
(c) in the $z = 0$ plane, if the eye is placed at (5, 5, –5)

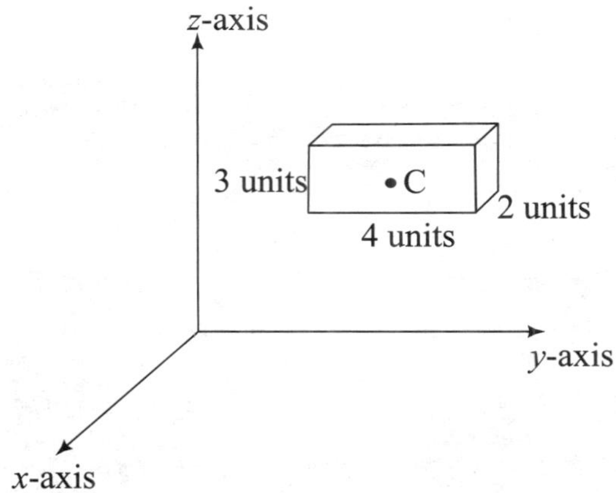

7. Find the vanishing points in the following paintings:
   (a) *School of Athens* by Raphael

Source: [C4].

   (b) *The Last Supper* by Leonardo da Vinci

Source: [C11].

(c) *Christ Handing the Keys to St. Peter* by Perugino [C19].

8. Explain how the following photographs illustrate various rules of perspective:

(a)

Source: [C29].

(b)

Source: [C29].

7.3 Linear Perspective

As mentioned earlier, perspective in art and painting can be achieved by various techniques, like varying the tone of colours or using blurring at far edges, etc. However, the aspect of perspective that we shall focus on is linear perspective, where geometrical concepts are used to convey three-dimensionality on a two-dimensional canvas. We introduce some terms that will help us understand and analyse linear perspective.

Horizon: The **horizon line** or **eye-level line** is a horizontal line drawn across the page, representing the eye level of the viewer who is creating the image.

This line is an important ingredient in ensuring linear perspective. All horizontal lines leading away from the picture plane have their vanishing points on the horizon. Artists generally avoid having the horizon line in the centre of their canvas (Figure 7.3.1).

Picture Plane: The flat surface on which the image is projected.

Station Point: The point where the viewer stands to observe the object.

Cone of Vision: An imaginary cone whose circular base lies on the picture plane. The vertex of the cone is the point of view, the point from where the viewer's eye observes the subject. Because of the limitations of the human eye, the angle of the cone of vision is usually kept under 60 degrees.

Vanishing Point: A point in the picture plane where a collection of parallel lines, which are not parallel to the picture plane, appear to meet.

Depending upon the number of vanishing points used in a perspective drawing or image, one can have **one-point perspective, two-point perspective** or **three-point perspective.**

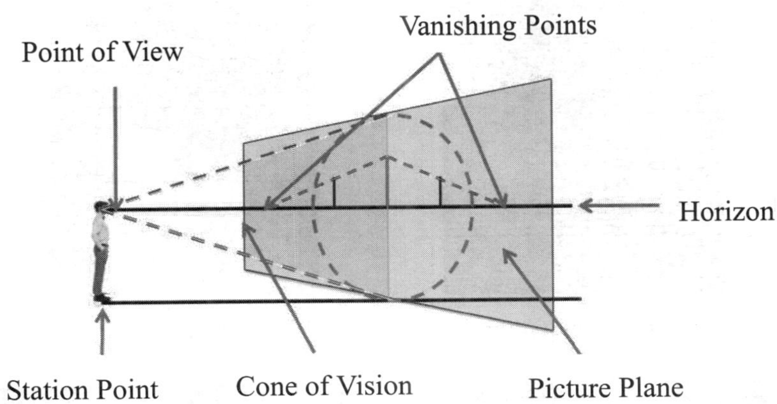

Figure 7.3.1 Diagrammatical representation of terms of linear perspective.

One-point Perspective

One-point perspective is the most common type of linear perspective. In this case, there is only one vanishing point in the painting or photograph. Typically, this occurs when you draw an image of a road, or railway tracks extending in front of you. This is also the view that emerges when a corridor is directly in front of the viewer (Figures 7.3.2 and 7.3.3).

In an image with one-point perspective, there is only one family of parallel lines receding from the picture plane. Other lines are kept parallel to the picture plane so that they do not create any more vanishing points.

Figure 7.3.2 The fifteenth-century painting *Banquet of Herod* by Masolino uses one-point perspective. [C22]

(a) Arches in a church

(b) Pillars

(c) Railway tracks

(d) Road

Figure 7.3.3 Examples of one-point perspective. [C29]

Two-point Perspective

An image that contains two vanishing points is said to have two-point perspective. An example of two-point perspective can be found when an image is drawn of a building with the station point being in front of the corner. The walls on either side lead towards two different vanishing points. In two-point perspective, the picture plane is kept perpendicular to the ground, so that vertical lines do not converge to vanishing points (Figures 7.3.4 and 7.3.5).

Figure 7.3.4 Each side wall has its own vanishing point, leading to two-point perspective. [C29]

Figure 7.3.5 A building with the front facade parallel to the picture plane and the side walls in two-point perspective. [C29]

There are relatively few paintings that illustrate two-point perspective. One of them is *Paris Street; Rainy Day* by Gustave Caillebotte that shows a building similar to the one in Figure 7.3.6 in two-point perspective.

Figure 7.3.6 Paris Street; Rainy Day. [C20]

Three-point Perspective

In three-point perspective, the picture plane is not perpendicular to the ground. Therefore, vertical lines also recede from the plane and appear to converge to vanishing points. It is generally achieved when a building is viewed either from the top or below and from a corner, so that none of the families of horizontal and vertical lines describing the building are parallel to the picture plane (Figure 7.3.7).

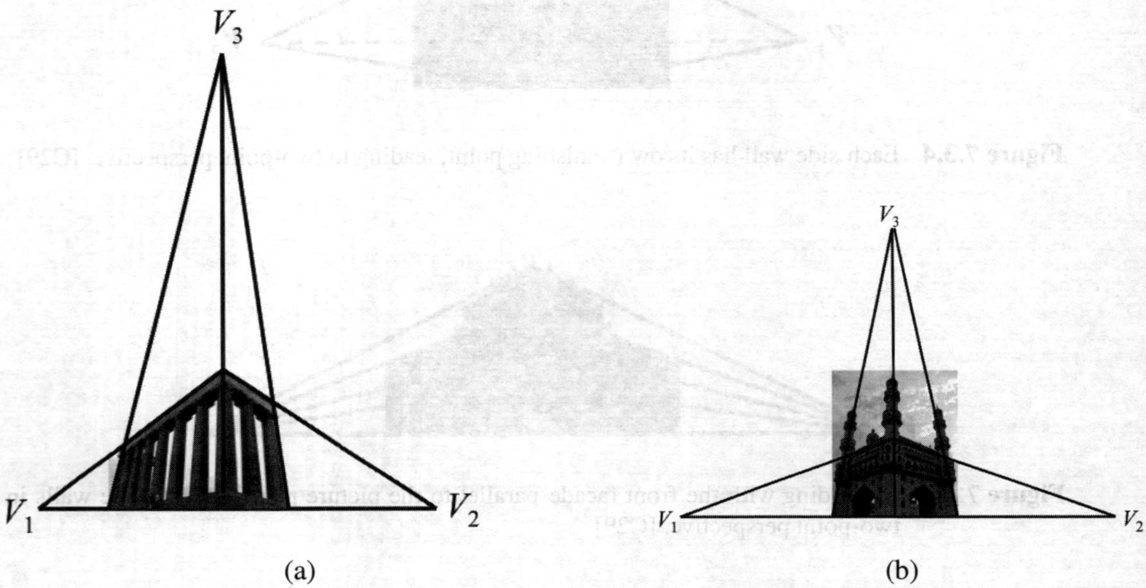

Figure 7.3.7 Examples of three-point perspective. [C29]

Quick Review

After reading this section, you should be able to:

- Use the concepts of vanishing points and horizon line to study the use of perspective by artists.
- Identify if a picture is in one-point, two-point or three-point perspective.
- Understand what an artist wants to highlight by his use of perspective.

EXERCISE 7.3

1. Determine whether the following pictures are in one-, two- or three-point perspective. In each case, find the vanishing point(s) and the horizon line.

(a)

(b)

(c)

(d)

Source: [C29].

2. Discuss how the artist has used linear perspective in the following paintings.

(a) *The Last Supper* by Leonardo da Vinci

(b) *School of Athens* by Raphael

7.4 Doing It Yourself

Our study of perspective so far has focused on understanding its basic elements, and recognising its use (or misuse) in paintings and drawings. In this section, you will learn some basic techniques for incorporating perspective in your own drawings. First, we will take up the use of repeating patterns to indicate depth. Just how should we shrink a shape as it moves into the distance, so that the eye correctly interprets it as the same shape being repeated? The second, and related, question is how do we incorporate the location of the viewer into the drawing?

The basic shape we need to master is the rectangle. If we know how to make repeating rectangles, we can handle other shapes by embedding them in rectangular grids.

Duplicating Simple Geometrical Figures

Before we begin duplicating simple geometrical figures in perspective, let us look at how we would duplicate them normally in a plane. We need to do this while keeping the following in mind, so that the techniques can be easily adapted to perspective drawings:

1. The duplication should not involve measuring any distance, since distance cannot be directly marked off in a perspective image due to foreshortening.
2. Parallel lines can be used, as parallel lines are easily represented in perspective due to the common vanishing point.
3. Intersections of lines are useful, as the perspective images would also intersect at the corresponding point.

Duplicating a Rectangle

Recall that a rectangle is a geometrical figure with opposite sides equal and parallel, and each angle a right angle. The diagonals of a rectangle are equal and bisect each other at the centre of the rectangle.

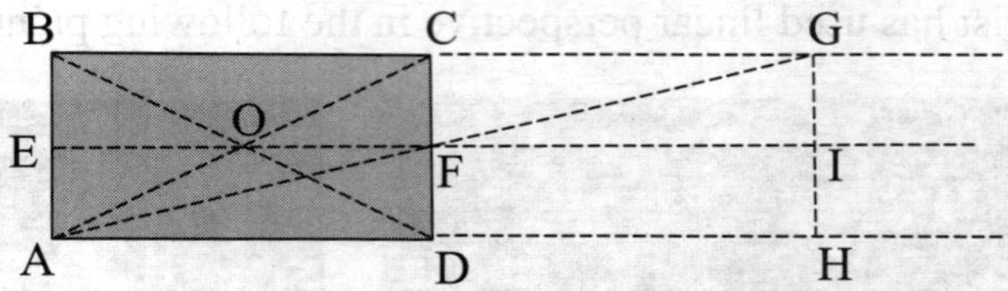

Figure 7.4.1 Duplicating a rectangle.

In order to duplicate a rectangle (Figure 7.4.1), we first find the centre O of the rectangle ABCD by joining the diagonals AC and BD. Through O we draw a line parallel to AD (and BC) that intersects the sides AB and CD at the points E and F. E and F are the mid-points of the sides AB and CD, respectively. If CGHD is to be the duplicated rectangle, then F would be the mid-point of the bigger rectangle BGHA, and AF would be part of the diagonal AG. Therefore, we simply extend AF to meet the extended line BC and call the intersection point G. Similarly, we extend BF to meet the extended line AD at H. The method can be repeated to obtain rectangles that are congruent to the rectangle ABCD.

Duplicating an Isosceles Triangle

Recall that an isosceles triangle has two equal sides. To duplicate an isosceles triangle ABC we first draw a rectangle ABDE (Figure 7.4.2) with the side AB equal to the base of the triangle ABC and the side AE equal to the height CF of the triangle. (The sides AC and BC of the triangle are equal). The perpendicular CF divides the base AB into equal halves, and hence also divides the rectangle ABDE into two congruent rectangles AFCE and FCDB. The rectangle FCDB is duplicated to BDJG and GJIH using the steps used for duplicating a rectangle. The diagonals BJ, JH and the side BH form the required duplicate triangle congruent to the triangle ABC.

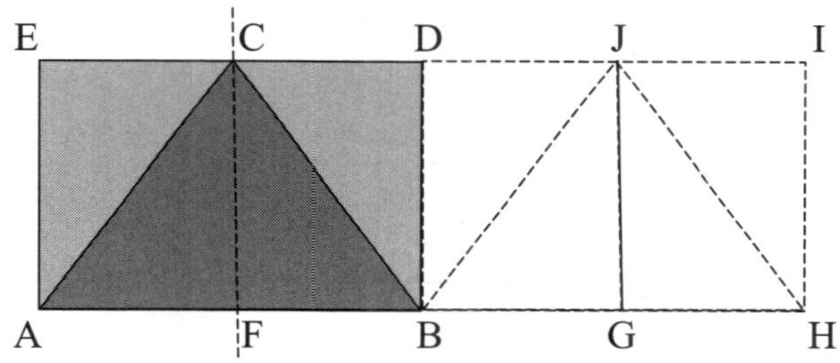

Figure 7.4.2 Duplicating an isosceles triangle.

The method described above can also be used to duplicate any regular polygon by inscribing it in a rectangle.

We now turn to creating repeated shapes in a perspective drawing.

Duplicating a Rectangle in Perspective

ABCD is a rectangle in one-point perspective (Figure 7.4.3). The edges AB and CD are parallel to the picture plane, while the edges AD and BC recede from it and meet at the vanishing point at V.

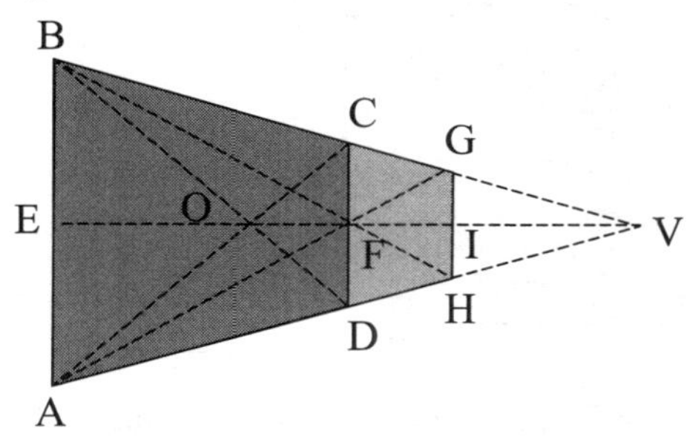

Figure 7.4.3 Duplicating a rectangle in perspective.

The diagonals AC and BD intersect at the centre O of the rectangle. The line OV intersects the sides AB and CD at E and F, respectively. E and F are the mid-points of the respective sides. Join AF and extend it to meet the line BV at G. Join BF and extend it to meet AV at H. DCGH is the required rectangle in perspective that would be congruent to ABCD in real space.

For more examples of such constructions, and especially to see case studies of how living artists use perspective, we highly recommend the book *Viewpoints: Mathematical Perspective and Fractal Geometry in Art* by Marc Frantz and Annalisa Crannnell [3]. An example from that book is presented below.

Example 7.4.1 A gift wrapped in the shape of a cube is to be drawn in one-point perspective. We wish to show a string tied around it so that it passes through the centre of each face.

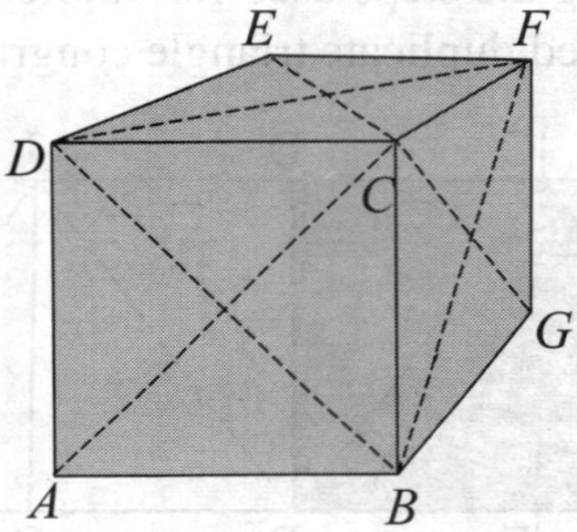

Figure 7.4.4 Midpoints of faces of a cube.

To keep the cube in one-point perspective, the front face ABCD is kept parallel to the picture plane. Hence, the visible edges EF and FG are also parallel to the picture plane. The edges DE, CF and BG are then perpendicular to the picture plane and so have a common vanishing point V,

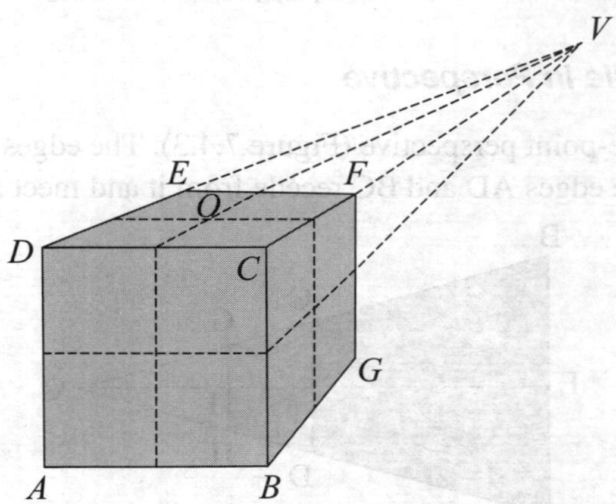

Figure 7.4.5 Lines through centres of faces of a cube.

which is in fact the origin of the picture plane. These aspects are illustrated in Figures 7.4.4 and 7.4.5. We locate the centre of each visible face by drawing the diagonals.

Figure 7.4.5 shows how to complete the task. Since face ABCD is parallel to the picture plane, we draw lines parallel to its sides and passing through its centre. For the faces CDEF and BCFG, we draw lines passing through their centres and the vanishing point V. These lines intersect the edges at their mid-points, and the string can be drawn on the cube along these lines. □

Viewing Distance

In order to appreciate a painting that has been drawn in perspective, we need to know the correct distance to view it. If a painting is drawn in one-point perspective, it is easy to find the correct viewing distance.

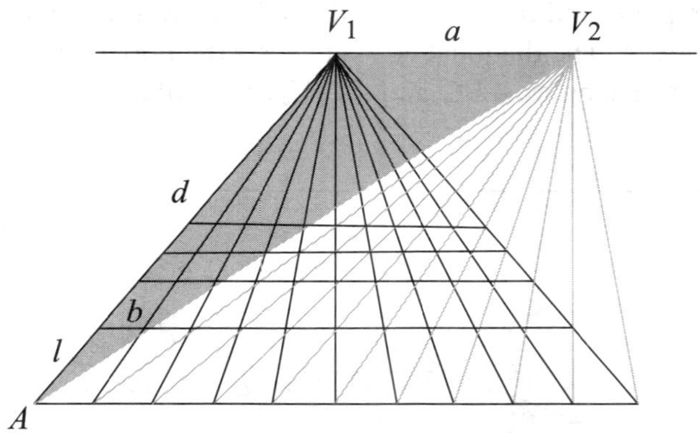

Figure 7.4.6 A tiled pavement in perspective.

Let us imagine a pavement of square tiles. Its one-point perspective image is shown in Figure 7.4.6. The grid created by the tiles has two sets of perpendicular lines. One set of lines is parallel to the picture plane, and hence their images are lines parallel to the horizon. The other set of lines is perpendicular to the picture plane, and hence their images converge to the vanishing point V_1 which is also the origin.

The diagonals of each square in the pavement are parallel, hence their images converge to another vanishing point V_2 on the horizon. A convenient fact which we shall use without proving it, is that the distance a between V_1 and V_2 is also the viewing distance, that is, the distance between the artist and the picture plane. If we now view the painting from the same distance in front of the origin, we shall get the exact experience the artist had planned. From any other position, objects will appear distorted.

Figure 7.4.6 also shows how to obtain the viewing distance without drawing the diagonal lines that meet at V_2. The smaller triangle with vertex A and edges of length l and b is similar to the triangle AV_1V_2. Hence, we have

$$\frac{a}{d} = \frac{b}{l} \implies a = \frac{b \cdot d}{l}$$

The numbers b, d and l can be read off from the image of the pavement.

Of course, we do not need the full pavement to find the viewing distance in this manner. The image of a single correctly oriented square will do. However, Figure 7.4.6 also shows us how to create the pavement's image, if the viewing distance is known. The grid corners will lie on the intersections of the lines to V_1 and V_2.

Finally, let us consider the case of a pavement made of rectangles which are not squares. Suppose, for instance, that the side perpendicular to the picture plane is twice as long as the side parallel to it. In the context of Figure 7.4.6, the dimensions l and b would become l and $b/2$, respectively. By similar triangles, the diagonal for the new rectangle would have its vanishing point at a distance $a/2$ from V_1. In other words, the viewing distance is 2 times the distance of V_1 from the vanishing point of the diagonal.

Example 7.4.2 Figure 7.4.7 illustrates a cuboid drawn in one-point perspective. The front square is parallel to the picture plane and the perpendicular sides converge to the vanishing point V_1 at a distance d from the vertex A. The diagonal AC intersects the horizontal line through V_1 at V_2, V_1 and V_2 are distant a apart. Triangles ABC and AV_1V_2 are similar.

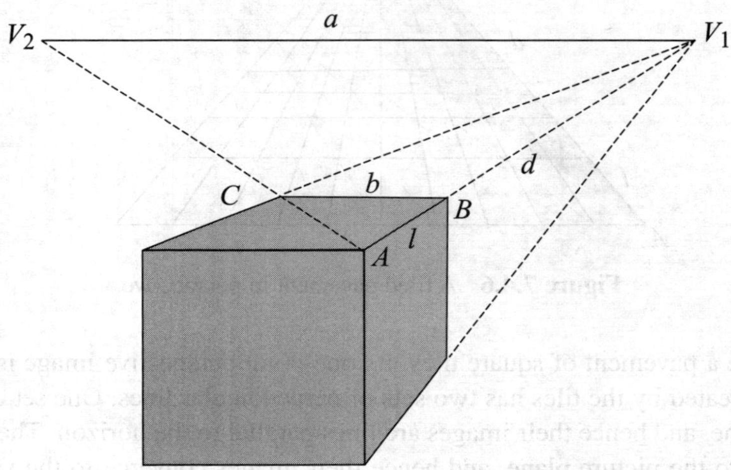

Figure 7.4.7 Measuring the viewing distance.

So we have the viewing distance $a = \dfrac{b \cdot d}{l}$. Try placing your eye at the viewing distance from V_1, and then look at the image of the cube for the best impression of its three dimensionality. □

Quick Review

After reading this section, you should be able to:

- Duplicate simple geometrical figures to create repetitive patterns in a perspective image.
- Work out the viewing point for a picture drawn in linear perspective.
- Draw squares with the correct foreshortening for a given viewing distance.

EXERCISE 7.4

1. Work out the steps to duplicate a hexagon inscribed in a rectangle (Figure 7.4.8).

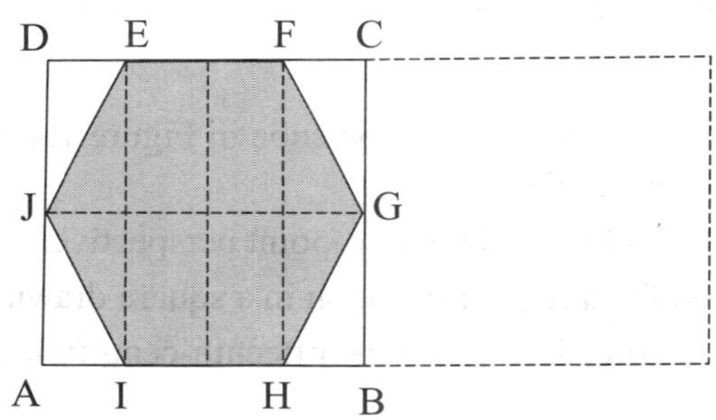

Figure 7.4.8 Duplicating a hexagon.

2. Describe how the artist should complete the drawing of the fence.

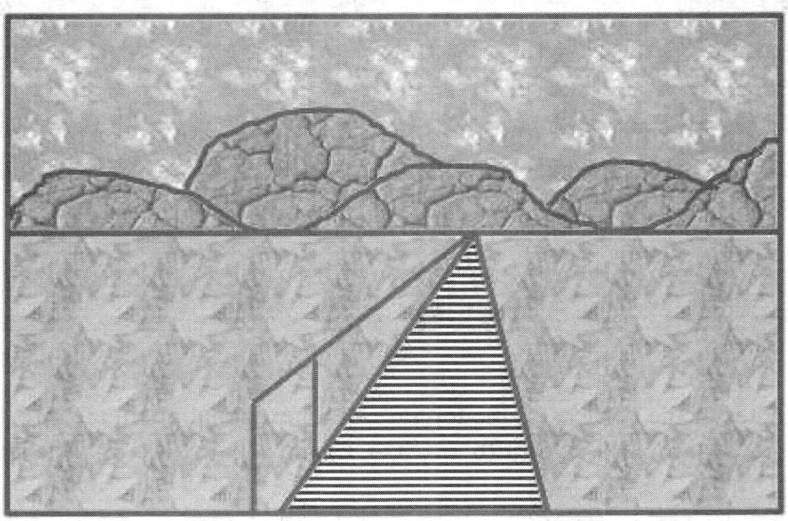

3. Inscribe an ellipse in the rectangle ABCD in Figure 7.4.3. Explain how to duplicate the ellipse in the rectangle DCGH.
4. (a) Find the centre of each face for the cube in two-point perspective (Figure 7.4.9).

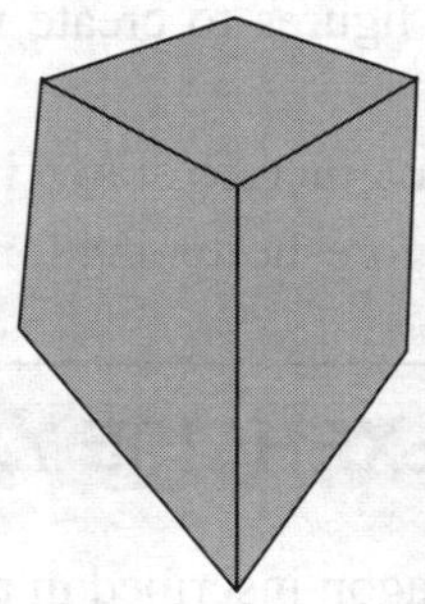

Figure 7.4.9 Centre of each face for a cube in two-point perspective.

   (b) An artist wishes to divide each face of the cube in Figure 7.4.9 into a 4×4 grid. Explain how he can draw the grid in the cube.
5. Explain how to draw a rectangle ABCD in two-point perspective.
6. Describe how you can inscribe a regular hexagon in a square drawn in two-point perspective.
7. Explain how to find the viewing distance for the fifteenth-century Italian painting *The Ideal City* (Figure 7.4.10).

Figure 7.4.10 *The Ideal City*. [C12]

Review Exercises

1. The coordinates of the vertices of the top face of a hexagonal prism are A(6, 7, 6), B($6-\sqrt{3}$, 6, 6), C($6-\sqrt{3}$, 4, 6), D(6, 3, 6), E($6+\sqrt{3}$, 4, 6) and F($6+\sqrt{3}$,6, 6). The height of the prism is 3 units.
   (a) What are the coordinates of the vertices of the bottom face of the prism?

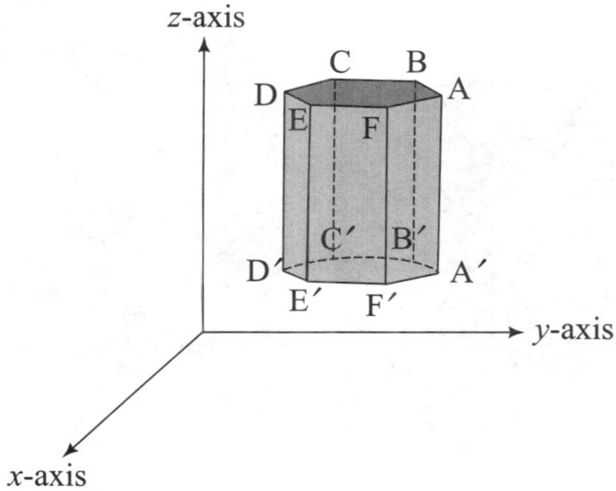

Figure 7.4.11 Hexagonal prism.

(b) What will be the shape of the projection of the prism, if it is projected in the (i) xy-plane? (ii) xz-plane? (iii) yz-plane?

2. Describe how perspective has been (or has not been) used in the following paintings (Figure 7.4.12).

(a)
Source: [C7].

(b)
Source: [C6].

(c)

Source: [C26].

(d)

Source: [C27].

Figure 7.4.12 Exercises on perspective.

3. The following figures consist of some blocks (Figure 7.4.13). What will the perpendicular projection of each figure look like if it is projected in the

   (a) xy-plane? (b) xz-plane? (c) yz-plane?

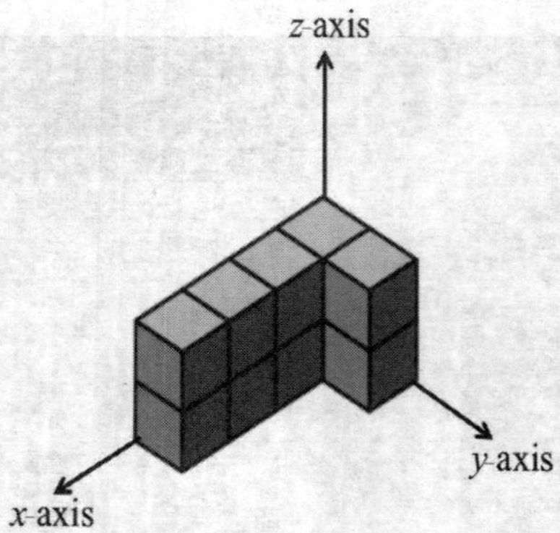

(a)

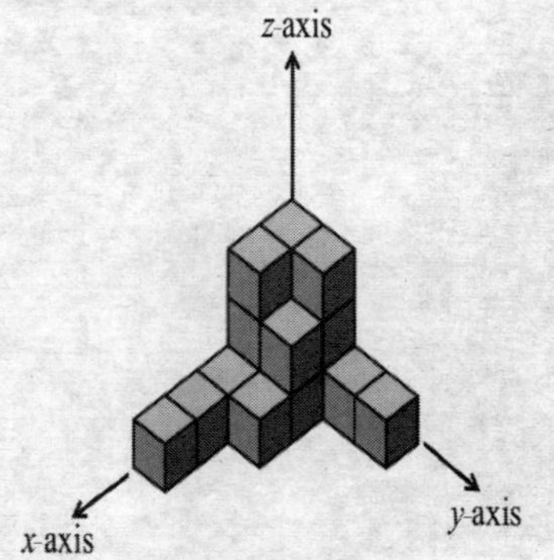

(b)

Figure 7.4.13 Exercises on perpendicular projection.

4. Determine whether the following drawings (or photographs) are in one-, two-, or three-point perspective. In each case find the vanishing point and the horizon line (Figure 7.4.14).

(a)

(b)

(c)

(d)

(e)

(f)

Figure 7.4.14 Exercises on perspective. Source: [C29].

5. A source of light, placed at a distance of 3 units from an object produces its shadow on a parallel plane placed at a distance of 6 units from the object. If the height of the object is 4 units, what will be the height of the shadow (Figure 7.4.15)?

Figure 7.4.15 Exercise on shadow.

6. Using the principles of perspective complete the following drawings (Figure 7.4.16).

(a) Pavement alongside a road

(b) Fence

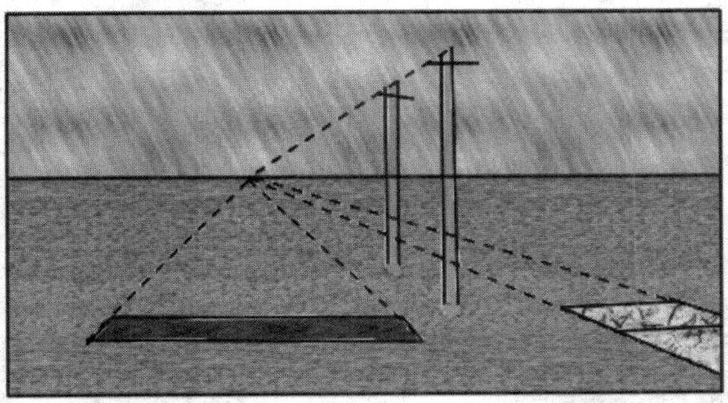

(c) Railway tracks with electric poles and parallel road

Figure 7.4.16 Exercise on extending patterns.

7. Draw a cube in three-point perspective. Find the mid-points of each of the faces of the cube.
8. (a) An octagon is inscribed in a square. Write the steps needed to duplicate it (Figure 7.4.17).

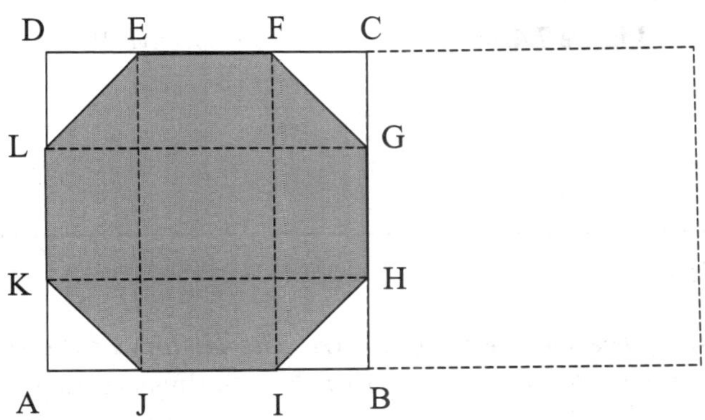

Figure 7.4.17 Duplicating an octagon.

(b) How will you duplicate the octagon in one-point perspective?
(c) How will you duplicate the octagon in two-point perspective?

9. *Satire on False Perspective* was engraved by William Hogarth in 1754 (Figure 7.4.18). The subtitle *Whoever makes a design without the Knowledge of perspective will be liable to such Absurdities as are shewn in this Frontispiece* clearly describes the purpose of the work. How many absurdities can you point out in the work?

Figure 7.4.18 *Satire on False Perspective.* [C23]

Bibliography

Books and Articles

[1] K. Andersen. 2006. *The Geometry of an Art: The History of the Mathematical Theory of Perspective from Alberti to Monge. New York*, NY: Springer-Verlag.

[2] J. Elkins. 1996. *The Poetics of Perspective*. Baltimore, MD: Cornell University Press.

[3] Marc Frantz and Annelisa Crannell. 2011. *Viewpoints: Mathematical Perspective and Fractal Geometry in Art*. Princeton University Press.

[4] Morris Kline. 2007. *Mathematics for the Nonmathematician*. Mineola, New York, NY: Dover Publications.

[5] E. R. Norling. 1999. *Perspective Made Easy*. New York, NY: Dover Publications.

[6] Dan Pedoe. 2000. *Geometry and the Visual Arts*. New York, NY: Dover Publications.

Website

[W1] C. Tyler and M. Kubovy. *Perspective: Science and Art of Perspective*. http://www.webexhibits.org/sciartperspective/ (accessed 28 January 2017).

Image Credits

[C1] Rameessos/Wikimedia Commons. Public domain. https://commons.wikimedia.org/wiki/File:AltamiraBison.jpg (accessed 12 January 2017).

[C2] Carla Hufstedler/Wikimedia Commons. CC BY-SA 2.0 license. https://commons.wikimedia.org/wiki/File:20,000_Year_Old_Cave_Paintings_Hyena.png (accessed 12 January 2017).

[C3] Wikimedia Commons.Public domain. https://commons.wikimedia.org/wiki/File:Mughal_painting_1.jpg (accessed 12 January 2017).

[C4] Paul_012/Wikimedia Commons. Public domain. https://commons.wikimedia.org/wiki/File:Sanzio_01.jpg (accessed 12 January 2017).

[C5] JarektUploadBot/Wikimedia Commons. Public domain. https://commons.wikimedia.org/wiki/File3ADuccio_di_Buoninsegna_-_Last_Supper_-_WGA06786.jpg (accessed 28 January 2017).

[C6] The Yorck Project/Wikimedia Commons. Public domain. https://commons.wikimedia.org/wiki/File:Indischer_Maler_um_1580_001.jpg (accessed 12 January 2017).

[C7] The Yorck Project/Wikimedia Commons. Public domain. https://commons.wikimedia.org/wiki/File:Meister_des_RasikapriyC3%A2-Manuskripts_001.jpg (accessed 12 January 2017).

[C8] Ekabhishek/Wikimedia Commons. Public domain. https://commons.wikimedia.org/wiki/File:Radha_celebrating_Holi,_c1788.jpg (accessed 12 January 2017).

[C9] The Yorck Project/Wikimedia Commons. Public domain. https://commons.wikimedia.org/wiki/File:NihC3A2l_Chand_001.jpg (accessed 12 January 2017).

[C10] Richardfabi∼commonswiki/Wikimedia Commons. Public domain. https://commons.wikimedia.org/wiki/File:Bunelleschi.jpg (accessed 12 January 2017).

[C11] Alonso de Mendoza/Wikimedia Commons. Public domain. https://commons.wikimedia.org/wiki/File:C3%9Altima_Cena_-_Da_Vinci_5.jpg (accessed 12 January 2017).

[C12] Oursana / Wikimedia Commons. Public domain. https://commons.wikimedia.org/wiki/File:Formerly_Piero_della_Francesca_-_Ideal_City_-_Galleria_Nazionale_delle_Marche_Urbino_2.jpg (accessed 12 January 2017).

[C13] Miaow Miaow / Wikimedia Commons. Public domain. https://commons.wikimedia.org/wiki/File:Loreto_Fresko.jpg (accessed 12 January 2017).

[C14] Jordi.2~commonswiki/Wikimedia Commons. Public domain. https://commons.wikimedia.org/wiki/File:GiottoMadonna.jpg (accessed 12 January 2017).

[C15] Emok / Wikimedia Commons. CC BY-SA 3.0 license. https://commons.wikimedia.org/wiki/File:First_angle_projection.svg (accessed 30 January 2017).

[C16] Ldo/Wikimedia Commons. CC BY-SA 3.0 license. https://commons.wikimedia.org/wiki/File:PerspectiveProjection.svg (accessed 30 January 2017).

[C17] Bob Mellish / Wikimedia Commons. CC BY-SA 3.0 license. https://commons.wikimedia.org/wiki/File:Pinhole-camera.png (accessed 30 January 2017).

[C18] SreeBot/Wikimedia Commons. Public domain. https://commons.wikimedia.org/wiki/File%3AStPaul'sCross.jpg (accessed 12 January 2017).

[C19] Alonso de Mendoza/Wikimedia Commons. Public domain. https://commons.wikimedia.org/wiki/File:Entrega_de_las_llaves_a_San_Pedro_(Perugino).jpg (accessed 12 January 2017).

[C20] DcoetzeeBot/Wikimedia Commons. Public domain. https://commons.wikimedia.org/wiki/File:Gustave_Caillebotte_-_Paris_Street;_Rainy_Day_-_Google_Art_Project.jpg (accessed 12 January 2017).

[C21] Oxag/Wikimedia Commons. Public domain. https://commons.wikimedia.org/wiki/File:Reconstruction_of_the_temple_of_Jerusalem.jpg (accessed 12 January 2017).

[C22] JarektUploadBot/Wikimedia Commons. Public domain. https://commons.wikimedia.org/wiki/File:Masolino_-_Banquet_of_Herod_-_WGA14245.jpg (accessed 12 January 2017).

[C23] Churchh/Wikimedia Commons. Public domain. https://commons.wikimedia.org/wiki/File:Hogarth-satire-on-false-pespective-1753.jpg (accessed 12 January 2017).

[C24] DcoetzeeBot/Wikimedia Commons. Public Domain. https://commons.wikimedia.org/wiki/File%3AAttributed_to_Hiranand_-_Illustration_from_a_Dictionary_(unidentified)-_Da'ud_Receives_a_Robe_of_Honor_from_Mun'im_Khan_-_Google_Art_Project.jpg (accessed 9 July 2017).

[C25] Ghirlanadajo/Wikimedia Commons. Public Domain. https://upload.wikimedia.org/wikipedia/commons/6/66/Yaroslavl_gospel.jpg (accessed 9 July 2017).

[C26] The Yorck Project/Wikimedia Commons. Public Domain. https://upload.wikimedia.org/wikipedia/commons/c/ca/Dieric_Bouts_004.jpg (accessed 9 July 2017).

[C27] The Yorck Project/Wikimedia Commons. Public Domain. https://upload.wikimedia.org/wikipedia/commons/8/8d/Dieric_Bouts_009.jpg (accessed 9 July 2017).

[C28] Redrawn by Shobha Bagai. Original picture at https://in.pinterest.com/pin/132152570289832015/

[C29] Photo by Shobha Bagai.

Solutions to Selected Exercises

Solutions to Exercise 2.1

1. (a) Not a set as *intelligent* is subjective.
   (b) Set. { 7, 14, 21, . . . }
   (c) Set. {Tokyo, Jakarta, Delhi, Manila, Seoul, Shanghai, Karachi, Beijing, New York City, Guangzhou} as of 2016.
   (d) Set as all the rivers of India can be listed.
   (e) Not a set as *top* is subjective

3. (a) $G = \{$M, A, T, H, E, I, C, S$\}$
   (b) $H = \{1\}$
   (c) $I = \{$Pratibha Patil$\}$
   (d) $J = \{2, 3, 5, 17, 101\}$
   (e) $K = \{2, 5, 10, 17, \ldots\}$

5. (a) Equal. $A = B = \{$F, L, O, W$\}$
   (b) Not equal. $A = \{-2, -3\}$, $B = \{-2, 3\}$
   (c) Equal. $A = B = \{$Blue, Red, White$\}$
   (d) Not equal. $A = \{1, 2, 3, 5\}$, $B = \{2, 3, 5\}$

Solutions to Exercise 2.2

1. (a) Finite. Cardinality is 12.
   (b) Infinite.
   (c) Infinite.
   (d) Finite. Cardinality is 4.
   (e) Finite. Cardinality is 8.

3. (a) True. Set J has elements that are prime.
   (b) False. Set K has elements that are not prime.
   (c) False. Cardinality of set C is 9, whereas the cardinality of set J is 5.
   (d) False. Cardinality of set B is 7, whereas the cardinality of set D is 6.

5. (a) $B \subsetneq A$ (b) $A \subsetneq B$ (c) $A \subsetneq B$ (d) $B \subsetneq A$ (e) $B \subsetneq A$

Solutions to Exercise 2.3

1. (a) $A \cup B = \{1, 2, 3, 4, 5\}$
   (b) $A \cap D = \emptyset$
   (c) $(B \cup C) \cup D = \{3, 4, 5, 6, 7\}$
   (d) $B \cup (B \cap D) = \{3, 4, 5\}$
   (e) $(B \cap C) \cup (B \cap D) = \{4, 5\}$
   (f) $B \cap (C \cup D) = \{4, 5\}$

3. (a) $A' = \{\text{C, P, L, E}\}$
   (b) $B' = \{\text{C, O, P, L, E}\}$
   (c) $(A \cap B)' = \{\text{C, O, P, L, E}\}$
   (d) $(B \cup C)' = \{\text{P, L, E}\}$
   (e) $(A')' = \{\text{S, T, A, I, O, N}\}$
   (f) $(A \cup B)' \cap C' = \{\text{P, L, E}\}$

Solutions to Exercise 2.4

1. (a) $A \subsetneq B$, $A \subsetneq C$, $A \subsetneq B \cup C$, $A \subsetneq B \cap C$
   (b) $B \subsetneq A$, $B \subsetneq A \cup C$
   (c) $A \subsetneq C$, $B \subsetneq C$, $A \cap B = \emptyset$

3. 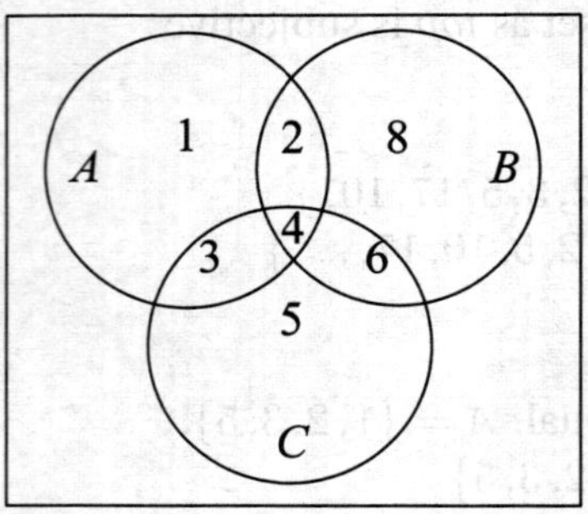

5. (a)
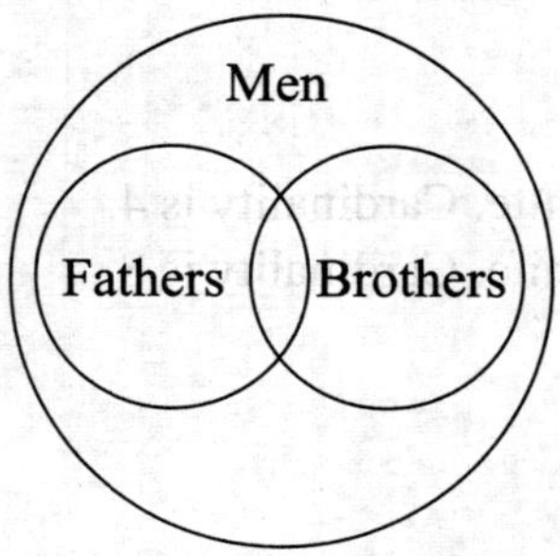

(b)
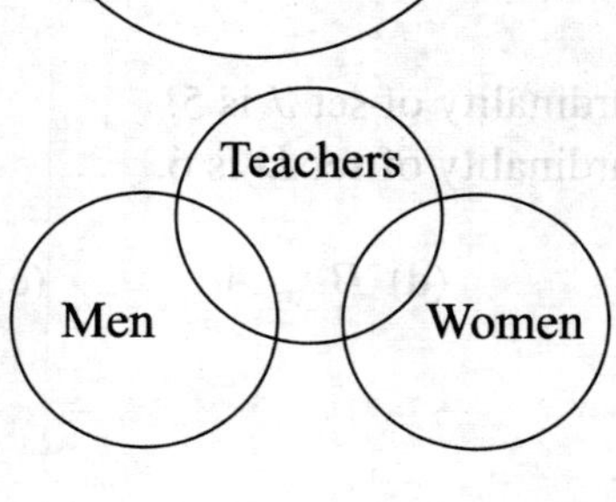

(c)
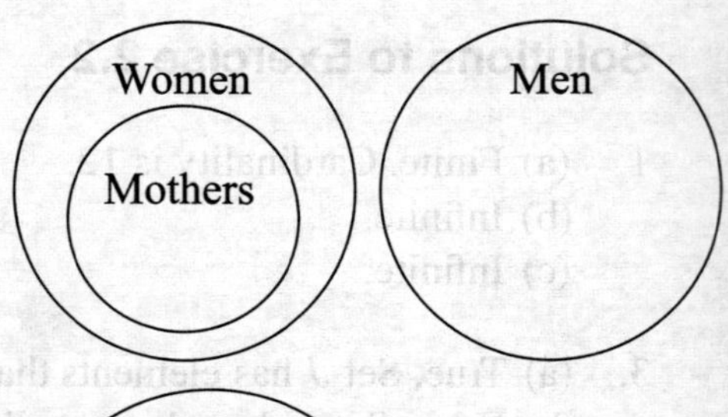

(d)
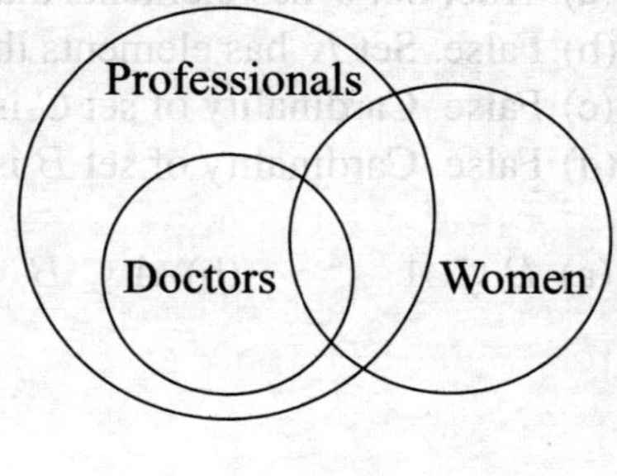

(e)

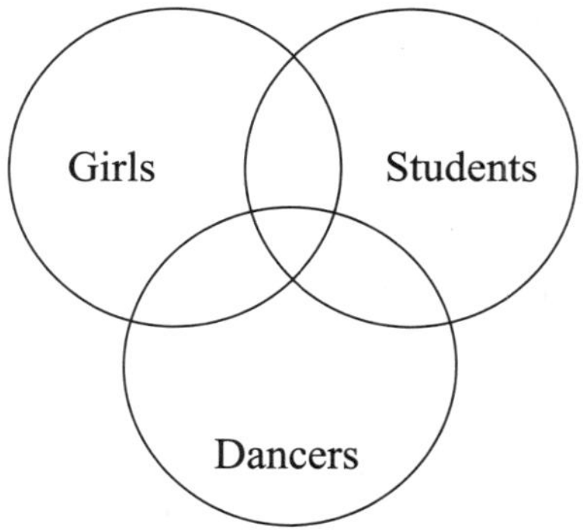

Solutions to Exercise 2.5

1. (a)

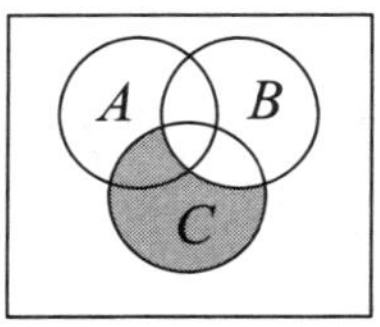

(b)

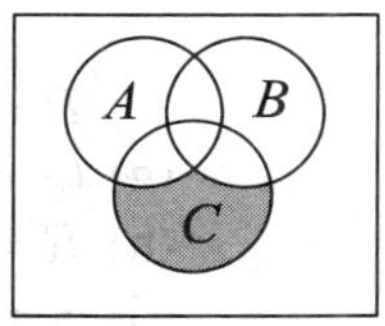

(c)

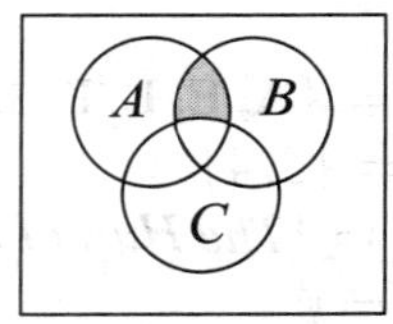

(d)

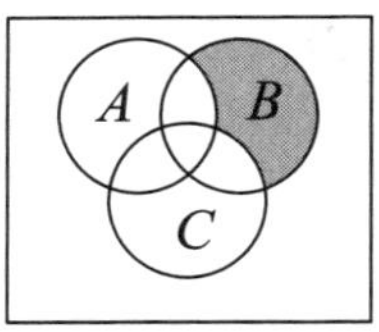

(e)

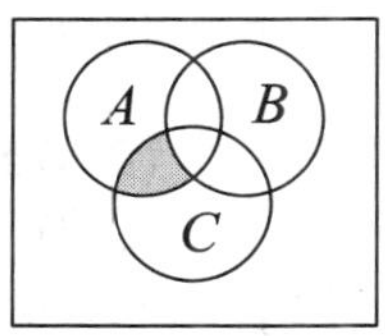

(f)

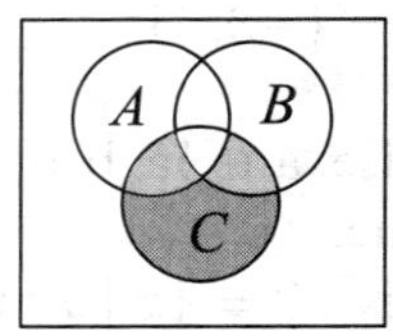

3. (a)

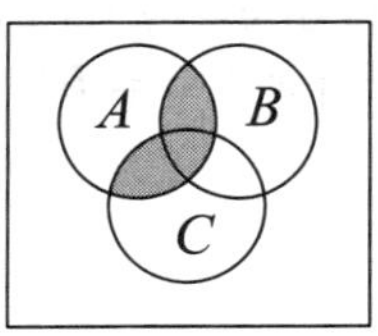

$A \cap (B \cup C)$

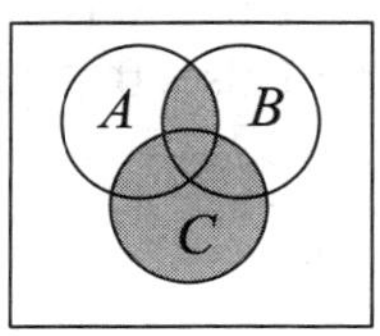

$(A \cap B) \cup C$

(b)

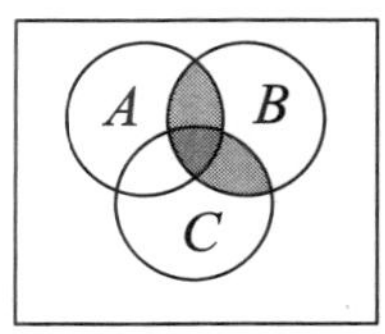

$(A \cap B) \cup (B \cap C)$

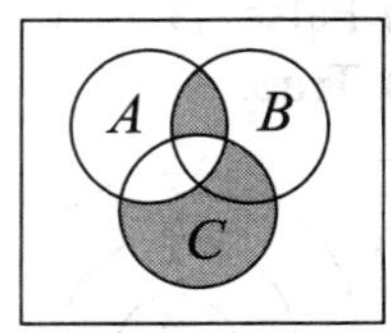

$(C \cap A') \cup (A \cap B)$

Solutions to Exercise 2.6

1. Number of students studying only mathematics is $12 - 5 = 7$
   Number of students studying only economics is $10 - 5 = 5$

   (a) 23 (b) 12

3. Percentage of students who passed in at least one paper $= 90\%$
   Percentage of students who passed in both the papers $70\% + 60\% - 90\% = 40\%$
   Total number of students $= 750$.

5. (a) 16 (b) 49 (c) 9 (d) 50

Solutions to Chapter 2 Review Exercises

1. (a) $A = \{x \mid x \text{ is a noble gas}\}$
   (b) $B = \{x \mid x \text{ is a letter of the word PETALS}\}$
   (c) $C = \{x \mid x \text{ is a month of the year having 31 days}\}$
   (d) $D = \{x \mid x \text{ is a Jungle Book character}\}$
   (e) $E = \{x \mid x = \frac{1}{n}, n \in \mathbb{N} \text{ and } n \leq 100\}$
   (f) $F = \{x \mid x = n^3, n \in \mathbb{N}\}$
   (g) $G = \{x \mid x \text{ is a holy book}\}$
   (h) $H = \{x \mid x \text{ is a prime factor of } 30\}$
   (i) $I = \{x \mid x = 2^n, n \in \mathbb{N}\}$
   (j) $J = \{x \mid x = 7 + 9n, n \in \mathbb{N}, 1 \leq n \leq 7\}$

3. (a) $K = \{\text{A, E, L, P, S, T}\}$
   (b) $L = \{3, 4\}$
   (c) $M = \{$*The Hurst Locker*$\}$
   (d) $N = \emptyset$
   (e) $O = \{-3, 5\}$
   (f) $P = \{1, 2, 5, 7, 10, 14, 35, 70\}$
   (g) $Q = \{-1, 1, 5, 13, \dots\}$
   (h) $R = \emptyset$
   (i) $S = \emptyset$
   (j) $T = \{\text{A, E, G, R, T}\}$

5. (a) False. $\text{a} \in \{\text{a, b, c}\}$ or $\{\text{a}\} \subseteq \{\text{a, b, c}\}$
   (b) True.
   (c) False. $\emptyset = \{x \mid x \text{ is an even factor of } 35\}$
   (d) False. $B \in A$
   (e) True.
   (f) True.
   (g) True.
   (h) False. $\mathcal{P}(A) = 2^8$
   (i) False. $\emptyset$ is a subset of every set.
   (j) False. $0 \in \{x \in \mathbb{R} \mid x + 5 = 5\}$

7. (a)
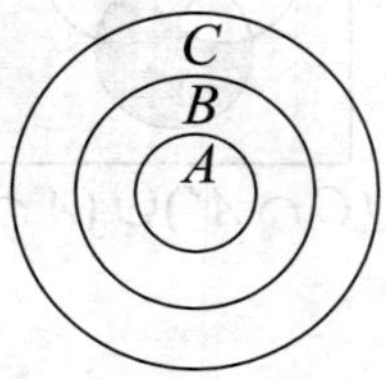

(b)
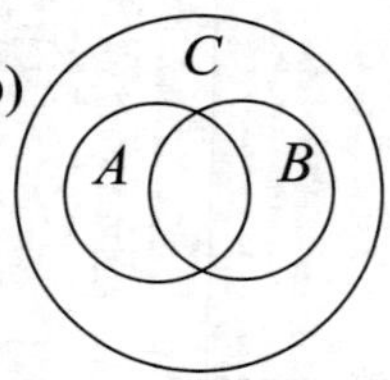

(c)
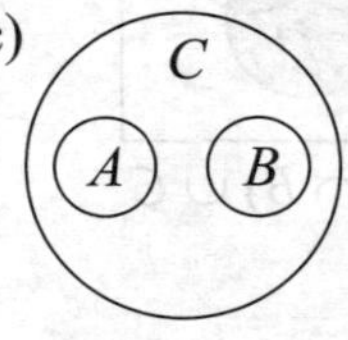

(d)
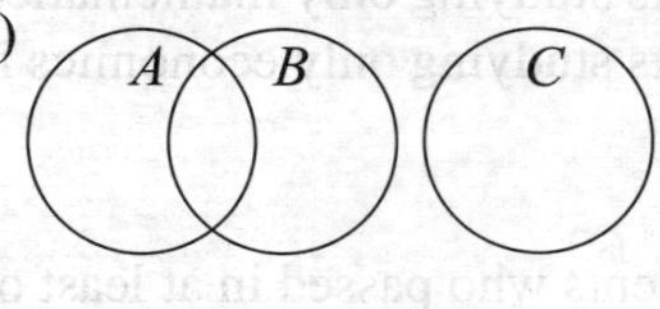

9. (a)
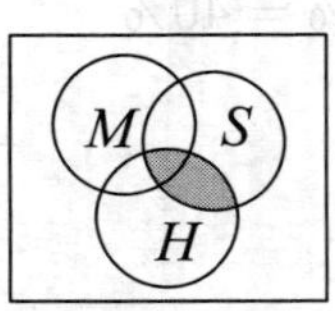

(b)
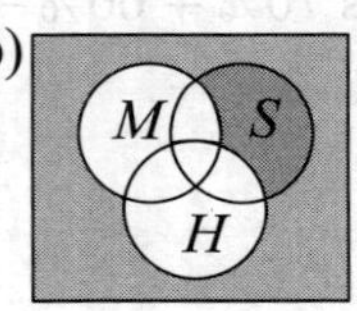

(c)
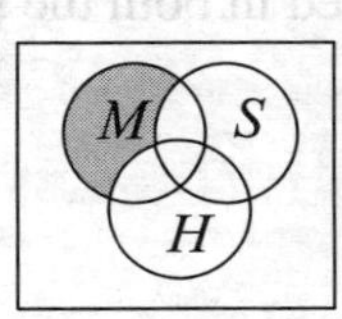

(d)
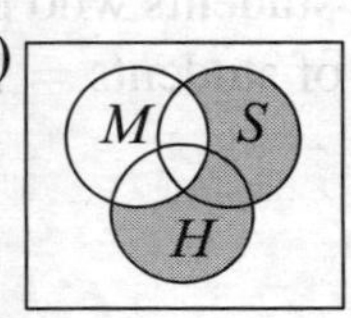

11. (a)

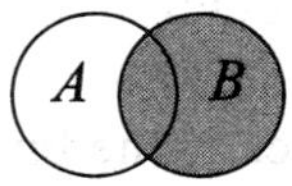

$(A \cup B) \cap B$

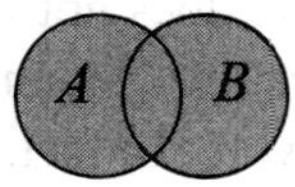

$(A \cup B)$

(b)

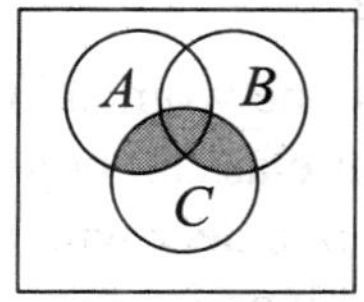

$(A \cup B) \cap C$

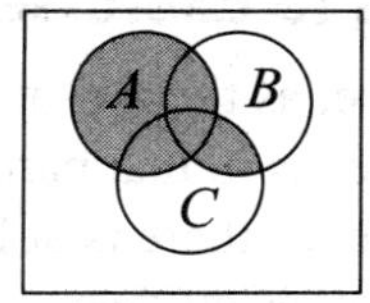

$A \cup (B \cap C)$

13. (a) Number of people who like only brand X is $860 - 600 = 260$
    (b) Number of people who like only brand Y is $725 - 600 = 125$
    (c) Number of people who did not like any brand is $1000 - 260 - 125 - 600 = 15$
15. The following Venn diagram represents the number of people in each society:

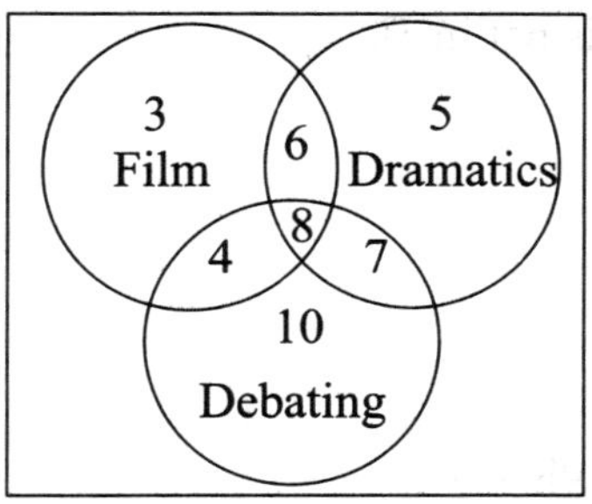

(a) Total number of members is 43.
(b) Number of members who are in exactly two societies is 17.
(c) Number of members who are in at least two societies is 25.
(d) Number of members who are in exactly one society is 18.

17. The following Venn diagram represents the given data:

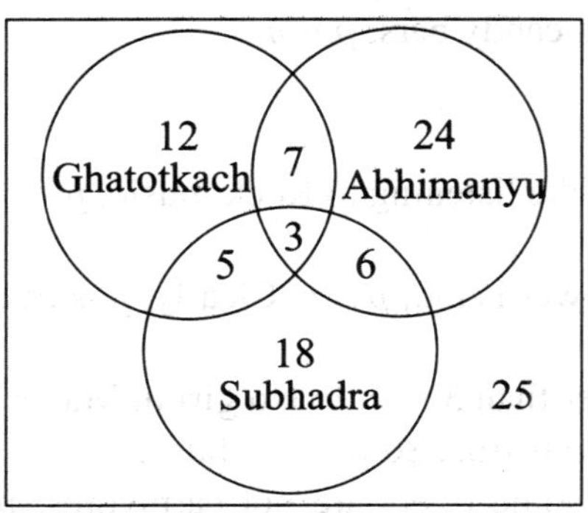

Therefore, 32 people knew about Subhadra.

Solutions to Exercise 3.1

1. (a) Statement.
   (b) Statement.
   (c) Not a statement as it is a command.
   (d) Statement.
   (e) Not a statement as it is a question.

Solutions to Exercise 3.2

1. (a) False, as there is at least one woman in India who does not wear a bindi.
   (b) True, as in that particular class all boys are over 3 feet in height.
   (c) True, as there is at least one deer who is a herbivore. In fact, all deer are herbivores.
   (d) False, as all deer are herbivores.
   (e) True, as there is no natural number lying strictly between 2 and 3. In fact, 3 is the next natural number after 2.
3. (a) $A \subseteq B$.
   (b) $A \cap B \neq \emptyset$. As there is at least one element of A which is also in B.
   (c) $A \cap B = \emptyset$. As A and B can have no element in common.
   (d) $A \setminus B \neq \emptyset$. As there is at least one element of A which is not in B.

Solutions to Exercise 3.3

1. (a) Some women in India do not wear Bindis.
   (b) No woman in India wears bindis.
   (c) Some women in India wear bindis.
   (d) All women in India wear bindis.
   (e) The set $\{x \in \mathbb{N} \mid 2 < x < 4\}$ does not have exactly one element.

Solutions to Exercise 3.4

1. (a) Conjunction. p: Today is a Sunday, q: I can sleep late, $p \wedge q$.
   (b) Conjunction. p: She can sketch, q: She cannot paint, $p \wedge q$.
   (c) Disjunction. p: He can walk, q: He can run, $p \vee q$.
   (d) Conjunction. p: Every tiger is a carnivore, q: Not all carnivores are tigers, $p \wedge q$.
   (e) Disjunction. p: He can eat chocolates, q: He can eat candy bars, $p \vee q$.

Solutions to Exercise 3.5

1. (a) Conjunction. p: I like watching Hindi movies, q: I like listening to Rock music, $p \wedge q$. For me, p is true and q is true. So $p \wedge q$ is true.
   (b) Conjunction. p: Every rose is a flower, q: Every flower is red, $p \wedge q$. Clearly, p is true and q is false. So $p \wedge q$ is false.
   (c) Conjunction. p: No boy in the Maths class is greater than 3 feet, q: No girl in Maths class is less than three feet, $p \wedge q$. Clearly, p is false and q is true. So $p \wedge q$ is false.
   (d) Disjunction. p: Some deer are not herbivores, q: some tigers are not carnivores, $p \vee q$. Clearly, p is false and q is also false. So $p \vee q$ is false.
   (e) Disjunction. p: All Indian women wear sarees, q: All Indian men wear dhotis, $p \vee q$. Here, p is false and q is also false. So $p \vee q$ is false.

Solutions to Exercise 3.6

1. (a) This is a tautology. It is therefore not a contradiction. The given statement is of the form $A \vee B$, where $A = p \vee r$ and $B = \sim (p \wedge q)$. But $A \vee B$ can be false only if both A and B are false. Note that A can be false only if p is false (and r is false). But if p is false, then B has to be true.

   (b) Not a tautology as when p is false and q is also false, the given statement will be false. Not a contradiction as when p is true, and q is also true, the given statement is true.

   (c) This is a tautology. It is therefore not a contradiction. The given statement is of the form $\sim A$, where
   $A = (p \wedge r) \wedge (\sim p \wedge \sim q)$. Now, A is of the form $C \wedge D$. We note that when p is true, D is false. Also, when p is false C is also false. Thus, A will always take the truth value false and hence $\sim A$ will always be true.

   (d) Not a tautology as when r is false and p is true, the given statement is false. Not a contradiction, as when r is true, the given statement is true.

   (e) This is a contradiction as whether p is true or false, the statement $\sim p \wedge (\sim q \wedge p)$ is false. It is therefore not a tautology.

1. (a) Not a tautology or a contradiction.
   Truth table for $\sim p \vee (\sim q \wedge \sim r)$

p	q	r	$\sim p$	$\sim q$	$\sim r$	$\sim q \wedge \sim r$	$\sim p \vee (\sim q \wedge \sim r)$
T	T	T	F	F	F	F	F
T	T	F	F	F	T	F	F
T	F	T	F	T	F	F	F
T	F	F	F	T	T	T	T
F	T	T	T	F	F	F	T
F	T	F	T	F	T	F	T
F	F	T	T	T	F	F	T
F	F	F	T	T	T	T	T

   (b) Not a tautology or a contradiction.

   Truth table for $\sim [\sim p \wedge (q \vee r)]$

p	q	r	$\sim p$	$q \vee r$	$\sim p \wedge (q \vee r)$	$\sim [\sim p \wedge (q \vee r)]$
T	T	T	F	T	F	T
T	T	F	F	T	F	T
T	F	T	F	T	F	T
T	F	F	F	F	F	T
F	T	T	T	T	T	F
F	T	F	T	T	T	F
F	F	T	T	T	T	F
F	F	F	T	F	F	T

(c) Not a tautology or a contradiction. The given statement is logically equivalent to the statement of part (a).

Truth table for $\sim [p \wedge (q \vee r)]$

p	q	r	$q \vee r$	$p \wedge (q \vee r)$	$\sim [p \wedge (q \vee r)]$
T	T	T	T	T	F
T	T	F	T	T	F
T	F	T	T	T	F
T	F	F	F	F	T
F	T	T	T	F	T
F	T	F	T	F	T
F	F	T	T	F	T
F	F	F	F	F	T

(d) This is a contradiction.

Truth table for $(\sim p \wedge (r \vee q)) \wedge (p \wedge (r \vee q))$

p	q	r	$\sim p$	$r \vee q$	$\sim p \wedge (r \vee q)$	$p \wedge (r \vee q)]$	$(\sim p \wedge (r \vee q)) \wedge (p \wedge (r \vee q))$
T	T	T	F	T	F	T	F
T	T	F	F	T	F	T	F
T	F	T	F	T	F	T	F
T	F	F	F	F	F	F	F
F	T	T	T	T	T	F	F
F	T	F	T	T	T	F	F
F	F	T	T	T	T	F	F
F	F	F	T	F	F	F	F

Solutions to Exercise 3.7

1. (a) $p \wedge q$ and $\sim (p \wedge q) = \sim p \vee \sim q$.
   p: No scoundrel is a ruffian, q: Some thieves are not scoundrels.
   $\sim p$: Some scoundrels are ruffians, $\sim q$: All thieves are scoundrels.
   $\sim p \vee \sim q$: Some scoundrels are ruffians or all thieves are scoundrels.
   (b) $p \vee q$ and $\sim (p \vee q) = \sim p \wedge \sim q$.
   p: None of us is out of breath, q: Some of us are not fat.
   $\sim p$: Some of us are out of breath, $\sim q$: All of us are fat.
   $\sim p \wedge \sim q$: Some of us are out of breath and all of us are fat.
   (c) $p \vee q$ and $\sim (p \vee q) = \sim p \wedge \sim q$.
   p: You wear a tie to the interview, q: You do not get hired. $\sim p$: You do not wear a tie to the interview, $\sim q$: You get hired.
   $\sim p \wedge \sim q$: You do not wear a tie to the interview and you get hired.
   (d) $p \wedge q$ and $\sim (p \wedge q) = \sim p \vee \sim q$.
   p: She is beautiful, q: She is modest.
   $\sim p$: She is not beautiful, $\sim q$: She is not modest.
   $\sim p \vee \sim q$: She is not beautiful or modest.
   (e) $p \wedge q$ and $\sim (p \wedge q) = \sim p \vee \sim q$. p: All rainy days are muggy, q: Some summer days are not hot.
   $\sim p$: No rainy day is muggy, $\sim q$: All summer days are hot.
   $\sim p \vee \sim q$: No rainy day is muggy or all summer days are hot.

Solutions to Exercise 3.8

1. (a) $\sim (p \to q) \equiv p \wedge \sim q$. So the required statement is 'You eat carrots, but you do not have good eyesight'.
   (b) $\sim p \to q$: If you commit a traffic violation, then your insurance premium does not go up.
   (c) $p \to \sim q$: If you don't carry an umbrella, then it will not rain.
   (d) $\sim p \to \sim q$: If you exercise regularly, then you stay healthy.
3. (a) p: We get a salary increase, q: We will be happy. Now $\sim (p \to q) \equiv p \wedge \sim q$. So the negation will be 'We get a salary increase, but we are not happy.
   (b) The dog wags its tail but it bites.
   (c) 'All tigers are predators' is the same as 'if you are a tiger, then you are a predator'. So the negation will be 'It is a tiger, but it is not a predator'.
   (d) 'No beggars are choosers' is the same as 'if you are a beggar, then you have no choice'. So the negation will be 'You are a beggar, but you have choice.
   (e) All scoundrels are ruffians, but no pirate is scoundrel.
5. (a) If p, s are true and q is false, then $(\sim s \wedge p)$ is false and $(\sim q \to s)$ is true. Therefore, $(\sim s \wedge p) \to (\sim q \to s)$ will be true.
   (b) When p is false but q, r and s are true, then $(p \to q)$ is true and $(r \wedge \sim s)$ is false. Therefore $(p \to q) \vee (r \wedge \sim s)$ is true.

(c) Since p is true we have that $\sim p$ is false. Therefore, $\sim p \rightarrow \{\sim [(q \vee \sim r) \wedge (w \wedge \sim q) \vee (u \vee \sim w)]\}$ has to be true.

Solutions to Exercise 3.9

1. (a) **Symbolising the argument:**
p: I want to be a lawyer.
q: I want to study logic.
r: I like to argue
Premises:
P_1: $p \rightarrow q$
P_2: $\sim p \rightarrow \sim r$
Conclusion:
C: $r \rightarrow q$
Truth table for the argument

Input		Variables	P_1	P_2	C
p	q	r	$p \rightarrow q$	$\sim p \rightarrow \sim r$	$r \rightarrow q$
T	T	T	T	T	T
T	T	F	T	T	T
T	F	T	F	T	F
T	F	F	F	T	T
F	T	T	T	F	T
F	T	F	T	T	T
F	F	T	T	F	F
F	F	F	T	T	T

This argument is valid as there is no 'bad row' in the truth table, that is there is no row in which both P_1 and P_2 are true but C is false.

(b) **Symbolising the argument:**
p: I buy cheap petrol.
q: My car runs badly.
r: I don't change the oil.
Premises:
P_1: $p \rightarrow q$
P_2: $r \rightarrow q$
Conclusion:
C: $p \rightarrow r$

Truth table for the argument

Input		Variables	P_1	P_2	C
p	q	r	$p \to q$	$r \to q$	$p \to r$
T	T	T	T	T	T
T	T	F	T	T	F
T	F	T	F	F	T
T	F	F	F	T	F
F	T	T	T	T	T
F	T	F	T	T	T
F	F	T	T	F	T
F	F	F	T	T	T

This argument is invalid as there is a 'bad row' in the truth table. In the '2nd row' of the truth-value entries, both P_1 and P_2 are true, but C is false.

(c) **Symbolising the argument:**

p: You are a clown.

q: You work in the circus.

r: You like cotton candy.

Premises:

P_1: $p \to q$

P_2: $\sim r \to \sim q$

Conclusion:

C: $p \to r$

Truth table for the argument

Input		Variables	P_1	P_2	C
p	q	r	$p \to q$	$\sim r \to \sim q$	$p \to r$
T	T	T	T	T	T
T	T	F	T	F	F
T	F	T	F	T	T
T	F	F	F	T	F
F	T	T	T	T	T
F	T	F	T	F	T
F	F	T	T	T	T
F	F	F	T	T	T

This argument is valid as there is no 'bad row' in the truth table.

(d) **Symbolising the argument:**
p: I do not pay my income taxes.
q: I file for an extension.
r: I am a felon.
Premises:
$P_1: p \rightarrow (q \vee r)$
$P_2: \sim q \wedge \sim r$
Conclusion:
$C: \sim p$
Truth table for the argument

Input		Variables	P_1	P_2	C
p	q	r	$p \rightarrow (q \vee r)$	$\sim q \wedge \sim r$	$\sim p$
T	T	T	T	F	F
T	T	F	T	F	F
T	F	T	T	F	F
T	F	F	F	T	F
F	T	T	T	F	T
F	T	F	T	F	T
F	F	T	T	F	T
F	F	F	T	T	T

This argument is valid as there is no 'bad row' in the truth table.

(e) All Koala bears are cuddly. No cat is a Koala bear. Thus, no cat is cuddly.

Symbolising the argument:
p: It is a Koala bear.
q: It is cuddly.
r: It is a cat.
Premises:
$P_1: p \rightarrow q$
$P_2: r \rightarrow \sim p$
Conclusion:
$C: r \rightarrow \sim q$

Truth table for the argument

Input		Variables	P_1	P_2	C
p	q	r	$p \to q$	$r \to \sim p$	$r \to \sim q$
T	T	T	T	F	F
T	T	F	T	T	T
T	F	T	F	F	T
T	F	F	F	T	T
F	T	T	T	T	F
F	T	F	T	T	T
F	F	T	T	T	T
F	F	F	T	T	T

This argument is invalid as there is a 'bad row' in the truth table. In the '5th row' of the truth-value entries, both P_1 and P_2 are true, but C is false.

Solutions to Chapter 3 Review Exercises.

1. (a) Statement
   (b) Not a statement as it is an appeal.
   (c) Not a statement as it is a command.
   (d) Statement.
   (e) Statement.

3. (a) Not necessarily. For example, 'Some dogs are not poodles' is true, but 'Some poodles are not dogs' is false.
   (b) The statement 'No A is B' is the same as saying $A \cap B = \emptyset$. Thus, 'No B is A' will also be true.
   (c) The statement 'Some A are B' is false only when the statement 'No A is B' is true. So the statement 'No B is A' will also be true. Thus, the statement 'Some B are A' will be false.
   (d) If the statement 'Some A are B' is false, then the statement 'No A is B' will be true.
   (e) The statement 'All A are B' is false only when the statement 'Some A are not B' is true and vice versa.

5. (a) Data is insufficient. It is only given that Reena does not teach on Wednesdays and on weekends. Nothing is said about which other weekdays Reena teaches.
   (b) True. Since Reena holds office hours in the University every Wednesday and there is at least one Saturday when she has come to the University.
   (c) True. A disjunction $p \vee q$ is true if either p or q is true. Here, both p and q are true.
   (d) False. Every Wednesday Reena is in the University.
   (e) True. It is given that on most Saturdays Reena stays home and so there is at least one Saturday when she is not in the University.

7. (a) Some deer are herbivores.
   (b) All men have long hair.
   (c) Some months have 30 days.
   (d) Today is a Saturday.
   (e) Some dogs are poodles.

9. (a) $p \wedge p$ is true only when p is true and $p \wedge p$ is false only when p is false. Thus, $p \wedge p \equiv p$.
   (b) $p \vee p$ is true only when p is true and $p \vee p$ is false only when p is false. Thus, $p \vee p \equiv p$.
   (c) Since $p \equiv q$, p and q are either both true or both false. Thus, $\sim p$ and $\sim q$ are both false or are both true. Hence $(\sim p) \equiv (\sim q)$.
   (d) Since $p \equiv (\sim q)$ then p and $\sim q$ are either both true or both false. Thus, $\sim p$ and q are both false or are both true. Hence, $\sim p \equiv q$.
   (e) $\sim (\sim p) \equiv p$, since p and $\sim p$ have opposite truth values, p and $\sim (\sim p)$ will either both be true or both be false.

11. (a) p: Some girls can sing, q: All men are mortal. p is true and q is true.
    $\sim p \wedge q$: No girl can sing and all men are mortal. This is false as $\sim p$ is false.
    $\sim p \vee q$: No girl can sing or all men are mortal. This is true as q is true.
    (b) p: No new-born human baby has teeth, q: All roses are red. p is true and q is false.
    $\sim p \wedge q$: Some new-born human babies have teeth and all roses are red. This is false as q is false.
    $\sim p \vee q$: Some new-born human babies have teeth or all roses are red. This is false as both $\sim p$ and q are false.
    (c) p: All men have moustaches, q: Some women wear skirts. p is false and q is true.
    $\sim p \wedge q$: Some men do not have moustaches and some women wear skirts. This is true as both $\sim p$ and q are true.
    $\sim p \vee q$: Some men do not have moustaches and some women wear skirts. This is true as $\sim p$ is true.
    (d) p: No boy likes to dance, q: All trains run on time. Both p and q are false.
    $\sim p \wedge q$: Some boys like to dance and all trains run on time. This is false as q is false.
    $\sim p \vee q$: Some boys like to dance or all trains run on time. This is true as $\sim p$ is true.

13. (a) Negation of $(\sim p \vee q) \wedge \sim (\sim q \wedge r)$.

$$\sim [(\sim p \vee q) \wedge \sim (\sim q \wedge r)] \equiv \sim (\sim p \vee q) \vee \sim [\sim (\sim q \wedge r)]$$
since $\sim (s \wedge t) \equiv \sim s \vee \sim t$
$$\equiv [\sim (\sim p) \wedge \sim q] \vee [\sim q \wedge r]$$
since $\sim (s \vee t) \equiv \sim s \wedge \sim t$ and $\sim\sim r \equiv r$
$$\equiv [p \wedge \sim q] \vee [\sim q \wedge r]$$
since $\sim\sim r \equiv r$

(b) Negation of $\sim p \vee \sim q$.

$$\sim [\sim p \vee \sim q] \equiv \sim (\sim p) \wedge \sim (\sim q)$$
since $\sim (s \vee t) \equiv \sim s \wedge \sim t$
$$\equiv p \wedge q$$
since $\sim\sim r \equiv r$

(c) Negation of $(p \vee q) \wedge r$.

$$\sim [(p \vee q) \wedge r] \equiv \sim (p \vee q) \vee \sim r \quad \text{since} \sim (s \wedge t) \equiv \sim s \vee \sim t$$
$$\equiv [\sim p \wedge \sim q] \vee \sim r \quad \text{since} \sim (s \vee t) \equiv \sim s \wedge \sim t$$

(d) Negation of $(p \vee r) \wedge (p \vee q)$.

$$\sim [(p \vee r) \wedge (p \vee q)] \equiv \sim (p \vee r) \vee \sim (p \vee q) \quad \text{since} \sim (s \wedge t) \equiv \sim s \vee \sim t$$
$$\equiv [\sim p \wedge \sim r] \vee [\sim p \wedge \sim q) \quad \text{since} \sim (s \vee t) \equiv \sim s \wedge \sim t$$

(e) Negation of $(p \wedge q) \vee (\sim p \wedge q)$.

$$\sim [(p \wedge q) \vee (\sim p \wedge q)] \equiv \sim (p \wedge q) \wedge \sim (\sim p \wedge q) \quad \text{since} \sim (s \vee t) \equiv \sim s \wedge \sim t$$
$$\equiv [\sim p \vee \sim q] \wedge [\sim (\sim p) \vee \sim q) \quad \text{since} \sim (s \wedge t) \equiv \sim s \vee \sim t$$
$$\equiv [\sim p \vee \sim q] \wedge [p \vee \sim q) \quad \text{since} \sim\sim r \equiv r$$

15. (a) $p \rightarrow q$ Here p: It is hot, q: I will sweat.
Converse: $q \rightarrow p$. If I sweat, then it is hot.
Inverse: $\sim p \rightarrow \sim q$. If it is not hot, then I will not sweat.
Contrapositive: $\sim q \rightarrow \sim p$. If I do not sweat, then it is not hot.

(b) $p \rightarrow q$. Here p: There is no coffee, q: I will drink tea.
Converse: $q \rightarrow p$. If I drink tea, then there is no coffee.
Inverse: $\sim p \rightarrow \sim q$. If there is coffee, then I will not drink tea.
Contrapositive: $\sim q \rightarrow \sim p$. If I do not drink tea, then there is coffee.

(c) $p \rightarrow q$. Here p: It is an Alsatian, q: It is a dog.
Converse: $q \rightarrow p$. If it is a dog, then it is an Alsatian.
Inverse: $\sim p \rightarrow \sim q$. If it is not an Alsatian, then it is not a dog.
Contrapositive: $\sim q \rightarrow \sim p$, If It is not a dog, then it is not an Alsatian.

(d) $p \rightarrow q$. Here p: It is a bat, q: It is not a bird.
Converse: $q \rightarrow p$. If it is not a bird, then it is a bat.
Inverse: $\sim p \rightarrow \sim q$. If it is not a bat, then it is a bird.
Contrapositive: $\sim q \rightarrow \sim p$. If It is a bird, then it is not a bat.

(e) $p \rightarrow q$. Here p: There is no rainfall, q: There will be a drought.
Converse: $q \rightarrow p$. If there is a drought, then there is no rainfall.
Inverse: $\sim p \rightarrow \sim q$. If there is rainfall, then there is no drought.
Contrapositive: $\sim q \rightarrow \sim p$. If there is no drought, then there is rainfall.

17. Test the validity of the following arguments.
   (a) p: You are not polite, q: You will not be treated with respect.
   $P_1: p \rightarrow q$
   $P_2: q$
   $C: p$
   This is not a valid argument by fallacy of the converse.
   (b) p: You are kind to a puppy, q: She will be your friend.
   $P_1: p \rightarrow q$
   $P_2: \sim p$
   $C: \sim q$ This is not a valid argument by fallacy of the inverse.
   (c) p: You are a sneak, q: You are a swindler, r: You are devious.
   $P_1: p \rightarrow r$
   $P_2: q \rightarrow p$
   $C: q \rightarrow r$
   This is a valid argument by transitive reasoning.
   (d) p: I am literate, q: I can read, r: I can write.
   $P_1: p \rightarrow (q \wedge r)$
   $P_2: q$ and $\sim r$
   $C: \sim p$
   C can be false only when p is true. When p is true, P_1 will be true only if $(q \wedge r)$ is true. This can happen only if q is true and r is true. But then P_2 will be false. Thus, we can never have both P_1 and P_2 true and C being false. Therefore, this argument is valid.
   (e) p: I will be a candidate for class representative elections, q: I will keep quiet.
   $P_1: p \vee q$
   $P_2: p$
   $C: \sim q$
   This is not a valid argument by disjunctive fallacy.

Solutions to Exercise 4.1

1. Let the initial price be 100.
   (a) When 100 increases by 50%, it becomes 150. When 150 decreases by 50%, it becomes 75. So overall there is a 25% decrease.
   (b) A 25% decrease.
3. We calculate and plot the amount of material required (in sq cm) for various allowed lengths and widths of the tray. If the length is L and the width is W, then the area of the required material is given by $2L + 2W + L \times W$.

Length	Width	Area
10	90	1,100
12	75	1,074
15	60	1,050
20	45	1,030
30	30	1,020
45	20	1,030
60	15	1,050
75	12	1,074
90	10	1,100

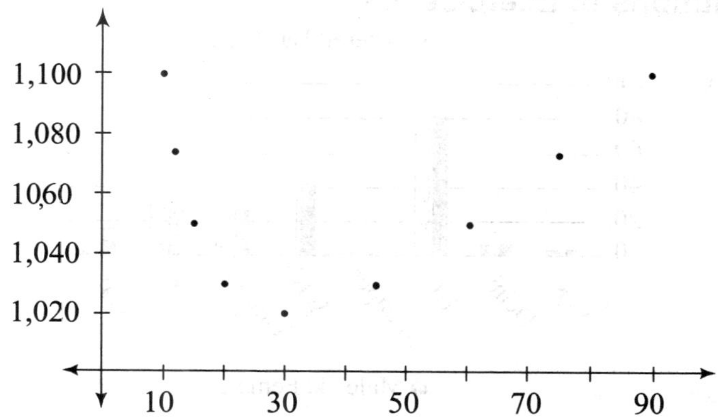

It appears that the optimum design is a square tray with sides of 30 cm. Can you establish this by a calculation?

5. We look at Table 4.1.2 where the same scores are listed in increasing order. Now, we can see that both 32 and 33 will work for us.

Solutions to Exercise 4.2

1. Remember to use a = before each of the following commands.
   (a) sqrt(10)^(1/3)
   (b) (2/3)/(3/5)
   (c) log(500) or log(500,10) or log10(500)
   (d) The sin command in Excel needs angles input in radians, so we use the radians command to convert 60° to radians: sin(radians(60)).
   (e) Fill the numbers 1–10 in cells A1–A10, and then in another cell use sum(A1:A10).
   (f) fact(100)
3. (a) 550 (b) 350 (c) 0
5. Among the class intervals, there should be one with 4 as an upper limit and another with 7 as a lower limit. The requirements 'less than' and 'at least' are consistent with the convention that upper limits are excluded from the class intervals, while the lower limits are included.

Solutions to Exercise 4.3

1. (a)

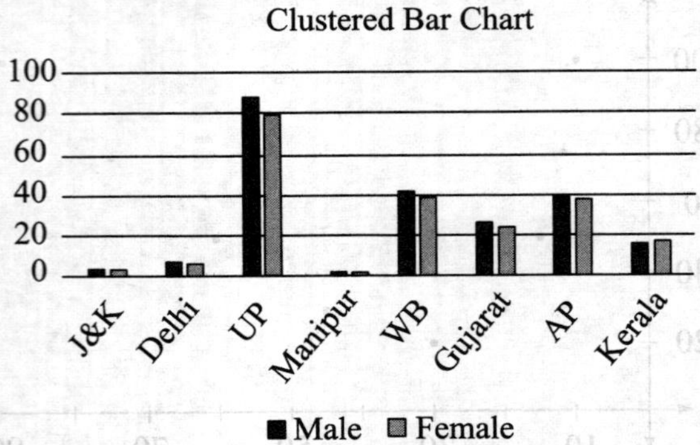

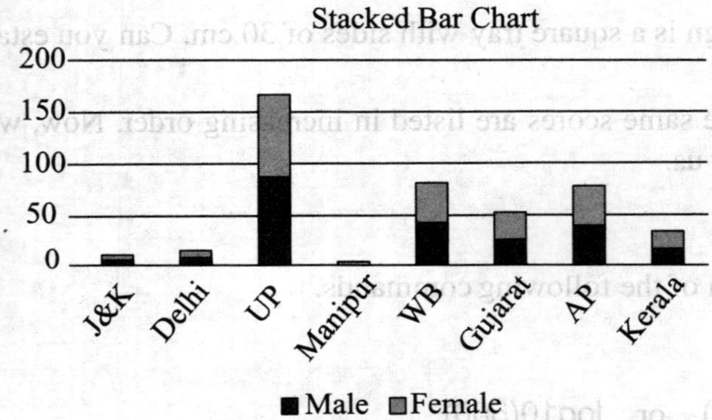

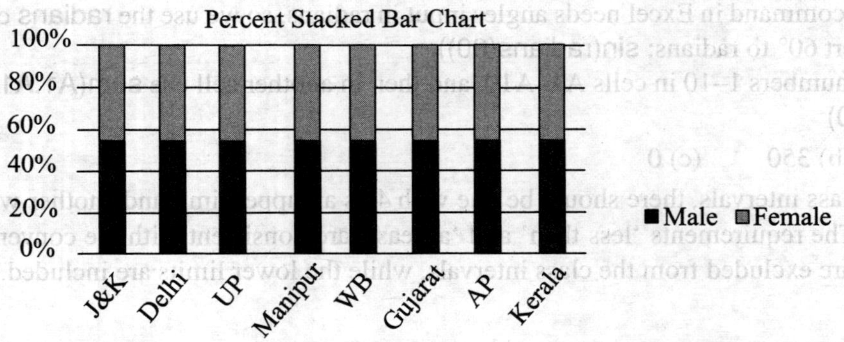

(b) The large variation in population among the states makes it hard to see the situation in the smaller states from the first two charts. The percent stacked bar chart makes it easy to see how the proportion varies.

3. In a typical bar chart, the categories could be placed in any order and so the shape of the chart has no significance.

Solutions to Exercise 4.4

1. (a) Cyclic
   (b) Increasing
   (c) Increasing
   (d) This has two plots, both are increasing (except China for a brief period around 1995).

3. 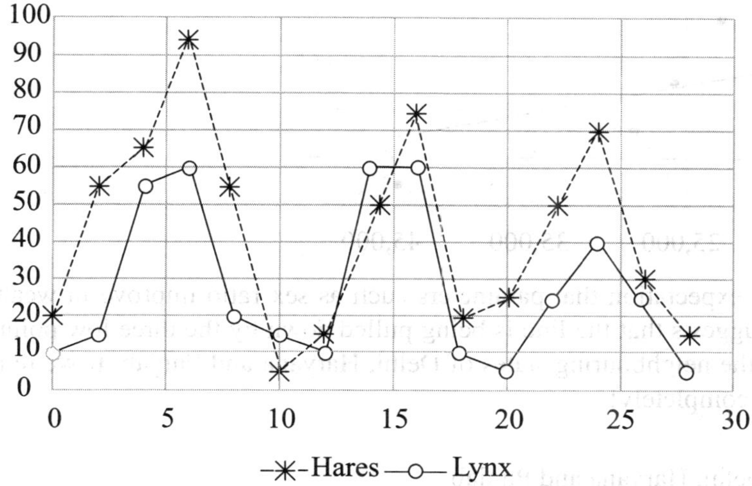

The lynx numbers rise and fall with the hare numbers.

Solutions to Exercise 4.5

1. 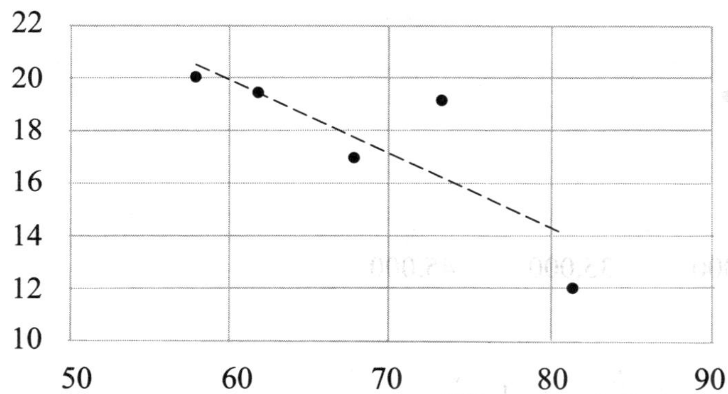

Suppose the X data are in cells A2:A6 and the Y data are in cells B2:B6. Then the interpolated value for $X = 70$ can be found with the Excel command forecast(70, B2:B6, A2:A6). It is 16.97.

3. Here is the scatter plot and line of best fit produced by Excel.

Sex Ratio vs GDP in Indian States

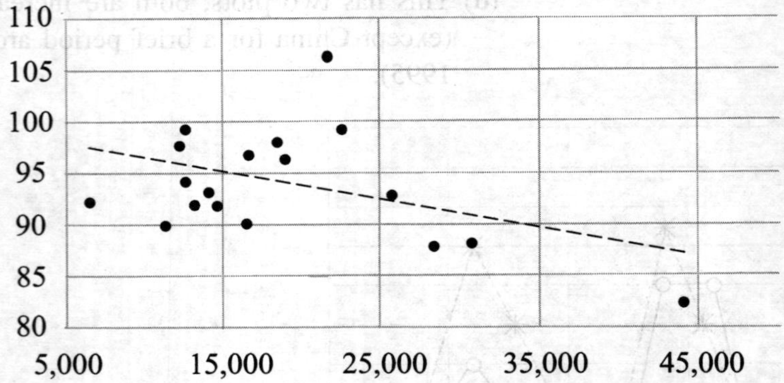

This plot belies the usual expectation that parameters such as sex ratio improve in wealthier societies. A closer look suggests that the line is being pulled down by the three low points on the right; these represent the neighbouring states of Delhi, Haryana and Punjab. If we remove them, the picture changes completely!

Without Delhi, Haryana and Punjab

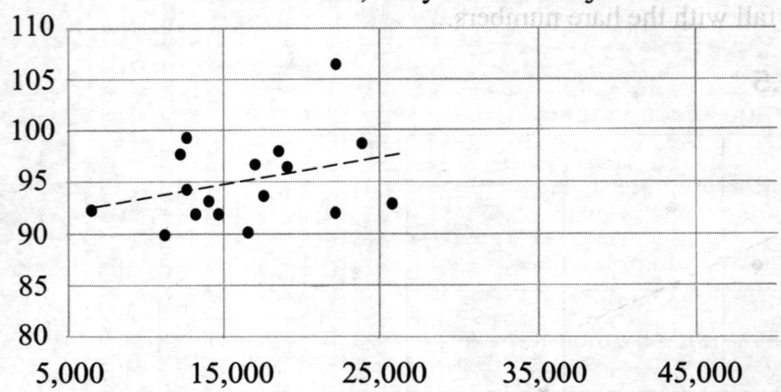

Solutions to Exercise 4.6

1. $\overline{z} = \frac{1}{N}\sum_{i=1}^{N} z_i = \frac{1}{N}\sum_{i=1}^{N}(x_i + y_i) = \frac{1}{N}\sum_{i=1}^{N} x_i + \frac{1}{N}\sum_{i=1}^{N} y_i = \overline{x} + \overline{y}$

3. If we put the 11 data points in cells A1:A11, we have the following calculations:

percentile(A1:A11,0.4)	4
percentile(A1:A11,0.45)	4
percentile(A1:A11,0.83)	7.9

Solutions to Exercise 4.7

1. Let $y_i = x_i + c$ be the shifted data. Then, the new mean is

$$\overline{y} = \frac{1}{N}\sum_{i=1}^{N}(x_i + c) = \frac{1}{N}\sum_{i=1}^{N} x_i + \frac{1}{N}\sum_{i=1}^{N} c = \overline{x} + c$$

Hence, the new variance is

$$\sigma_y^2 = \frac{1}{N}\sum_{i=1}^{N}(y_i - \overline{y})^2 = \frac{1}{N}\sum_{i=1}^{N}((x_i + c) - (\overline{x} + c))^2 = \frac{1}{N}\sum_{i=1}^{N}(x_i - \overline{x})^2 = \sigma_x^2$$

3. (a) A (b) B

Solutions to Chapter 4 Review Exercises

1. In each increase, the previous amount is multiplied by 1.1. So the final amount is $(1.1)^3 = 1.331$ times the original, showing an increase of 33.1%.
3. The mean is

$$\frac{1 + 2 + \cdots + 100}{100} = \frac{1 + 100}{100} + \frac{2 + 99}{100} + \cdots + \frac{50 + 51}{100}$$
$$= \underbrace{\frac{101}{100} + \frac{101}{100} + \cdots + \frac{101}{100}}_{\text{50 times}} = 50 \times \frac{101}{100} = \frac{101}{2}$$

5. (a) False. Since the Average function ignores blank cells, it will display the error message #DIV/0!.
   (b) False. Consider the following data: 1, 3, 4, 6, 6.
   (c) True.
   (d) True, as long as the distribution of the data has a simple shape like that in Figure 4.3.12.
7. The likely explanation is that the calculator treated blank cells as zeros, for example, by just summing over the cells and dividing by their total number.
9.

$$V(a) = \frac{1}{N}\sum_{i=1}^{N}(x_i - a)^2 = \frac{1}{N}\sum_{i=1}^{N}((x_i - \overline{x}) + (\overline{x} - a))^2$$
$$= \frac{1}{N}\sum_{i=1}^{N}((x_i - \overline{x})^2 + (\overline{x} - a)^2 + 2(x_i - \overline{x})(\overline{x} - a))$$
$$= \sigma^2 + \frac{1}{N}\sum_{i=1}^{N}(\overline{x} - a)^2 + \frac{2}{N}\sum_{i=1}^{N}(x_i - \overline{x})(\overline{x} - a)$$
$$= \sigma^2 + (\overline{x} - a)^2 + \frac{2(\overline{x} - a)}{N}\sum_{i=1}^{N}(x_i - \overline{x}) = \sigma^2 + (\overline{x} - a)^2$$

Solutions to Exercise 5.1

1. (a) $\mathcal{S} = \{HH,\ HT,\ TH,\ TT\}$
   (b) $\mathcal{S} = \{(H,1),\ (H,2),\ \ldots,\ (H,6),\ (T,1),\ (T,2),\ \ldots,\ (T,6)\}$
3. No. As discussed in this section, the three possibilities are not equally likely.
5. The experiment is equivalent to picking one ball from the original 5, and asking if is black. Hence, the probability is 2/5.
7. If we never picked a white bead, we would think its probability is 0. If we did pick it, we would think the probability is at least 1/1000. In this case, we cannot get a good estimate using only 1,000 repetitions.

Solutions to Exercise 5.2

1. Let the sample space have N outcomes, and let the event be $A = \{a_1, \ldots, a_n\}$. Then
$$P(A) = \frac{n}{N} = \underbrace{\frac{1}{N} + \cdots + \frac{1}{N}}_{n \text{ times}} = P(a_1) + \cdots + P(a_n)$$
3. We have $P(A \cup B) = P(A) + P(B) - P(A \cap B) \leq P(A) + P(B) = 0.6$. Hence the largest possible value of $P(A \cup B)$ is 0.6 and it will happen if $A \cap B = \emptyset$.
   Since $A \cap B$ is contained in both A and B, its probability cannot exceed their individual probabilities. Hence the largest possible value of $P(A \cap B)$ is 0.2 and it will happen if $B \subset A$.
5.
$$P(E) = \frac{16+20}{100} = 0.36 \qquad P(C) = \frac{35+20}{100} = 0.55$$
$$P(E \cup C) = \frac{16+20+35}{100} = 0.71 \qquad P(E^c) = \frac{35+29}{100} = 0.64$$

Solutions to Exercise 5.3

1. Let A be the event that the total is greater than 6.
   (a) Let B be the event that the first roll results in 2. Then
$$B = \{(2,1), \ldots, (2,6)\} \implies P(B) = \frac{6}{36} = \frac{1}{6}$$
$$A \cap B = \{(2,5), (2,6)\} \implies P(A \cap B) = \frac{2}{36} = \frac{1}{18}$$
   Hence, $P(A|B) = \dfrac{1/18}{1/6} = \dfrac{1}{3}$.

(b) Let B be the event that the total is less than 8.

$$B = \left\{ \begin{array}{l} (1,1),\ldots,(1,6),(2,1),\ldots,(2,5) \\ (3,1),\ldots,(3,4),(4,1),\ldots,(4,3) \\ (5,1),(5,2),(6,1) \end{array} \right\} \implies P(B) = \frac{6+5+\cdots+1}{36} = \frac{7}{12}$$

$$A \cap B = \{(1,6),(2,5),\ldots,(6,1)\} \implies P(A \cap B) = \frac{6}{36} = \frac{1}{6}$$

Hence, $P(A|B) = \dfrac{1/6}{7/12} = \dfrac{2}{7}$.

3. We have $P(E) = 0.36$, $P(C) = 0.55$ and $P(C \cap E) = 0.2$. Therefore,

$$P(C|E) = \frac{0.2}{0.36} = 0.56, \qquad P(E|C) = \frac{0.2}{0.55} = 0.36$$

5. (a) No. $P(A \cap B) = 0$ but $P(A)P(B) = 0.4 \cdot 0.4 = 0.16$.
   (b) No. $P(A \cap B) = 0.2$ but $P(A)P(B) = 0.5 \cdot 0.5 = 0.25$.

7. (a) True: Since $B = (A \cap B) \cup (A^c \cap B)$ is a disjoint union, we have $P(B) = P(A \cap B) + P(A^c \cap B)$. Hence,

$$P(A^c|B) = \frac{P(A^c \cap B)}{P(B)} = \frac{P(B) - P(A \cap B)}{P(B)} = 1 - P(A|B)$$

   (b) False: Let $\mathcal{S} = \{a,b,c\}$ with classical probability. Let $A = \{a\}$ and $B = \{a,b\}$. Then, $P(A|B^c) = 0$, but $1 - P(A|B) = 1 - 1/2 = 1/2$.
   (c) True: $P(A \cap B^c) = P(A) - P(A \cap B) = P(A) - P(A)P(B) = P(A)(1 - P(B)) = P(A)P(B^c)$.
   (d) True : $P(A^c \cap B^c) = P((A \cup B)^c) = 1 - P(A \cup B) = 1 - P(A) - P(B) + P(A \cap B) = 1 - P(A) - P(B) + P(A)P(B) = (1 - P(A))(1 - P(B)) = P(A^c)P(B^c)$.

9. The probability of not getting a pair of sixes in one throw of the dice is 35/36. Assume that the results of the throws are independent. Then, the probability of getting zero pairs of sixes in 24 throws is $(35/36)^{24}$. Hence,

$$P(\text{at least one pair}) = 1 - (35/36)^{24} = 0.4914$$

The bet is not advantageous.

11. Let $A = \{BB\}$ be the event that both children are boys.
   (a) Let E be the event that the elder one is a boy: $E = \{BG, BB\}$. Then, $P(A|E) = \dfrac{1/4}{1/2} = \dfrac{1}{2}$.
   (b) Let O be the event that one of them is a boy: $O = \{BG, BB, GB\}$. Then, $P(A|O) = \dfrac{1/4}{3/4} = \dfrac{1}{3}$.

Solutions to Exercise 5.4

1. The possible values of X are 0, 1, 2. The primes are 2, 3 and 4 and so the probability of one die showing a prime number is 1/2. The pdf of X is given by

$$P(X=0) = P(\text{first is not prime})P(\text{second is not prime}) = \frac{1}{2}\cdot\frac{1}{2} = \frac{1}{4}$$

$$P(X=1) = P(\text{first is prime})P(\text{second is not prime}) + P(\text{first is not prime})P(\text{second is prime}) = \frac{1}{2}\cdot\frac{1}{2}+\frac{1}{2}\cdot\frac{1}{2} = \frac{1}{2}$$

$$P(X=2) = P(\text{first is prime})P(\text{second is prime}) = \frac{1}{2}\cdot\frac{1}{2} = \frac{1}{4}$$

3. Mean and variance of X from problem 1:

$$\mu_X = \frac{1}{4}\cdot 0 + \frac{1}{2}\cdot 1 + \frac{1}{4}\cdot 2 = 1$$

$$\sigma_X^2 = \frac{1}{4}\cdot(0-1)^2 + \frac{1}{2}\cdot(1-1)^2 + \frac{1}{4}\cdot(2-1)^2 = \frac{1}{2}$$

5. The possible prices after a month are ₹110 and 90, with respective probabilities 0.6 and 0.4. So the expected price is $0.6\times 110 + 0.4\times 90 = 102$.

7. Let X take values $a_1,\ldots,a_n$ with probabilities $p_1,\ldots,p_n$, respectively. Then, $X+k$ takes values $a_1+k,\ldots,a_n+k$ with probabilities $p_1,\ldots,p_n$, respectively. First, we calculate the mean of $X+k$:

$$\mu_{X+k} = \sum_{i=1}^{n} p_i(a_i+k) = \sum_{i=1}^{n} p_i a_i + \sum_{i=1}^{n} p_i k = \mu_X + k\sum_{i=1}^{n} p_i = \mu_X + k$$

Now, we calculate the variance:

$$\sigma_{X+k}^2 = \sum_{i=1}^{n} p_i(a_i+k-\mu_{X+k})^2 = \sum_{i=1}^{n} p_i(a_i-\mu_X)^2 = \sigma_X^2$$

9. We first calculate the mean and variance:

$$\mu_X = p\cdot(-1) + (1-2p)\cdot 0 + p\cdot 1 = 0$$

$$\sigma_X^2 = p\cdot(-1)^2 + (1-2p)\cdot 0^2 + p\cdot 1^2 = 2p$$

Hence, the largest value of the variance is 1, achieved for $p = 1/2$.

Solutions to Exercise 5.5

1. (a) $P(XY=0) = P(X=0 \text{ or } Y=0) = 1/8+1/4+1/8+1/4 = 3/4$.

(b)

	$Y=-1$	$Y=0$	$Y=1$
f_Y	1/4	1/2	1/4

,

	$X=0$	$X=1$
f_X	1/2	1/2

(c) $E[X] = 1/2,\ E[Y] = 0,\ E[XY] = 0$; hence, $\text{Cov}[X,Y] = E[XY] - E[X]E[Y] = 0$. Therefore, $\rho_{X,Y} = 0$.

(d) Yes

3. (a) For example, $P(X+Y=0) = P(X=0, Y=0) + P(X=1, Y=-1)$ $= 1/2 + 1/4 = 3/4$.

(b)
$$E[X+Y] = (-1)\cdot 0 + 0\cdot 3/4 + 1\cdot 0 + 2\cdot 1/4 = 1/2$$
$$E[X] = 0\cdot 1/2 + 1\cdot 1/2 = 1/2$$
$$E[Y] = (-1)\cdot 1/4 + 0\cdot 1/2 + 1\cdot 1/4 = 0$$

(c)
$$\sigma^2_{X+Y} = (-1)^2\cdot 0 + 0^2\cdot 3/4 + 1^2\cdot 0 + 2^2\cdot 1/4 - (1/2)^2 = 3/4$$
$$\sigma^2_X = 0^2\cdot 1/2 + 1^2\cdot 1/2 - (1/2)^2 = 1/4$$
$$\sigma^2_Y = (-1)^2\cdot 1/4 + 0^2\cdot 1/2 + 1^2\cdot 1/4 - 0^2 = 1/2$$
$$\text{Cov}[X,Y] = E[XY] - E[X]E[Y] = E[XY] = (-1)\cdot 1/4 + 1\cdot 1/4 = 0$$

5.
$$\begin{aligned}\text{Cov}[X+Y, Z] &= E[(X+Y)Z] - E[X+Y]E[Z]\\ &= E[XZ] + E[YZ] - E[X]E[Z] - E[Y]E[Z]\\ &= E[XZ] - E[X]E[Z] + E[YZ] - E[Y]E[Z]\\ &= \text{Cov}[X,Z] + \text{Cov}[Y,Z]\end{aligned}$$

Solutions to Exercise 5.6

1. $P(35 < X < 65) = \mathsf{binom.dist}(64, 100, 0.5, 1) - \mathsf{binom.dist}(35, 100, 0.5, 1) = 0.996$
3. Let X be the number of correct answers to 10 questions. Then,
$$P(X \geq 6) = 1 - P(X \leq 5) = 1 - \mathsf{binom.dist}(5, 10, 0.5, 1) = 0.377$$
If there are 20 questions, the probability is 0.252. A longer test reduces the chances of luck playing a role.
5. We have $\mu_X = np = 5$ and $\sigma_X = \sqrt{np(1-p)} = 2$. Therefore,
$$P(\mu_X - \sigma_X \leq X \leq \mu_X + \sigma_X) = P(3 \leq X \leq 7) = P(X \leq 7) - P(X \leq 2) = 0.793$$
7. Y is a Binomial variable with parameters n and $1-p$.

Solutions to Exercise 5.7

1. We calculate the higher cutoff for a two-sided test: $c_2 = \mathsf{Binom.Inv}(1000, 0.5, 1 - 0.05/2) = 531$. Since $530 < c_2$, we cannot reject the hypothesis of a fair coin.
3. We calculate $P = \mathsf{Binom.Dist}(1.01 * N/2, N, 0.5, 1) - \mathsf{Binom.Dist}(0.99 * N/2, N, 0.5, 1)$ for different values of N. $N = 40{,}000$ gives $P = 0.954$.
5. The variance of a binomial random variable is $np(1-p)$ and hence is maximum for $p = 0.5$. Therefore, confidence intervals are widest when calculated assuming $p = 0.5$ and the corresponding sample size of 1,000 would be safe for any background p value.

Solutions to Chapter 5 Review Exercises

1. (a) We represent each outcome by an ordered pair in which the first entry identifies the selected urn (1 or 2) and the second entry identifies the colour of the picked ball (B for black, R for red and W for white). Then, the sample space is $\{(1,B),\ (1,R),\ (2,W),\ (2,R)\}$.
   (b) We represent each outcome by an ordered pair in which the first entry identifies the colour of the ball selected from Urn 1 and the second entry identifies the colour of the ball selected from Urn 2. Then, the sample space is $\{(B,W),\ (B,R),\ (R,W),\ (R,R)\}$.

3. $P(A\cap B)$ cannot exceed the probabilities $P(A)$ and $P(B)$. Hence, its maximum possible value is 0.5, and it will happen if $A \subset B$.
   Since $P(A\cup B)$ cannot exceed 1, we have $P(A\cap B) = P(A)+P(B)-P(A\cup B) \geq 0.5+0.7-1 = 0.2$. Thus, the minimum possible value of $P(A\cap B)$ is 0.2.

5. $$\begin{aligned} P(A\cup B\cup C) &= P(A\cup B)+P(C)-P((A\cup B)\cap C) \\ &= P(A)+P(B)+P(C)-P(A\cap B)-P((A\cap C)\cup(B\cap C)) \\ &= P(A)+P(B)+P(C)-P(A\cap B)-P(A\cap C)-P(B\cap C)+P(A\cap B\cap C) \end{aligned}$$

7. Let $H, T. I$ denote the events that a selected reader reads the *Hindustan Times*, *Times of India* or *Indian Express*, respectively. Note that there are a total of 100 readers. Hence:
   (a) $P(H\cap T') = \dfrac{23+5}{100} = 0.28$
   (b) $P((H\cup T)\cap I') = \dfrac{23+20+32}{100} = 0.75$
   (c) $P([(H\cap T)\cup(H\cap I)\cup(T\cap I)]\cap[H\cap T\cap I]') = \dfrac{20+5+7}{100} = 0.32$
   (d) $P(I) = \dfrac{9+5+4+7}{100} = 0.25$
   (e) $P(I|T) = \dfrac{P(I\cap T)}{P(T)} = \dfrac{4+7}{4+7+20+32} = \dfrac{11}{63} = 0.175$

9. (a) $P(B) = P(A\cup B)-P(A)+P(A\cap B) = 0.6-0.4 = 0.2$
   (b) $P(B) = P(A\cup B)-P(A)+P(A)P(B)$
   $$\Longrightarrow P(B) = \frac{P(A\cup B)-P(A)}{1-P(A)} = \frac{0.2}{0.6} = \frac{1}{3}$$

11. We have $P(X=k) = \alpha k$. Then, $\alpha(0+1+2+3+4+5) = 1$ implies $\alpha = 1/15$. Hence, $P(X=k) = k/15$, and
   $$\begin{aligned} E[X] &= (0^2+1^2+2^2+3^2+4^2+5^2)/15 = 55/15 = 11/3 \\ E[X^2] &= (0^3+1^3+2^3+3^3+4^3+5^3)/15 = 225/15 = 15 \\ \sigma_X^2 &= E[X^2]-E[X]^2 = 15-121/9 = 14/9 \end{aligned}$$
   We leave it to you to confirm that in the general case we have $E[X] = \dfrac{2n+1}{3}$ and

$$\sigma_X^2 = \frac{n(n+1)}{2} - \frac{(2n+1)^2}{9} = \frac{n^2+n-2}{18}.$$

13. When it can only take one value.
15. No. For example, when a coin is tossed once, the number of heads and that of tails have the same probability distribution but are never equal.

Solutions to Exercise 6.1

1. (a) Not a symmetry as the two figures are not coinciding.
   (b) It is a symmetry.
   (c) It is a symmetry.
   (d) It is a symmetry.
   (e) Not a symmetry as the two figures are not coinciding.
   (f) Not a symmetry as the two figures are not coinciding.
3. Denote the three rotational symmetries as R_0, R_{120} and R_{240}. These are the 0°, 120° and 240° rotations, respectively, in the anticlockwise direction about the rotocenter O. The effects, of these on the vertices are described as follows.

$$\begin{array}{llll} R_0: & A \to A & R_{120}: & A \to B & R_{240}: & A \to C \\ & B \to B & & B \to C & & B \to A \\ & C \to C & & C \to A & & C \to A \end{array}$$

   Denote the three reflection symmetries about the lines of reflection L_A, L_B and L_C as R_A, R_B and R_C respectively. The effects of these on the vertices are described as follows.

$$\begin{array}{llllll} R_A: & A \to A & R_B: & A \to C & R_C: & A \to B \\ & B \to C & & B \to B & & B \to A \\ & C \to B & & C \to A & & C \to C \end{array}$$

5. Mark the vertices of the rectangle in the anticlockwise direction as A, B, C and D, respectively. Let O denote the point of intersection of the diagonals AC and BD. The non-square rectangle has two rotational symmetries of angles 0° and 180° in the anticlockwise direction about the rotocenter O. Let us denote these by R_0 and R_{180} respectively. Also, let L_H be the line joining the mid-points of sides AB and CD. Let L_V be the line joining the mid-points of sides AD and BC. The non-square rectangle has two reflection symmetries R_H and R_V about the lines L_H and L_V, respectively. The effects of these symmetries on the vertices are described as follows.

$$\begin{array}{llllllll} R_0: & A \to A & R_{180}: & A \to C & R_H: & A \to B & R_V: & A \to D \\ & B \to B & & B \to D & & B \to A & & B \to C \\ & C \to C & & C \to A & & C \to D & & C \to B \\ & D \to D & & D \to B & & D \to C & & D \to A \end{array}$$

The non-square rectangle has fewer symmetries than a square since the square is the regular four-sided figure (equal sides and equal angles) and hence more 'symmetrical' compared to the rectangle.

Solutions to Exercise 6.2

1. (i)

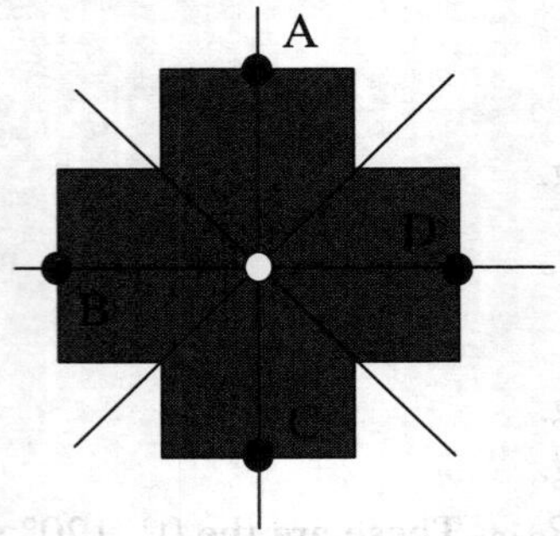

The white dot marks the rotocenter. The black lines are the lines of reflection. This figure has eight symmetries: four reflections and four rotations. It is of type D_4. The four reflections are about the horizontal, vertical and the two diagonal lines of reflection. The four rotations about the rotocenter in the anticlockwise direction are by angles of 0°, 90°, 180° and 270°. The vertices are marked in the figure. The effects on the vertices of each of these four rotations are exactly the same as in the case of a square with vertices marked similarly in an anticlockwise manner.

(ii)

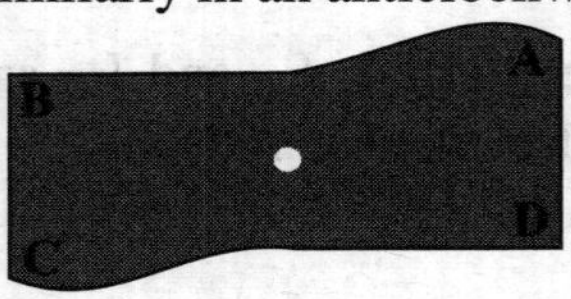

The white dot marks the rotocenter. The four corners are the marked vertices. This figure has no reflection symmetries but has two rotational symmetries. It is of type C_2. The two rotations about the rotocenter in the anticlockwise direction are by angles of 0° and 180°. The 0° rotation keeps every point in the figure fixed and hence each vertex stays fixed. The 180° rotation interchanges the vertices A and C and also B and D.

(iii)

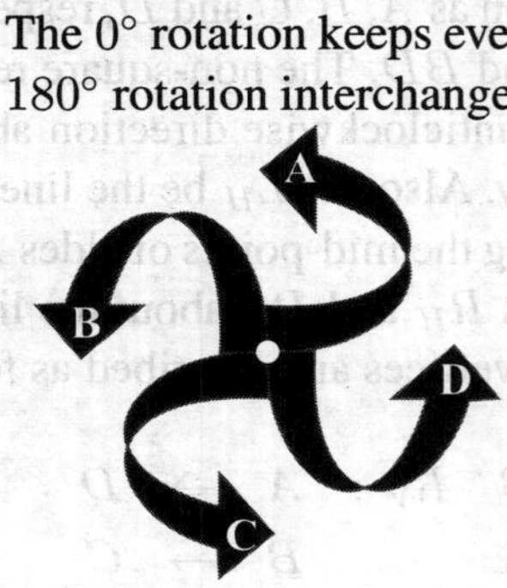

The four tips of the arrows are the suitably marked vertices. The white dot in the centre marks the rotocenter.This figure has no reflection symmetries but has four rotational symmetries. It is of type C_4. The four rotations about the rotocenter in the anticlockwise direction are by angles of 0°, 90°, 180° and 270°. The effects on the vertices of each of these four rotations are exactly the same as in the case of a square with vertices marked similarly in an anticlockwise manner.

(iv) 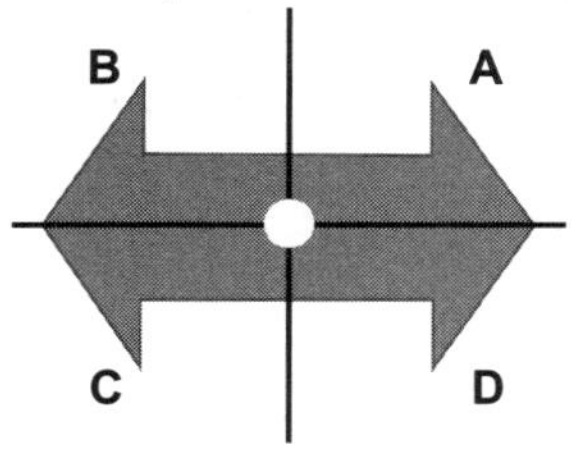

The white dot marks the rotocenter. The four vertices are marked in an anticlockwise direction about the upper and lower tips of the arrow–heads. This figure has the same symmetries as that of a non-square rectangle whose vertices are marked similarly in an anticlockwise direction. Thus, it is of type D_2.

(v) 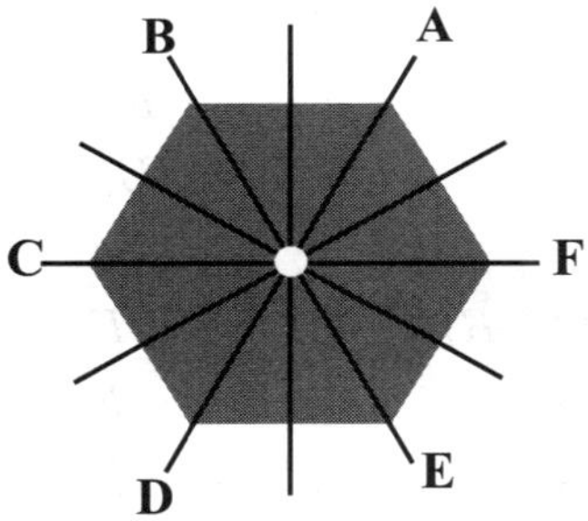

The vertices are marked in an anticlockwise direction with the centre, marked with the white dot being the rotocenter. This figure has six reflection symmetries and six rotational symmetries. It is of type D_6. The six rotations about the rotocenter in the anticlockwise direction are by angles of 0°, 60°, 120°, 180°, 240° and 300° and will be denoted by $R_0, R_{60}, R_{120}, R_{180}, R_{240}$ and R_{300}, respectively. The six reflection symmetries will be denoted as follows. Reflection about the horizontal and vertical lines of reflection, respectively, will be denoted by R_H and R_V. Reflections about the lines joining vertices A, D, B, E and C, F will be called R_{AD}, R_{BE} and R_{CF}, respectively. The two remaining reflections are as follows. The reflection about the line joining the mid-points of the sides BC and EF will be denoted as R_{D1}. The reflection about the line joining the mid-points of the sides CD and AF will be denoted as R_{D2}. The effects of these 12 symmetries on the 6 vertices are as follows.

$R_0:$		$R_{60}:$		$R_{120}:$	
$A \to A$		$A \to B$		$A \to C$	
$B \to B$		$B \to C$		$B \to D$	
$C \to C$		$C \to D$		$C \to E$	
$D \to D$		$D \to E$		$D \to F$	
$E \to E$		$E \to F$		$E \to A$	
$F \to F$		$F \to A$		$F \to B$	

R_{180}:		R_{240}:		R_{300}:	
$A \to D$		$A \to E$		$A \to F$	
$B \to E$		$B \to F$		$B \to A$	
$C \to F$		$C \to A$		$C \to B$	
$D \to A$		$D \to B$		$D \to C$	
$E \to B$		$E \to C$		$E \to D$	
$F \to C$		$F \to D$		$F \to E$	

R_H:	R_V:	R_{D1}:
$A \to E$	$A \to B$	$A \to D$
$B \to D$	$B \to A$	$B \to C$
$C \to C$	$C \to F$	$C \to B$
$D \to B$	$D \to E$	$D \to A$
$E \to A$	$E \to D$	$E \to F$
$F \to F$	$F \to C$	$F \to E$

R_{AD}:	R_{BE}:	R_{D2}:
$A \to A$	$A \to C$	$A \to F$
$B \to F$	$B \to B$	$B \to E$
$C \to E$	$C \to A$	$C \to D$
$D \to D$	$D \to F$	$D \to C$
$E \to C$	$E \to E$	$E \to B$
$F \to B$	$F \to D$	$F \to A$

3. (a) Let the ABC denote the equilateral triangle with vertices being marked in an anticlockwise manner. Let R_A denote the reflection about line joining vertex A to the mid-point of the opposite side (BC) and R_{120} denote the 120° rotation in the anticlockwise direction about the center of the equilateral triangle. Then, $R_{120} \circ R_A$ affects the vertices as follows: $A \to A \to B$, $B \to C \to A$ and $C \to B \to C$, where the first arrow denotes the action of R_A and the second arrow that of R_{120}. Thus, we see that $R_{120} \circ R_A$ takes $A \to B$, $B \to A$ and $C \to C$. Thus, it is the same as the reflection R_C, which denotes the reflection about line joining vertex C to the mid-point of the opposite side (AB).
   (b) Let the $ABCD$ denote the square with vertices being marked in an anticlockwise manner. Let R_{AC} be a reflection about the diagonal AC of the square. Then, it is easy to check that $R_{AC} \circ R_{AC} = R_0 = I$; thus the inverse of R_{AC} is itself.
   (c) $R_{180} \circ R_{180} = R_0 = I$.
   (d) $R_{90}{}^{-1} = R_{270}$.
   (e) $R_V \circ R_{180} = R_H$.

Solutions to Exercise 6.3

1. (i) This strip pattern has translation symmetries, reflection symmetries about vertical lines (two types), reflection symmetry about a horizontal line, 180 ° rotations about two different types of rotocentres (black and white dots) and glide reflections. The different types of lines of mirror symmetry, rotocentres and direction of translation symmetry are marked in the following figure. The glide reflection occurs about the horizontal mirror symmetry line with the translation as shown in the figure. This is of Pattern VII type.

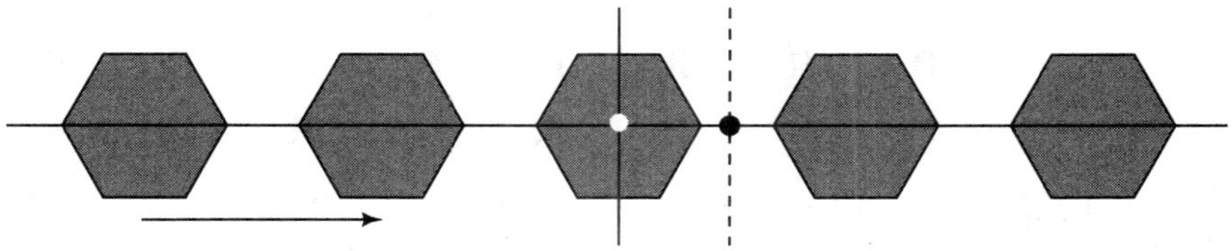

(ii) This strip pattern has translation symmetries, no reflection symmetries, 180° rotations about two different types of rotocentres (black and white dots) and no glide reflections. The different types of rotocentres and direction of translation symmetry are marked in the following figure. This is of Pattern IV type.

(iii) This strip pattern has translation symmetries, reflection symmetries about vertical lines (two types), no reflection symmetry about a horizontal line, no 180 ° rotations and no glide reflections. The different types of lines of vertical mirror symmetry and direction of translation symmetry are marked in the following figure. This is of Pattern III type.

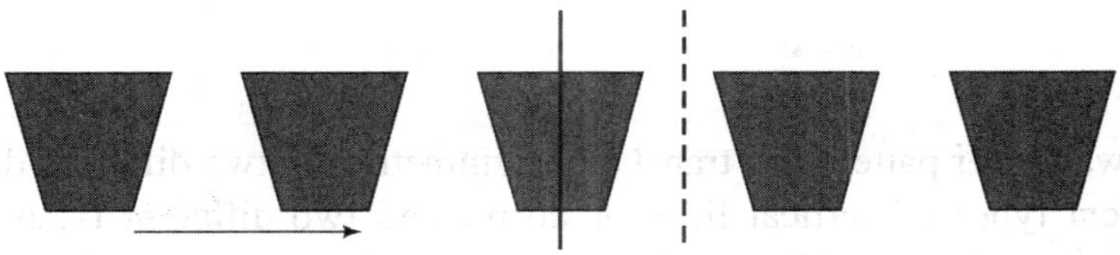

3. (a) $\rho \circ \rho = I$, where I is the identity symmetry or the 0° rotational symmetry of the strip pattern.
   (b) $\rho \circ \sigma = T^{2n}$, where n is the number of units of 'translation distance **t**' between the two different rotocenters that are of the same type and T denotes the translation symmetry of moving **t** units to the right in the strip pattern.
   (c) Inverse a reflection R in a strip pattern is itself, that is, $R^{-1} = R$.
   (d) If the translation involves moving n units of length **t** (the basic translation length) towards the right in a strip pattern, then its inverse is the translation which involves moving n units of length **t** towards the left. So $T^{n^{-1}} = T^{-n}$.
   (e) Inverse of a 180° rotation S in a strip pattern is itself, that is, $S^{-1} = S$.

Solutions to Exercise 6.4

1. (i) This wallpaper pattern has translation symmetries in two different directions. No reflection symmetry, no glide reflections and no rotation symmetry other than the do-nothing or 0° rotation. The translations in two different directions are marked in the following figure.

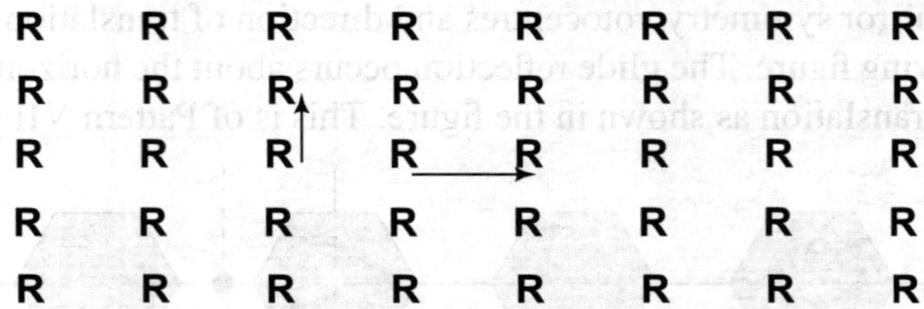

(ii) This wallpaper pattern has translation symmetries in two different directions and has two different types of vertical lines of mirror symmetry. It has glide reflections along the two types of vertical lines of mirror symmetry with the translation being the same as the unit length in the vertical direction. It has no rotation symmetry other than the 0° rotation. The translations in two different directions and the two different types of vertical mirror symmetry are marked in the following figure.

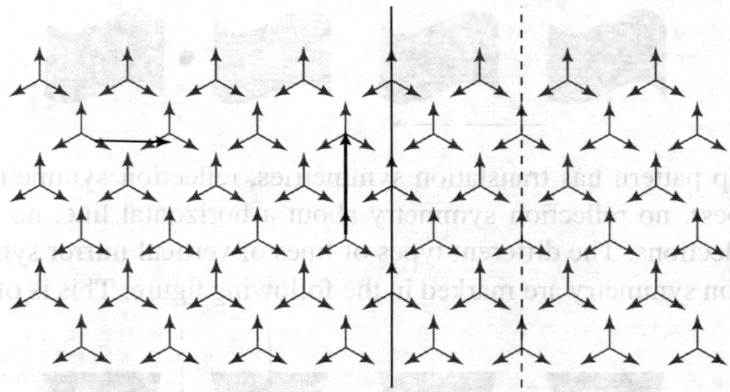

(iii) This wallpaper pattern has translation symmetries in two different directions and has two different types of vertical lines of mirror and two different types of horizontal mirror symmetries. It has 180° rotations about two different type of rotocenters. It has glide reflections along the vertical lines of mirror symmetry with the translation being the same as the unit length in the vertical direction. Similarly, it also has glide reflections along the horizontal lines of mirror symmetry with the translation being the same as the unit length in the horizontal direction. The translations in two different directions, the two different types of vertical mirror symmetry, the two different types of horizontal lines of mirror symmetry and the two different types of rotocenter are marked in the following figure.

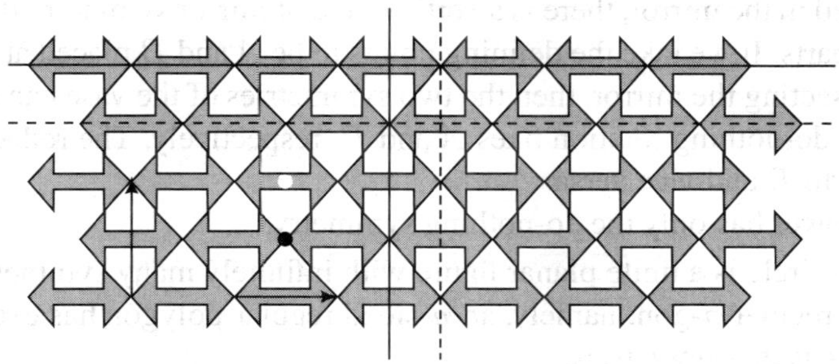

3. In Figure 6.4.4, glide reflection occurs along both the vertical lines of mirror reflection and the horizontal lines of mirror reflection. The translation length is the basic unit of translation. So the lines of reflection for the glide reflections are the same as the lines of reflection.
   Notice that Figure 6.4.5 is constructed by fitting together regular hexagons. The oblique lines of reflection will occur along the diagonals of the hexagons and lines joining the mid-points of opposite sides. But these are not new types of reflections as the nonzero rotations move the oblique lines of reflection to the horizontal and vertical lines of mirror symmetry. The reflection lines of the glide reflections are the same as the lines of reflection. There are essentially three types of rotocenters: centre of the hexagons, the mid-points of the edges of the hexagons (rotations of 180° happen about these) and the vertices of the hexagons.
   In Figure 6.4.6, the lines of glide reflection occur along the lines of reflection.

Solutions to Exercise 6.5

1. Figure 6.5.8 is of Pattern VII as it has translations, vertical reflections, horizontal reflection, 180° rotation and glide reflections. For both the wallpaper patterns, in Figures 6.5.9 and 6.5.11, the lines of glide reflection are the same as the lines of mirror reflection, and the translation is the basic unit between motifs along the line of mirror reflections.

Solutions to Exercise 6.7

1. (a) D_2 as it has exactly two symmetries, one rotation of 0° and a mirror reflection about the vertical line joining the top vertex to the mid-point of the opposite side.
   (b) Only symmetry is the 0° rotation. So it is of C_1 type.
   (c) Only symmetry is the 0° rotation. So it is of C_1 type.
   (d) Symmetries are the same as a square. So it is of D_4 type.

Solutions to Chapter 6 Review Exercises

1. (i) Only rotational symmetry is the do-nothing symmetry denoted by R_0. If the lateral veins on the leaf are ignored, then the leaf has a line of mirror symmetry, namely, the central spine. If we mark two defining points say A and B at the end of an imaginary line bisecting the central spine and perpendicular to it, then the two symmetries of the leaf can be described as follows: R_0 the do-nothing rotation fixes A and B, respectively. The reflection symmetry R_V will take A to B and vice versa.

(ii) Only rotational symmetry is the do-nothing symmetry denoted by R_0. For the flower vase depicted in the mirror, there is a vertical line of mirror symmetry dividing the vase into two equal parts. If we take the defining points to be A and B placed at each end of a horizontal line bisecting the mirror, then the two symmetries of the vase can be described as follows: R_0 the do-nothing rotation fixes A and B, respectively. The reflection symmetry R_V will take A to B and vice versa.

(iii) This flower has only the do-nothing symmetry.

3. (a) False. A circle is a finite planar figure with infinitely many symmetries.
   (b) True. A regular n-gon, namely, an n-sided regular polygon has exactly n reflections and n rotations as its symmetries.
   (c) False. A scalene triangle does not have a reflection symmetry.
   (d) True. Every planar figure has the do-nothing rotational symmetry.
   (e) True. See Figure 6.2.2.
5. (a) True.
   (b) False. See strip with Pattern VI in Figure 6.3.12
   (c) True. See strip with Pattern I in Figure 6.3.11
   (d) True. Every strip pattern will have infinitely translation symmetries.
   (e) False. Every strip pattern has infinitely translation symmetries.
7. (a) Since the rotations are about rotocentres of two different types, the distance between any two such rotocenters will be $1/2n$ where n is a natural number and $1/2n$ is the number of units of 'translation distance **t**' between the two different rotocenters of different types. If T denotes the translation symmetry of moving **t** units to the right in the strip pattern then $\rho \circ \sigma = T^n$.
   (b) Let ρ be the reflection of the strip pattern about a horizontal line of reflection and σ be a 180° rotational symmetry of a strip pattern. Then $\rho \circ \sigma$ is the same as a reflection symmetry about a vertical line of reflection passing through the rotocenter for the 180° rotation σ.
   (c) A glide reflection g in a strip pattern can be represented by $\rho \circ \sigma$ where σ is a translation along the horizontal axis say by s units to the right followed by ρ, the reflection along the horizontal line of mirror symmetry. Inverse of g is again a glide reflection given by $\sigma^{-1} \circ \rho$ which is the symmetry obtained by applying ρ first and then followed by translation of s units to the left.
   (d) Inverse of a reflection is always itself.
9. (a) False. There are wallpaper patterns of 'cmm' type which have rotocenters about which rotations of 180 degree can be done and such that they do not lie on lines of reflection. For an example of 'cmm' type wallpaper pattern see https://en.wikipedia.org/wiki/Wallpaper_group#/media/File:SymBlend_cmm.svg.
   (b) False. See Figure 6.4.5.
   (c) False. See Figure 6.4.3.
   (d) False. See Figure 6.4.3.

Solutions to Exercise 7.1

1. (a) Top (b) Side (c) Front

3. Drawing damsels standing behind, higher up in the painting brings out the sense of depth. Similarly, the trees in the background are drawn along a diagonal line to emphasise depth. Another instance that the artist has tried to adopt is by drawing the canopy using diagonal lines. The diagonal lines of the canopy and the trees meet at a point outside the painting. The stairs have been drawn in diminishing order as an incorrect attempt at perspective. Another flaw is that subjects—damsels or trees—do not diminish in size with distance.

5. The feeling of depth in the painting is created by the diminishing sizes of the arches and pillars that converge to the central figure in the painting. The figures in the foreground are larger in size than the figures in the background. The design on the floor is also so drawn that the diagonal lines all meet at the vanishing point. Also, these lines emerge from the corner points of the rectangles in the front.

7. The main architectural lines on the front facade of the cathedral converge to different points. The people sitting in the courtyard are drawn between two parallel diagonal lines that give a wrong perspective of depth. The noblemen (or the bishops) sitting in the lower part of the cathedral are drawn larger than the people in the courtyard, although they are further away from the viewing point.

Solutions to Exercise 7.2

1. (a) (1, 3, 7)
   (b) (1, 6, 2)
   (c) (3, 3, 2)
   (d) (−3, 3, 2)
   (e) (1, 2, 2)
   (f) (1, 3, −5)

3. (a) A′(3.5, 1.4, 0), B′(3.5, −1.4, 0)
   (b) 4 units, 2.8 units
   (c) A′(0.35, 0.14, 0), B′(0.35, −0.14, 0), 0.28 units
   (d) A′(0.035, 0.014, 0), B′(0.035, −0.014, 0), 0.028 units
   (e) Our everyday experience shows that as an object moves farther away, the image formed becomes smaller. This is exactly what we obtain mathematically.

5. In Figure 7.2.17, $A'B' \parallel AB$. Therefore, the triangles $EB'A'$ and EBA are similar and $A'B'$ is proportional to AB.

7. (a) Centre of the painting.
   (b) Jesus.
   (c) Door opening in the central dome.

Solutions to Exercise 7.3

1. (a) Two-point perspective.
   (b) Two-point perspective.
   (c) One-point perspective.
   (d) One-point perspective.

Solutions to Exercise 7.4

1. Duplicate the smaller rectangles as described in the text. Join the corresponding vertices to get the hexagon.
3. Draw a line through O parallel to AB and DC, cutting the lines BC and AD at J and K, respectively. O is the centre of the ellipse, and the sides of the rectangle are tangent to the ellipse at the points E, J, F and K. To duplicate the ellipse in CDHG, find the centre of the rectangle and the mid-points of the sides of CDHG. Use this to duplicate the ellipse.
5. In two-point perspective, there are two vanishing points. A rectangle has two sets of parallel lines. If we draw a rectangle in two-point perspective, these two set of parallel lines should meet in two different vanishing points.

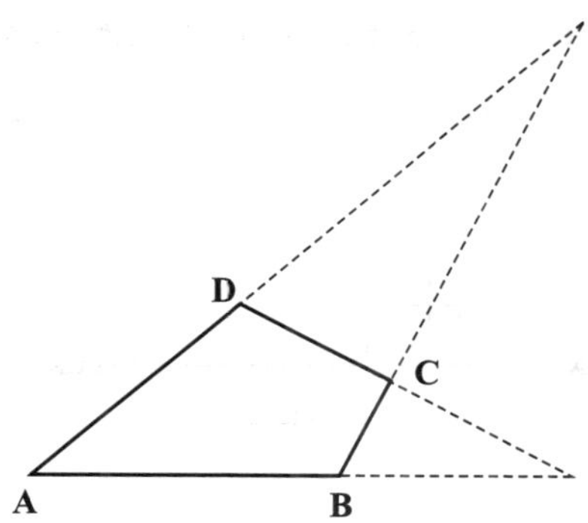

7. The vanishing point V_1 at the origin is found by the intersection of lines joining the side walls (black lines). The vanishing point V_2 of the diagonal lines of the arches on the side wall (black dotted lines) of the building on the left will lie vertically above V_1.
   We assume that the arches on the side wall have the same proportions as the ones on the front wall, whose base is about half the height. Therefore, the viewing distance is also about half the distance between V_1 and V_2.

Solutions to Chapter 7 Review Exercises

1. (a) A′(6, 7, 3), B′(6 – $\sqrt{3}$, 6, 3), C′(6 – $\sqrt{3}$, 4, 3), D′(6, 3, 3), E′(6 + $\sqrt{3}$, 4, 3), F′(6 + $\sqrt{3}$,6, 3).

   (b) (i) A hexagon with vertices (6, 7, 0), (6 – $\sqrt{3}$, 6, 0), (6 – $\sqrt{3}$, 4, 0), (6, 3, 0), (6 + $\sqrt{3}$, 4, 0), (6 + $\sqrt{3}$,6, 0)

   (ii) A rectangle with vertices (6 – $\sqrt{3}$, 0, 6), (6 + $\sqrt{3}$, 0, 6), (6 + $\sqrt{3}$,0, 3) and (6 – $\sqrt{3}$, 0, 3)

   (iii) A rectangle with vertices (0, 3, 6), (0, 6, 6), (0,6, 3) and (0, 3, 3)

3. (a)

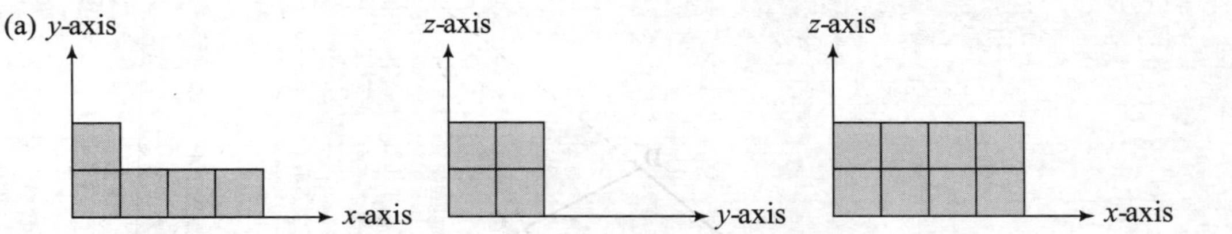

(a) *xy*-plane (b) *yz*-plane (c) *xz*-plane

(b)

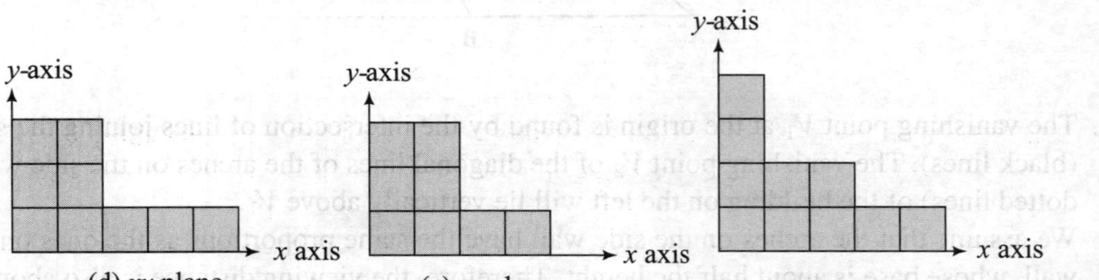

(d) *xy*-plane (e) *yz*-plane (f) *xz*-plane

5. $\dfrac{9}{h} = \dfrac{3}{4}$ which gives $h = 12$ units.

7. Recall that a cube is a symmetric solid and in three-point perspective it will have three vanishing points. Place the three vanishing points symmetrically in the plane. Draw two lines from each of the vanishing points. A third line bisecting the angle (dotted line) is drawn from each of the

vanishing point. The points of intersection of three lines will give the vertices of the required cube.

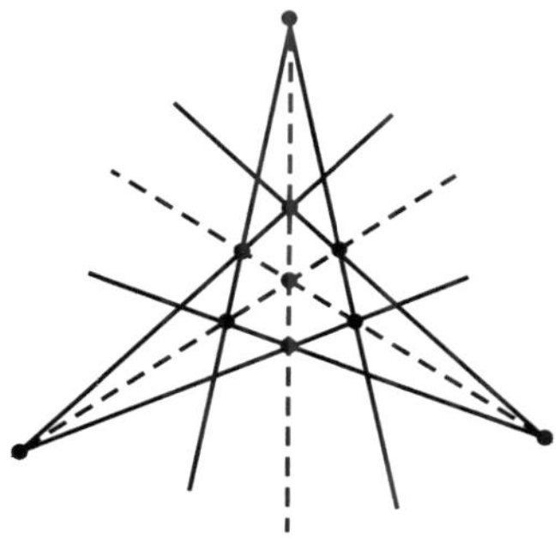

9. The list of absurdities are listed at https://en.wikipedia.org/wiki/Satire_on_False_Perspective.

Index